INTEGRATION OF FUNDAMENTAL POLYMER
SCIENCE AND TECHNOLOGY—3

The proceedings of the international meeting on polymer science and technology, Rolduc Polymer Meeting—3 held at Rolduc Abbey, Limburg, The Netherlands, 24–28 April 1988

INTEGRATION OF FUNDAMENTAL POLYMER SCIENCE AND TECHNOLOGY—3

Edited by

P. J. LEMSTRA

Eindhoven University of Technology, Eindhoven, The Netherlands

and

L. A. KLEINTJENS

DSM-Research, Geleen, The Netherlands

ELSEVIER APPLIED SCIENCE
LONDON and NEW YORK

ELSEVIER SCIENCE PUBLISHERS LTD
Crown House, Linton Road, Barking, Essex IG11 8JU, England

Sole Distributor in the USA and Canada
ELSEVIER SCIENCE PUBLISHING CO., INC.
655 Avenue of the Americas, New York, NY 10010, USA

WITH 39 TABLES AND 252 ILLUSTRATIONS

© 1989 ELSEVIER SCIENCE PUBLISHERS LTD
Softcover reprint of the hardcover 1st edition 1989

British Library Cataloguing in Publication Data

Integration of fundamental polymer science
and technology—3
1. Polymer science
I. Lemstra, P. J.   II. Kleintjens, L. A.
547.7

ISBN 978-94-010-6993-9   ISBN 978-94-009-1115-4 (eBook)
DOI 10.1007/978-94-009-1115-4

Library of Congress CIP Data applied for

FOREWORD

The Rolduc Polymer Meetings, of which the contents of this volume represent the third, are already on their way to occupying a unique place in the crowded calendar of symposia on every aspect of polymer science and engineering. They combine manageable meeting size with a theme, 'Integration of Fundamental Polymer Science and Technology', which is often discussed but seldom realized in practice.

The technological, or applied, areas of polymers have perhaps received more emphasis historically than those of other allied disciplines. Indeed, various plastic and rubber materials were successful items of commerce long before the macromolecular concept itself was firmly established. The more fundamental aspects of the field were also largely developed in industrial laboratories. The early work of Mark and Meyer at IG Farben, and that of Carrothers and Flory at Du Pont, are good examples of this. The present situation, in which polymers are being applied to more and more demanding end uses, from high performance materials on the one hand to the biomedical and electronics fields on the other, calls for an ever greater understanding of the basic scientific principles governing their behavior. It is evident, therefore, that interactions between those engaged in the 'pure' and 'applied' parts of the field must be promoted effectively. The Rolduc Polymer Meetings contribute significantly to such interactions, not only by interweaving technological and scientific presentations, but also by providing a forum for the participants to discuss problems of mutual interest in all their complexity.

The interdisciplinary nature of polymer science and engineering makes it necessary for successful practitioners to be broadly aware of aspects of many fields, from chemistry and physics to materials science to engineering. No one individual can hope to master all these disciplines in depth. It is necessary to possess the ability to recognize what is required to address the problem at hand and to have sufficient acquaintance with the relevant areas to communicate effectively with the appropriate specialists. The contents of this volume represent contributions from experts in a wide variety of fields and exemplify the opportunities the conference represents for the participants to expand their technical horizons. A glance at the table of contents reveals that 'Rolduc 3' dealt with most of the current areas of interest in polymeric materials, from liquid crystalline polymers through engineering thermoplastics to the science and technology of high performance composites. There was also emphasis on polymers in the electronics field, novel polymer chemistry, and polymer blends. In all cases, basic science and technology were judiciously balanced.

It may be that a 'well-behaved' analytical integration of science and technology is not possible in the polymer field, but approximate numerical integration methods are essential to progress, and the Rolduc Meeting represents the development of a rather successful example of one such method. The organizers, Piet Lemstra and Ludo Kleintjens, are to be congratulated for their efforts and it is to be confidently expected that future meetings will lead to increasingly better approximations to the value of the integral.

WILLIAM J. MACKNIGHT
Polymer Science and Engineering
University of Massachusetts
Amherst, Massachusetts 01003
USA

PREFACE

The polymer industry is doing better than ever. Well-known bulk polymers, optimized to customers' demands, are taking an ever increasing part of the materials market. New polymeric materials, including polymer alloys, high modulus fibres, molecular composites, high temperature resistant thermoplastics, liquid crystalline polymers and electrically conductive polymers, are being studied worldwide and some of these products are finding their first successful applications. Fascinating aspects are announced, at least by the manufacturers of these materials.

Although the turnover of such high-tech polymers is still a small percentage of the total polymer market, the automotive, aeronautical, printing and electronics industries seem keen to use the possibilities of these polymer materials. It is well accepted nowadays that successful development and application of new polymer materials can only be obtained using an integrated approach. Polymer chemists, processing engineers, mechanical engineers and designers have to work together and understand each others' possibilities (and impossibilities). Major projects even call for a multinational approach, and joint ventures and projects in which competitors are working together are daily practice (Carmat, Brite, etc.). The aim of the Rolduc Polymer Meetings was and still is to serve this goal, to stimulate multidisciplinary discussions between academic and industrial polymer scientists and engineers. Rolduc Polymer Meeting—3 brought together producers, processing researchers and end-users of new polymer materials. The prospects for such materials, and examples of successes and failures, were presented and discussed. All active participants were requested to submit a manuscript for this book. Contrary to the usual meeting proceedings, a substantial part of this book contains industrial contributions which sometimes give controversial opinions. We hope the reader will enjoy this up-to-date report on the broad field of polymer materials.

The reader may look for an integrated presentation of all material presented at Rolduc Polymer Meeting—3. Unfortunately, this is an impossible task in such proceedings. We are sure, however, that the 250 participants of RPM—3 have experienced the integrating atmosphere during the meeting and the many lively discussions in Rolduc Abbey.

The editors wish to thank all contributors to this volume.

P.J.L.
L.A.K.

ix

CONTENTS

Part 3: Fibres/Composites

xii

Part 4: Morphology/Structure

xiii

Part 5: Speciality Polymers

Part 6: Techniques

Part 1

PLENARY PAPERS

INTEGRATION OF POLYMER SCIENCE AND TECHNOLOGY?

L.C.E. Struik
Plastics and Rubber Research Institute TNO
P.O. Box 71
2600 AB   Delft
The Netherlands

1. INTRODUCTION

One of the aims of these Rolduc Conferences is to stimulate integration of/in Polymer Science and Technology. The organisers asked for a critical review of the present-day status of this integration. Using a number of examples, it will be shown that the goal is far from being reached; these examples will refer to fracture, yielding, curing-reactions, UV-degradation and finite-element calculations. It will appear that some essential knowledge, available from material science, physics, etc. or even from certain branches of polymer science is not effectively used in (other branches of) the polymer field. As a consequence, research projects are sometimes ineffective, too expensive, based on wrong or questionable assumptions, unworkable or even unneeded.

2. AN EXAMPLE OF AN UNWORKABLE/UNNEEDED PROJECT

Sometime ago, the author became aware of a research proposal aiming at generalisation of a conventional, linear-elastic finite-element computer code (see Figure 1) to visco-elastic materials; the costs amounted to several millions of guilders. The proposers asked the author's opinion which was as follows:

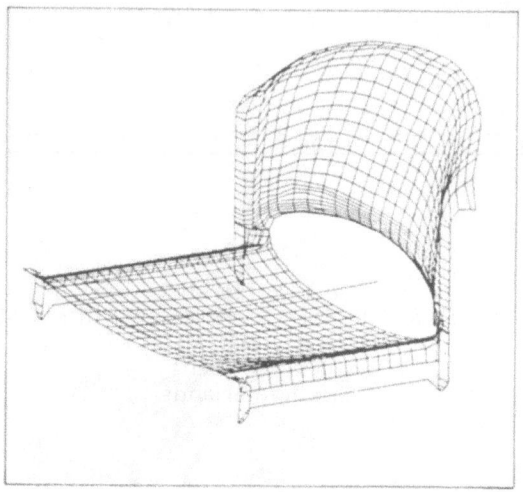

Figure 1. Finite-element mesh for describing the stress-strain behaviour of a seat for a chair; reproduced with permission of De Constructeur 25 (1987) page 6.

For small strains the material will be linearly visco-elastic. According to the correspondence principle, derived 40 years ago [1-7] the linear-elastic solution, found with the existing finite-element codes can be used more or less straightforwardly for linear visco-elastic materials in quite many cases. So it is unneeded to complicate the finite-element calculation by giving the elements visco-elastic properties. For larger strains, the material becomes non-linear visco-elastic. But for this strain-regime, we don't have a reliable constitutive equation, so the problem is of a physical instead of a computational character. Some constitutive equations are described in [8]: Shapery's reduced time theory is easy to use but cannot explain the strain-softening occurring in cyclic loading [9]. The Green-Rivlin method is more complicated but ignores physical ageing; this is questionable because there are strong indications that the non-linear visco-elastic behaviour of plastics arises from the effect of deformation on the ageing process [10]. So, as a conclusion, the proposed project is unneeded (small strains) or unworkable (large strains). A more specialized example of the same kind is given in [11].

3. <u>RESEARCH BASED ON WRONG ASSUMPTIONS</u>

Sometimes, projects are (have been) based on assumptions which from "an integrated point of view" would immediately be classified as wrong. The main result of such projects is frustration. Examples are the following:

3.1 PREDICTION OF THE DEGRADATION OF POLYOLEFINES DUE TO THE UV COMPONENTS OF THE SUNLIGHT

The correlation between outdoor UV degradation and accelerated ageing in the laboratory has been an important research topic for many years [12]. Several projects ended with the conclusion that there is no unique acceleration factor characteristic for a particular laboratory tester (e.g. a Xenon-tester 1200). This conclusion might have been reached beforehand; i.e. without testing [13-14]. For polyolefines, the UV-stimulated oxidation can be described by the classical scheme of Bolland [15]. There is a prime production rate $[\dot{R}]_+$ of radicals ([R] is the radical concentration, a dot means differentiation with respect to time) which is proportional to the UV-light intensity I. However, the breakdown reaction rate $[\dot{R}]_-$ depends on the stabilization system. If the radicals recombine with each other, $[\dot{R}]_-$ will be proportional to $[R]^2$. If the radicals are scavenged by, e.g., an excess of stabilizer, the rate $[\dot{R}]_-$ will be proportional to [R]. This implies that the steady-state radical concentration [R] will be proportional to $\sqrt{I}$ or I depending on the stabilizer; in fact dependences in between $\sqrt{I}$ and I are also possible. Since the degradation rate is proportional to the radical concentration, the degradation rate cannot be expected to be proportional to I in general.
Such proportionality, however, is implicitely assumed if one thinks of a unique acceleration factor for a particular lab-tester; research aiming at finding this factor will not be very fruitful.

3.2 INCREASE OF THE HARDNESS OF ACRYLIC SHEET BY CHEMICAL MODIFICATION OF THE POLYMER [18]

In the fifties and sixties, a lot of research has been done aiming at a substantial increase of the hardness of unoriented glassy acrylic sheets.

The goal was to penetrate into the huge window-glass market. The goal has not been reached and this frustrating result might have been foreseen. Already at that time it was known that the hardness is more or less proportional to the elastic modulus E or to a power of E [19]. For the classical case of an elastic indentation (Hertz, 1881) this is self-evident. For "plastic" indendations, Tabor (1947) [16] showed that the hardness will be proportional to the yield stress σ_y. But for unoriented glassy amorphous polymers, the strain at yield does not vary over wide ranges; so, in contrast with e.g. metals, the yield stress must be more or less proportional to the modulus [20] and the same thus holds for hardness H.

Consequently, the question whether the hardness of acrylic sheet can be raised to values comparable to those of inorganic glasses reduces to the question whether the modulus of an unoriented glassy polymer can be raised to that of inorganic glass. Already in the forties, hardly nobody would have believed in such a possibility; the modulus of all unoriented glassy polymers prepared at that time remained far below that of inorganic glass. In retrospect, the key factor overlooked in the studies mentioned above was the fair constancy of the σ_y/E ratio; the same applies to many present-day studies on the yielding- and fracture behaviour of unoriented polymers (see Section 5).

4. RESEARCH IN WHICH ESSENTIAL KNOWLEGDE IS NEGLECTED

Sometimes research projects in one area of polymer science (e.g. dealing with the curing of coatings or thermosets) ignore essential knowledge built-up in other areas (e.g. the formation of inorganic or organic glasses by (rapid) cooling from the melt). The reason is a lack of "horizontal" integration. Elaborating the example indicated above, we can conclude the following: Already in the thirties, it was known that organic and inorganic glasses prepared by solidification of a melt or a liquid are unstable [21-23]. Solidification is not completed during the brief period (minutes to hours) of cooling, evaporation of solvent, etc., but continues during the whole life time of the glassy material and so induces large changes in properties. In the fifties to seventies, this phenomenon (physical ageing) has been extensively studied for thermoplastics solidified by rapid cooling from the melt [24-28]. The ageing phenomenon, however, is much more general and also occurs in glasses prepared by solvent evaporation, polymerization or cross-linking reactions or by a combination of such mechanisms [29]. Yet physical ageing is hardly considered in studies on the curing of thermosets, composites or organic coatings, or on the preparation of fibres by spinning from solutions.

5. RESEARCH BASED ON QUESTIONABLE ASSUMPTIONS

As an example [30] of current research based on questionable assumptions we now consider the application of linear-elastic fracture mechanics (LEFM) [31-32] and classical yield criteria (von Mises, etc. [8]) to unoriented glassy polymers. That there actually is a basic problem needs some clarification:

Since Grüneisen's work (1908) we know that the elastic moduli of crystalline solids are determined by the intermolecular forces. Taking a Mie-potential for the pair-interaction energy

$$U(r) = \frac{m}{n-m} U_0 [(\frac{r_0}{r})^n - \frac{n}{m} (\frac{r_0}{r})^m]$$ (1)

in which n and m are constants whilst U_0 and r_0 are explained by Fig. 2, one obtains [17] Grüneisens first rule:

$$KV_0 = \frac{mn}{9} U_0$$ (2)

where K denotes the bulk modulus and V_0 the molar volume.

Eq. (2) predicts that the modulus K is proportional to the intermolecular energy (lattice energy U_0); strictly speaking Eq. (2) only applies at 0 K; at higher temperatures the modulus slightly decreases due to thermal expansion.

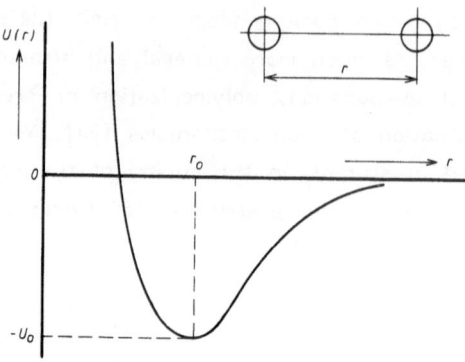

Figure 2. Interaction energy U(r) as a function of the distance r of the two molecules; r_0 is the (equilibrium) distance at which U(r) attains the minimum value $-U_0$.

The Grüneisen theory can be applied to unoriented glassy polymers [19] [33] [43]. We then face the problem that in such materials different types of binding forces are active. Along the chains, we have strong, stiff, covalent bonds; perpendicular to the chains we have weak van der Waals bonds. It can, however, be argued that the elastic modulus is mainly determined by the weak van der Waals forces. Using a Lennard-Jones potential (n = 12, m = 6) and taking into account that there is hardly no deformation in the chain direction (stiff bonds), we find instead of Eq. (2):

$$KV_0 = 18 \ U_0 \qquad\qquad (3)$$

This equation can be verified experimentally. For amorphous glassy polymers, the "lattice" energy may be equated to the Cohesive Energy Density (CED) tabulated in [19]. The bulk modulus K may be converted to the shear modulus G or Youngs modulus E by means of the classical formulas:

$$G = \frac{3}{2} \ K \ (1-2v)/(1+v) \qquad\qquad (4A)$$

$$E = 3 \ K \ (1-2v) \qquad\qquad (4B)$$

where v denotes the Poisson ratio.

For unoriented glassy polymers, we have $v \sim 1/3$ at low temperatures. Consequently, $G \sim 0.375\ K$ and $E = K$. Writing E^*_{coh} for the cohesive energy density per unit volume (U_o/V_o), we finally obtain:

$$G \sim 6.75\ E^*_{coh} \quad ; \quad E \sim 18\ E^*_{coh} \qquad (5)$$

A check of these formulae against experimental data is presented in Fig. 3. Obviously the order of magnitude of the low-temperature modulus is well predicted from E^*_{coh}, i.e. from a property measured on polymers swollen in liquids! The same conclusion has been drawn earlier (see e.g. Willbourn [34]).

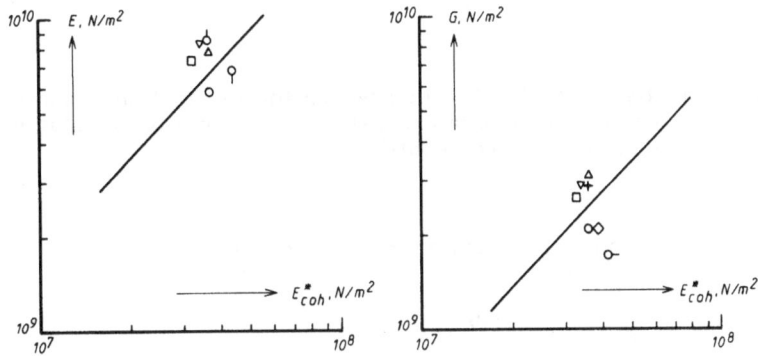

Figure 3. Youngs Modulus E and Shear Modulus G at very low temperatures (few K) vs cohesive energy density per unit volume E^*_{coh}. All quantities are given in N/m². The E^*_{coh}-data are from [19]. The symbols mean:

o : PS [35] [39]; ϙ : PVC [37]; Δ : PMMA [37] [39]; ∇ : PEtMA [35] [37] [39]; □ : poly(isobutyl MA) [37] [39]; ◇ : PVAc [35]; + : poly (vinyl-propionate) [35]; o- : PET [38] [39]; ⌀ : poly(d.l.propylene oxide) [37]; the modulus data were obtained from the references indicated. The straight lines obey Eq. (5).

The results shown above strongly support the view that the modulus of and coherence in an unoriented glassy polymer is mainly determined by the weak van der Waals forces. This conclusion has far-reaching consequences on the yield- and fracture behaviour of such materials [33]. An illustration is given in Fig. 4, which shows a replot of Fig. 2, but now for the intermolecular force $f(r) = -dU/dr$. At a distance $r = r* = r_0[(n+1)/(m+1)]^{1/(n-m)}$, force $f(r)$ passes the maximum value f_{max} and this is the maximum tensile force which the bond can survive.

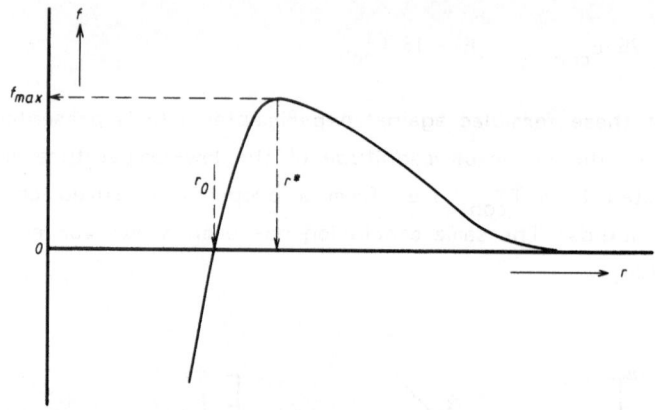

Figure 4. Replot of Fig. 2, but now for the intermolecular force $f = -dU/dr$ versus the separation distance r. A positive f-value means attraction, a negative repulsion.

With $n = 12$ and $m = 6$ (Lennard-Jones) we find:

$$\varepsilon_{max} = \frac{r*}{r_0} - 1 = 0.109 \qquad (6A)$$

$$f_{max} = 2.69 \frac{U_0}{r_0} \qquad (6B)$$

Quantity f_{max} is proportional to the so called theoretical strength σ_{max}, i.e. the strength as calculated from the strength of the molecular bonds. Eq. (6B) thus tells that σ_{max} will be proportional to "lattice" energy U_0 but Eqs. (3)-(5) tell that the same holds for modulus E. Consequently, σ_{max} will be proportional to modulus E too. A detailed calculation shows that, for unoriented amorphous polymers at room temperature, we will have

$$\sigma_{max} \sim E/30 \qquad (7)$$

For unoriented polymers the theoretical strength, according to Eq. (7), must be identified as the yield stress. It is well known that it are the van der Waals bonds which are broken at the yield point; moreover Eq. (7) reasonably agrees with experiment (Fig. 5).

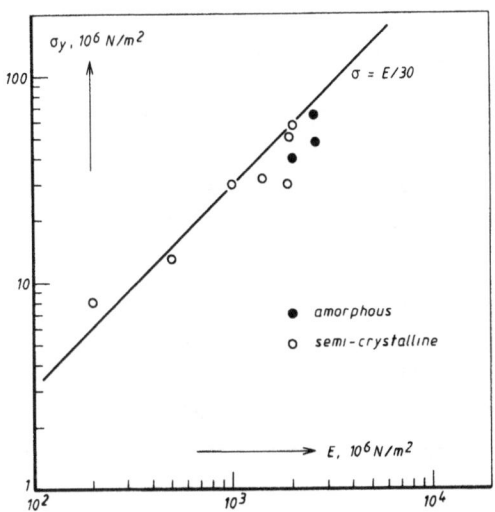

Figure 5. Tensile yield stress σ_y vs Youngs modulus E for various polymers; data from Table 13.10 of [19].

The above results show that unoriented polymers which don't break before yield reach their theoretical strength [40]. This is remarkable because classical materials such as metals or glasses yield or fracture at stress levels far below σ_{max} = E/30. Some data, collected in Table 1, show that unoriented polymers fail at their theoretical strength (ductile materials) or at a substantial fraction of σ_{max} (PS, PVC or Epoxy); for classical materials (glass, concrete, steel) the σ/σ_{max} ratio at failure is much lower. This remarkable conclusion has already been drawn by Buchdahl [42] in 1958, see also [40].

TABLE 1

Comparison of theoretical strength σ_{max} and practical strength σ for some materials. The data for glass, concrete and steel were taken from [41], the other data are from Table 13.10 of [19]. All stresses are given in MPa = 10^6 N/m².

Material	σ_{max} = E/30	Actual strength σ (yield or fracture)	Ratio σ/σ_{max}
glass	1700-3400	30-90	0.02-0.03
concrete	1300	40-60	0.03-0.045
Chr-Ni steel	6500	550-750	0.08-0.12
LDPE	7	8	1.1
HDPE	33	30	0.9
Rigid PVC	87	48	0.6
PS	113	50	0.4
Epoxy	80	55	0.7

The implications of these results are the following:

(a)   The low fracture stress (compared with σ_{max}) of most classical materials has been the basis for the development of fracture mechanics [31-32]. The "premature" fracture is explained from pre-existing flaws which act as stress concentrators. In view of the high σ/σ_{max} ratios for polymers, the straightforward use of LEFM is highly questionable. We do not deny that LEFM can be a useful tool for describing the fracture behaviour of artificially pre-cracked laboratory specimens; what is questionable is whether tests on such pre-cracked samples have any relevance to actual products made without artificial cracks.

(b)   The low yield stress of most metals ($\sigma_y \ll \sigma_{max}$) has led to the dislocation theory [17] and to yield criteria (von Mises, etc.) which assume that yielding is due to the deviatoric (shearing) components of the stress tensor [8] [17]. The hydrostatic component of the stress tensor is assumed to have no or only an indirect influence. Such criteria cannot be true for materials failing at their theoretical strength. What now counts are the bond elongations and critical bond elongations can be achieved in shear, tension or negative hydrostatic pressure as well.

Notwithstanding the fact that all things mentioned above have been published many years ago [42], most workers in the field seem unaware of these problems.

6. DISCUSSION

Turning back to the question put forward in the introduction, we must conclude that integration of knowledge has certainly not been reached in the polymer field. Some reasons might be the following:

(1) Polymer science and technology is a relatively young and rapidly growing field. New people with other backgrounds continuously "flow in" and they sometimes have difficulties with the "connections". This certainly explains the problem of Section 2 and to some extent that of Section 5 (fracture and yielding studies on polymers have mainly been taken up by workers with a background of metallurgy; furthermore in metallurgy the subject is much further developed than in the polymer field).

(2) The field is very wide; people working with organic coatings, adhesives or glass-reinforced epoxies have little connection with those working on the effect of processing or the properties of thermoplastics. This lack of horizontal integration might explain the problem of Section 4.

(3) The field is stratified; practice oriented research has sometimes little connection with more fundamental studies. This might explain the problems of Section 3.

I don't think that these problems are exceptional and a blame for our field. However, attempts such as these Rolduc Meetings and the post-academic TOP and PTN coarses presently organised in this country and

which all aim at more integration of knowledge should be highly wel-
comed.

7. ACKNOWLEDGEMENTS

The author is indebted to Dr P. Vink for the example of Section 3.1 and
Dr Ir J. Heijboer for valuable discussions about the problem discussed in
Section 3.2.

REFERENCES

1.   W. Flügge, "Viscoelasticity", Springer, 1975.

2.   W.N. Findley, J.S. Lai, K. Onaran, "Creep and Relaxation of
     non-linear viscoelastic materials", North-Holland, 1976.

3.   T. Alfrey, Qu. Appl. Math. $\underline{2}$ (1944), 113.

4.   W.T. Read, J. Appl. Phys. $\underline{21}$ (1950), 671.

5.   M.A. Biot, J. Appl. Phys. $\underline{25}$ (1954), 1385.

6.   E.H. Lee, Qu. Appl. Math. $\underline{13}$ (1955), 183.

7.   L.C.E. Struik, "Het tijdsafhankelijk gedrag van kunststoffen", part
     of College Kunststoffen, Un. Twente, code nr. 135020. KRITNO
     publication nr. P 1/'88.

8.   I.M. Ward, "Mechanical Properties of Solid Polymers", Wiley, 1983,
     Chapter 9.

9.   L.C.E. Struik, "Physical Aging of Amorphous Polymers and Other
     Materials", Elsevier, Amsterdam, 1978, Chapter 8.

10. L.C.E. Struik, "Physical Aging: Influence on the Deformation
     Behaviour of Amorphous Polymers", in "Mechanical Failure of
     Plastics", Brostow & Corneliussen, Eds., Hanser, Munnich-Vienna,
     1986.

11. L.C.E. Struik, in "Kunststoffen 1986, Terugblik en Toekomst",
     H.M. Brüggeman, Ed. published by KRITNO, 1986, page 280-281.

12. J.D.M. Wisse, book of Ref. 11, pp 203-210.

13. P. Vink, in "Developments in polymer stabilisation" - 3; Ed.,
     G. Schott, Applied Science Publishers, Ltd., London, Chapter 4
     (1980).

14. P. Vink and J.D.M. Wisse, Polymer Degradation and Stability, 4 (1982), 51-57.

15. J.L. Bolland, Proc. Roy. Soc. (London), A 186 (1946) 218, Trans. Faraday Soc. 42 (1946) 236, 244; 43 (1947) 201; 44 (1948) 669; 45 (1949) 93; 46 (1950) 358.

16. D. Tabor, Proc. Roy. Soc (London) 192 (1947) 247.

17. C. Zwikker, "Physical Properties of Solid Materials", Pergamon Press, London 1954.

18. J. Heijboer, "Technological Properties of hard methacrylate Polymers", Central Laboratory TNO, Report no. CL 55/45; September 1955.

19. D.W. van Krevelen, "Properties of Polymers", Elsevier, Amsterdam, 1976.

20. N. Brown, Mat. Sci. and Eng. 8 (1971), 69-73.

21. F. Simon, Z. Anorg. Chemie 203 (1931), 219.

22. H.R. Lillie, J. Amer. Ceram. Soc. 16 (1933), 619.

23. E. Jenckel, Z. Elektrochem. 43 (1937) 769, 45 (1939) 202.

24. A.J. Kovacs, Thesis, Fac. Sci. Paris, France, 1954.

25. A.J. Kovacs, Fortschr. Hochpolym. Forschung, 3 (1964) 394.

26. L.C.E. Struik, "Physical Aging of Amorphous Polymers and other Materials", Elsevier, Amsterdam, 1978.

27. L.C.E. Struik, Chapter 11 in "Failure of Plastics", Brostow and Corneliussen, Eds., Hanser, Munnich-Vienna, 1986.

28. L.C.E. Struik, "The mechanical behaviour and physical aging of semi-crystalline polymers", Sequence of 4 papers in Polymer. Parts 1 and 2 have been published: Polymer 28 (1987) 1521, 1534. Parts 3 and 4 have been issued to Polymer. See also Plastics and Rubber Process. and Appl. 2 (1982) 41.

29. L.C.E. Struik, "Aging-Physical" in the Mark-Bikales-Overberger-Menges "Encyclopedia of Polymer Science and Engineering", Vol. 1, Second Ed., Wiley, New York, 1985.

30. L.C.E. Struik, "De Fysika van polymere materialen", Oratie, Un. Twente, 22-01-1987.

31. D. Broek, "Elementary Engineering Fracture Mechanics", Noordhoff, Leyden, 1974.

32. J.G. Williams, "Fracture Mechanics of Polymers", Wiley, New York, 1984.

33. L.C.E. Struik, College Kunststoffen Code nr. 135020, Un. Twente, 1986-87.

34. A.H. Willbourn, Polymer 17 (1976) 965.

35. V. Frosini and A.E. Woodward, J. Polym. Sci., A-2, 7 (1969) 525.

36. J.M. Crissman, A.E. Woodward and J.A. Sauer, J. Polym. Sci. A 3 (1965) 2693.

37. J.M. Crissman, J.A. Sauer and A.E. Woodward, J. Polym. Sci A 2 (1964) 5075.

38. J.M. Roe and E. Baer, Int. J. Polymeric Mat. 1 (1972) 11.

39. J.A. Sauer and R.G. Saba, J. Macromol. Sci. (Chem.) A 3 (1969) 1217.

40. N. Brown, Chapter 6 of the book of Ref. 27.

41. F. Kohlrausch, "Praktische Physik", Bd 2, Table 16.

42. R. Buchdahl, J. Polym. Sci. 28 (1958) 239.

43. A.V. Tobolsky, "Properties and Structure of Polymers", Wiley, New York, 1960.

Polymers for Electronics and Photonics

"Conducting, ferromagnetic, and nonlinear optical properties of polymers"

E.W. Meijer*, S. Nijhuis, and E.E. Havinga

Philips Research Laboratories, P.O. Box 80000, 5600 JA Eindhoven,
The Netherlands

ABSTRACT

Recently, polymers show promising prospects as active constituents of optical and electronic devices, especially polymers for nonlinear optics and conducting polymers. In a remote future even organic ferromagnets and molecular electronics may become more than wishful thinking. For this reason several fundamental aspects of these materials are the subject of intensive research. In this paper we will review the present state of the art in these new fields of polymers for electronics and photonics. Special emphasis will be laid on recent results from our laboratories.

1 INTRODUCTION

It is well-known that polymers are important materials in various electronic and photonic applications[1]. For example, packaging and interconnection of electronic components, coatings for optical fibers, optical recording media, polymeric waveguides and lenses, and microlithography[2]. Recently, polymers show also prospects as active constituents of optical and electronic devices[3]. With these applications in mind, the important properties of the polymers studied are physical phenomena like electrical conductivity, ferromagnetism, and nonlinear optical properties as optical switching or frequency doubling. In order to be really useful, the polymers should retain their typical polymer properties like mechanical strength and ease of processing. A combination of properties will afford a new class of active materials in devices. For this reason several fundamental aspects of these materials are the subject of intensive research. High conductivity is measured in a variety of doped polymers and the nonlinear optical properties of polymers exceed those of the best inorganic materials, whilst very recently an organic polymer is claimed to show ferromagnetism. Although these interesting properties are found in polymers, they show quite a number of disagreeable properties too, such as insolubility and nonprocessability or instability in time. No wonder that a large part of the research going on in this field is aimed at restoring, by means of chemical modifications, some of the properties that make polymers in general such valuable materials. In this paper we will review the present state of the art in these new fields of polymers for electronics and photonics. Successively, we will discuss the conducting, ferromagnetic, and nonlinear optical properties of polymers. Special emphasis will be laid on recent results from our laboratories.

2 CONDUCTING POLYMERS

Since the seminal experiments in 1977 on polyacetylene[4] showing that this polymer could be made electrically conductive upon doping, an intensive research activity has been unfold world-wide[5]. In the years following these first disclosures, polyacetylene has been studied in great detail, eventually leading to improved syntheses and a conductivity of 1.5 x 10^5 S/cm, which equals that of copper at room temperature[6].

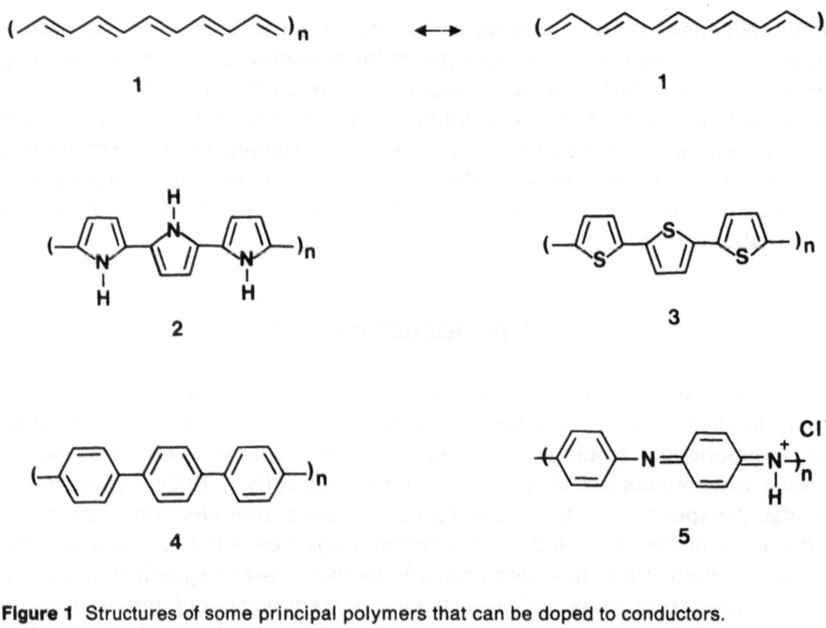

Figure 1 Structures of some principal polymers that can be doped to conductors.

The all-trans structure of polyacetylene (**1**) is shown in figure 1, together with some other well-studied polymers like: polypyrrole (**2**), polythiophene (**3**), poly-p-phenylene (**4**), and polyaniline (**5**). They represent the key structures of conducting polymers. The polymers as given in figure 1 are insulators with bandgaps in the order of 1.5 eV or more. In order to become electrically conductive the polymers have to be doped, that is to be oxidized (to p-type conductors) or reduced (to n-type conductors), yielding positively charged backbones or negatively charged backbones, respectively. Not in all cases both types of doping can be realized. For instance, the reduction potential of neutral polypyrrole is so high, that only oxidative doping can occur. With respect to environmental stability this oxidative doping is preferred for all polymers. The doping occurs either simultaneously with the preparation of the polymers (e.g. the electrochemical polymerization of pyrrole and thiophene) or by separate doping with strongly oxidizing agents (e.g. gaseous I_2, AsF_5, etc.).

Several experiments have been performed on conducting polymers to establish the mechanism of conduction in these polymers[7]. Conducting polymers are often divided into two classes, namely those having a degenerated ground state and those where the ground state degeneracy is lifted. Polyacetylene (**1**) is an example of the first class, where the interchange of double and single bonds involves no loss of energy. Here solitons are considered to be the important excitations [5]. In the other class, comprising compounds like polypyrrole (**2**), polythiophene (**3**), and poly-p-phenylene (**4**), polarons and bipolarons are the dominant charge-storage configurations. Most of these polymers have a nondegenerated ground state that is aromatic , while the quinoid form is of higher energy[8].

Despite numerous reports on the mechanism and theory of conduction in these polymers, exact descriptions are still hard to give, let alone to prove. Only for the very high conductivities recently measured in pure, well-characterized and ordered polymers, the assumptions made in theoretical models may be realistic. For a good review on these aspects we like to refer to several papers in the Handbook of Conducting Polymers[9].

The topics of special interest for polymer chemists are the search for new and better polymers or new procedures to the well-known polymers. In order to improve the properties of conducting polymers emphasis is given to higher conductivity, better environmental stability, and processability. A fundamental approach to increase the conductivity is based on decreasing the bandgap of the undoped polymer[10]. A "pure" intrinsically metallic polymer would be reached at a bandgap of 0 eV. However, such a low bandgap in a simple conjugated polymer is hardly to be expected, as theory gives many reasons for lifting the degeneracy of the extended π-electron groundstate (correlation energy, Peierls effect because of onedimensionality). Up to now the polymer with the lowest bandgap shows a value around 1 eV [10]. However, the search for polymers that really show intrinsic metallic properties remains challenging.

To the contrary, in the area of processability much progress has been made. Specially designed polymer synthesis has been carried out leading to either soluble precursors or even soluble conductors [11] .

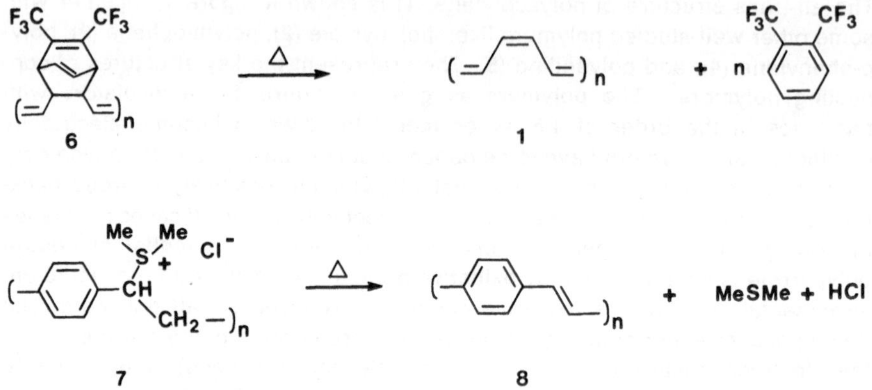

Scheme 1

Feast and Edwards introduced this concept of soluble precursors with their polyacetylene synthesis as outlined in scheme 1 [12]. The soluble precursor **6**, obtained by metathesis polymerization, yielded the socalled Durham polyacetylene in a thermal retro Diels Alder reaction. Another illustrative example is the synthesis of poly-*p*-phenylene-vinylidene **8** from the poly-sulfonium salt **7** by thermal elimination (scheme 1)[13]. When the polymer is stretched during this elimination a highly oriented film is obtained with a high anisotropy in its conductivity after doping (for instance with I_2). The conductivities can become as high as 10^4 S/cm. In these and many other cases the precursor polymers are made processable[11]. Another breakthrough in this area of research was achieved by synthesizing soluble conjugated polymers. This aspect will be discussed in paragraph 2.2 as introduction to self-doped conducting polymers.

2.1 POLYMERS OBTAINED BY THE STETTER REACTION

Since new strategies for polymer synthesis are at the basis of the progress made in the area of processable conducting polymers, we have conducted an investigation to a new synthetic scheme for polyheterocyclic polymers[14]. For heterocyclic polymers, all syntheses known today are based on the coupling of pyrrole or thiophene monomers, either electrochemically[15], by $FeCl_3$ oxidation of pyrrole[16] or by Grignard-type reactions of 2,5-dibromo-thiophene[17]. To date no schemes have been proposed in which the heterocycle is formed during a reaction performed on a preformed polymer. This is surprising, since the synthesis of pyrroles and thiophenes by ring closure of the appropriate precursor is a well-known reaction[18]. The reason that this procedure has never been used in polymer chemistry is probably due to the lack of suitable precursor polymers.

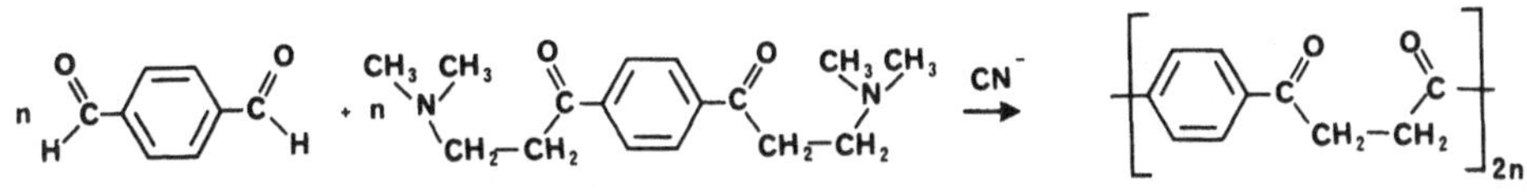

Scheme 2 The Stetter polymerization

We have synthesized such a precursor polymer by using the Stetter reaction (scheme 2). Poly-1,4-phenylene-1′,4′-butanedione **9** was formed in the reaction of terephthalic-dicarboxaldehyde **10** and the bis-Mannich base of 1,4-diacetylbenzene **11** in a yield of 81%. The conversion of **9** into the alternating copolymers of p-phenylene-2,5-pyrrole **12** and p-phenylene-2,5-thiophene **13** is performed with liquid NH_3 and Lawesson's reagent, respectively (scheme 3). Upon doping with either I_2 or AsF_5 both polymers **12** and **13** become electrically conducting with specific conductivities up to 0.1 S/cm.

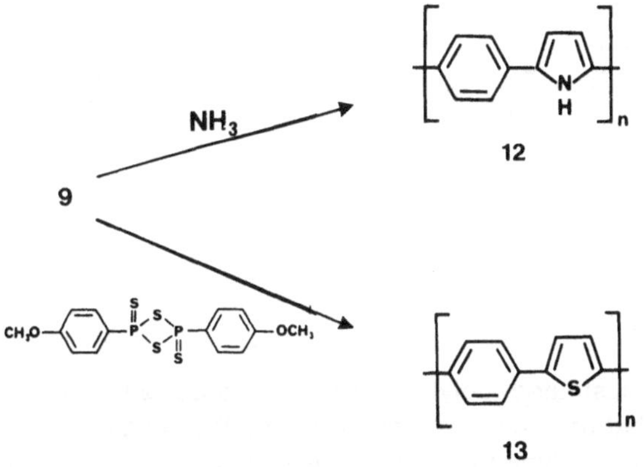

Scheme 3

In order to increase the conductivity to still higher values we will optimize the reactions used in this new strategy. Since it is expected that polymer **9** can be substituted and that also other aromatics than benzene can undergo the Stetter polymerization, we foresee a broad scope for this reaction scheme in the area of conducting polymers.

2.2 SELF-DOPED WATER-SOLUBLE CONDUCTING POLYMERS

Polyheterocyclic polymers that are soluble in the highly doped state were made for the first time by Frommer et al.[19]. However the solvents used were the toxic and environmentally unstable AsF_3-AsF_5 or liquid I_2, being both solvent and dopant simultaneously[19]. More recently, several groups, with seeding work from Elsenbaumer et al., prepared soluble and highly conducting polymers based on polythiophenes substituted with alkyl chains (like **14**, figure 2)[20]. The solubility is achieved through the addition of appropriate side-chains. Elsenbaumer prepared polymers like **14** by coupling of substituted 2,5-diiodo-thiophenes, but later on similar polymers have also been made by electrochemical polymerization of the corresponding monomers[21]. Conductivities in the range of 10-100 S/cm were obtained. Recognized drawbacks of the electrochemical polymerization of thiophene is the relatively high oxidation potential (1.65 eV versus 0.7 eV for pyrrole) and the susceptibility to the reaction for oxygen, leading to sulfon formation. It is a prerequisite to perform the polymerizations in inert media, whilst substituted α-terthienyls can be used to lower the oxidation potential to values of 0.7 eV [22]. In the latter case polymers (like **15**) are obtained with only one substituent per three thiophene units (figure 2).

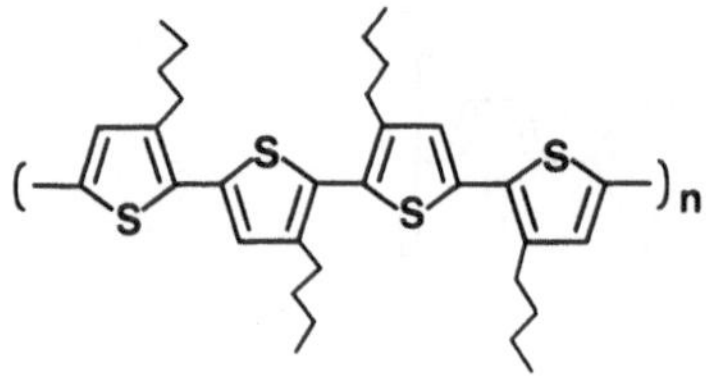

| 14 | 15 |

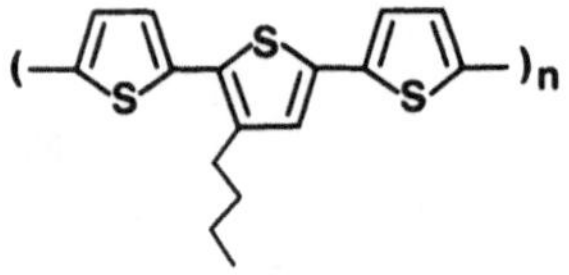

Figure 2

These alkyl substituted polythiophenes initiated our idea to synthesize polythiophenes that are so-called self-doped by the covalent attachment of the counter anion to the polymer chain[23]. The first successful example is given in figure 3, based on a sulfonate anion substituted poly(alkyl)thiophene (**16**). This polymer is made by electrochemical polymerization of the corresponding α-terthienyl **17** in acetonitrile *without the addition of a conducting salt*. The use of a conduction salt is generally considered as obligatory in order to have a sufficient conducting solution. Part of the anions is incorporated in the polymer during its electropolymerization, providing the dopant. In the case of **16** the monomer itself is a salt and addition of another conduction salt can be omitted. This experimental procedure gives a solid proof that the conduction in the polymer is due to doping with the counterions of the sidechains. The conducting polymer, with conductivities up to 0.01 S/cm, proved to be soluble in water. However, the stability of this doped polymer in aqueous solutions is very limited while in methanol the polymer is already undoped upon solvation. Similar results were found by Wudl et al. by preparing and studying self-doped polythiophene **18** as made via an indirect route[24]. In order to increase the stability of the self-doped polymers we directed our attention to polypyrroles[23,25].

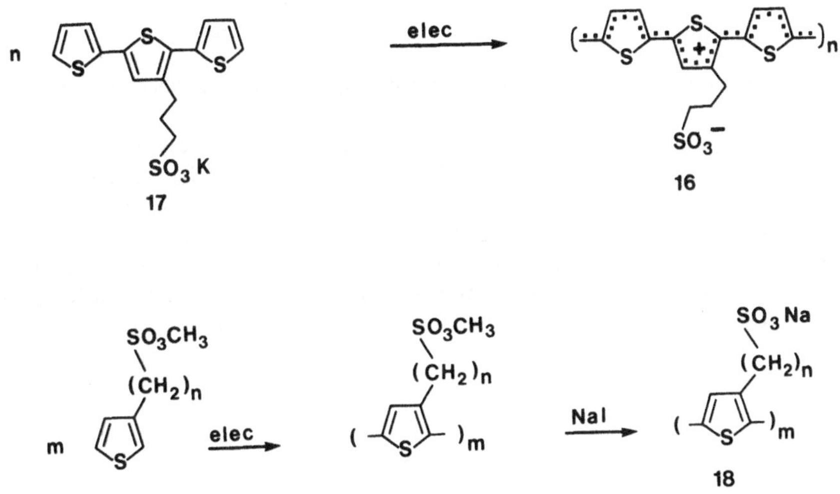

Figure 3

Upon electrochemical polymerization of 3-substituted pyrroles **19a-c** (again without additional conduction salts) an acetonitrile-insoluble polymer is formed as a blue, changing into black, precipitate at the (Pt or ITO) anode. In these polypyrroles part of the sulfonate groups act as a dopant and form "Zwitter-ions" having delocalized positive charges. The remaining sulfonate ions not participating in the self-doping are neutralized with either sodium ions or H_3O^+ (figure 4).

The electronic spectra of solid films of the polypyrroles, prepared on an ITO-glass electrode are compared with those of freshly prepared aqueous solutions of the same polymer (figure 5). Both spectra exhibit low-energy humps that are characteristic for doped conjugated polymers[26]. Contrary to the

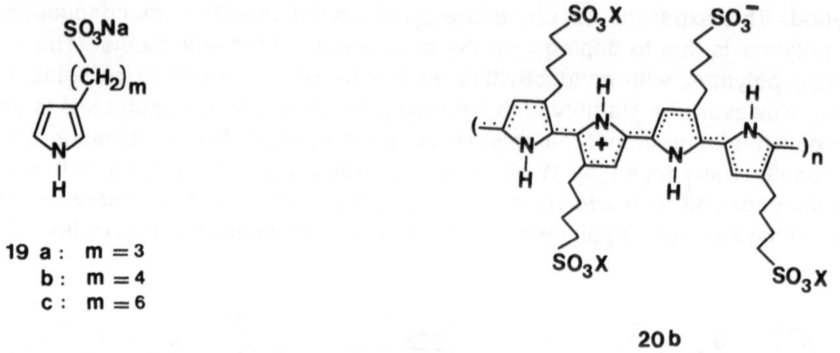

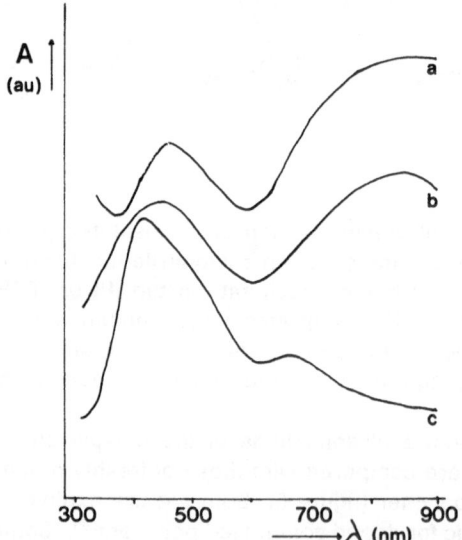

Figure 4 Monomers 19a-c and a sketch of self-doped polypyrrole 20b (X is either H_3O or Na).

self-doped polythiophenes **16**, the aqueous solutions of the polypyrroles **20a-c** are stable (hardly any change in the UV-vis spectrum could be observed after several months). Pressed pellets of the self-doped material showed conductivities of the order of 0.1 to 0.5 S/cm. Films obtained from these solutions by evaporating the water at about 70 °C exhibited a low conductivity of about 10^{-8} S/cm. However, slow evaporation of the solvent at ambient temperature afforded films with a conductivity of 0.01 S/cm.

Figure 5 Electronic spectra of the self-doped polypyrrole 20b. a) Solid film grown on ITO-glass electrodes in acetonitril, b) Aqueous solutions obtained from these films c) Film after the evaporation of water at 70 °C.

These first examples of self-doped conducting polymers that are stable as aqueous solutions will no doubt lead to new applications as well as to new insights in the behaviour of charge carriers in conducting polymers.

2.3 POLY-1,2-AZEPINES BY THE PHOTOPOLYMERIZATION OF PHENYL AZIDES

Up to now most of the effort to achieve processability of conducting polymers is directed to arrive at soluble (precursor) polymers. This allows films to be cast from solutions of these polymers. In order to produce microscopic structures as may be used in the electronic industry, these films have to be processed using conventional techniques like photolithography, etching etc. In another approach, Wrigthon et al. used patterns of gold electrodes on a substrate to form electrochemically polypyrrole or polyaniline on the gold-patterns only, leading to the production of polymer-based diodes and transistors[27]. We were interested in methods by which conducting polymers could be formed pattern-wise on a substrate via a direct one-step method.

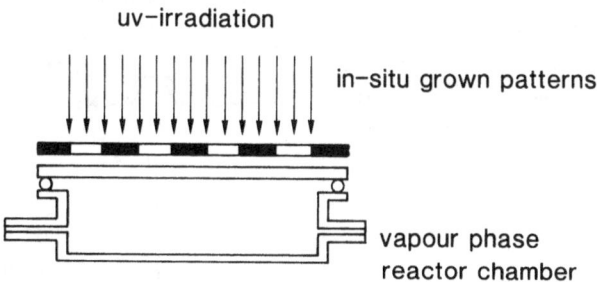

Figure 6   Reaction vessel for poly-1,2-azepine 24 film formation.

In our studies of the photochemistry of phenylazides we first thought that it could possibly provide a new route towards polyaniline. The results showed that a different polymer was formed instead; poly-1,2-azepines. We found that upon irradiation of phenylazides in the gasphase the poly-1,2-azepines are grown on the surface of the reaction vessel[26]. When the irradiation is performed through a photomask, as illustrated in figure 6, high resolution patterns are formed (figures 7 and 8). Upon doping with I_2 or AsF_5 these films (or patterns) become electrically conductive with conductivities up to 0.01 S/cm.

Figures 7 and 8 SEM photographs of poly-1,2-azepine as grown pattern-wise on fused silica using contact illumination. The bars represent 100 μm (fig.7) and 10 μm (fig.8).

Although the photochemistry of phenylazide **21** has been the subject of intensive research since its discovery by Wolff in 1912 [29], only a few contradictory notes have been made concerning the composition of the "tar" and its possible mechanism of formation[30]. In a variety of other reactions both singlet (**22**) and triplet (**23**) phenylnitrenes act as intermediate. Several reactions can be distinguished (scheme 5)[31].

Scheme 5 Phenylazide photochemistry

We have found that the primary product of the photochemistry of phenylazides is the formation of poly-1,2-azepines **24** (Scheme 6)[28]. This polymer is formed via the ring-enlarged aza-cycloheptatetraene (**25**). The structure elucidation of the polymer is based on a variety of spectroscopic data and chemical analysis. The polymerization succeeded with a series of substituted phenylazides (figure 9).

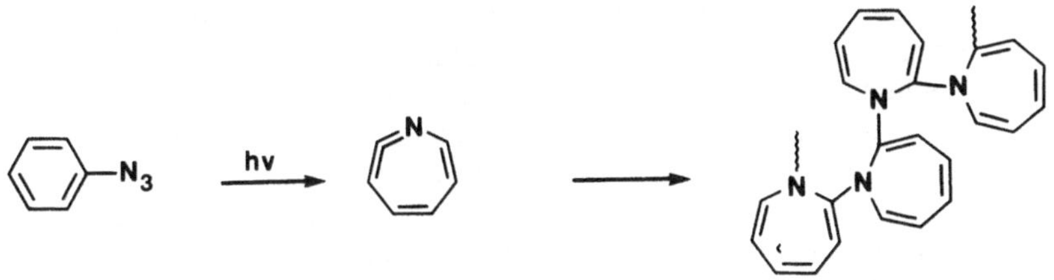

Scheme 6 Photopolymerization of phenylazide 21

The virgin poly-1,2-azepines (**24**) are very susceptible to oxidation by air, leading to charged species. When the polymers are subjected to strong oxidants like I$_2$ or AsF$_5$, conducting polymers are formed.

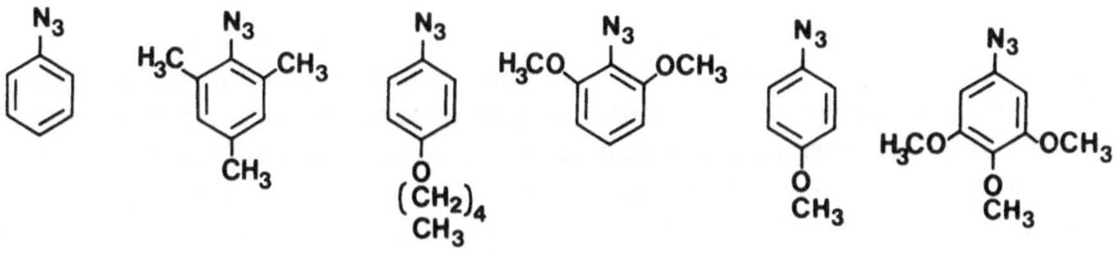

Figure 9 Monomers that are used in the photopolymerization

The formation of conducting poly-1,2-azepines suggests that the charged species formed are stabilized by conjugation or even aromaticity. The species to be expected are radical cations and dications of azepines. Delocalization of these species and 6-π electron aromaticity of the dications is evident both from theoretical studies and from experiments[32]. However, extended conjugation is excluded owing to steric hindrance between individual azepines (similar to ortho-substituted phenyls). Hence a moderate conductivity is expected and found. In order to increase the conductivity from poly-azepines, it seems necessary to synthesize 1,4-azepines. However, no design for such a synthesis is available at this time.

2.4 CONCLUSIONS AND PROSPECTS

The field of conducting polymers is still full of activity, mainly due to progress made in the synthesis of new polymers and the design of new synthetic schemes for the well-known polymers. In the first section of this paper we have given an anthology of our own recent results together with some of the landmarks in the field worldwide. Higher conductivities are reached and processability is improved. What can we expect in the near future? A further improvement in all areas of conducting polymers can be foreseen, especially since at the present time scientists from several disciplines are cooperating.

One of the reasons for an increasing interest in conducting polymers is the expectation of their use in electronic devices[33]. Several applications are announced and prototype devices have been demonstrated[34]. By far the most studied application is the polymer battery, in which polyaniline (5) and polypyrrole (2) are the important candidates for the electrodes. Other applications are based on the semi-conducting properties of the polymers; e.g. transistors for Liquid Crystal Displays based on processable polythiophenes. Another announced application is found in conducting adhesives. In the future we will probably see more of these types of applications.

3 FERROMAGNETIC POLYMERS

In 1987 the first claims on ferromagnetic properties found in organic polymers were published[35,36]. Together with the comprehensive study of ferromagnetic charge transfer complexes from ferrocene compounds [37] it represents the first experimental counterpart of the theoretical studies published since 1963. In that year, McConnell presented his proposals[38] for high spin charge-transfer complexes from the combination of a triplet and singlet ground state donor and acceptor. In those systems a parallel alignment of all spins is expected. Polymer structures with such a high spin multiplicity were proposed in 1968 by Mataga[39]. Several investigations along the lines of these proposals have been performed.

The experiments following the McConnell model were concentrated on model studies and organometallic stacks 26 (figure 10).

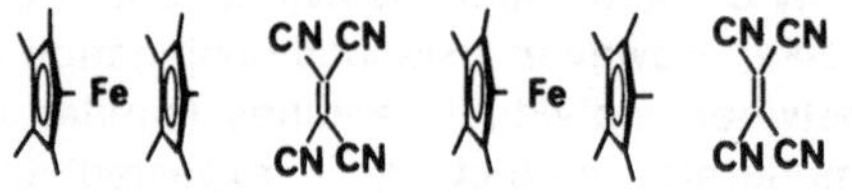

26

Figure 10

These low-molecular weight charge-transfer complexes exhibit bulk ferromagnetism below the Curie temperature of 4.8 K, as observed by Miller and Epstein[37]. A model of configuration mixing of the lowest charge-transfer excited state with the ground state was developed to understand the magnetic coupling as a function of electron configuration and direction of charge transfer. In a series of studies by Breslow et al. pure organic charge transfer complexes were made[40]. However no bulk ferromagnetism has been observed.

From hard-to-reproduce experiments on the polymer network from 1,3,5-triaminobenzene and iodine, Torrance et al. concluded that it represents the first polymer with bulk ferromagnetism[36]. Despite the possible presence of impurities the work is stimulating for further research. In an elegant study on oligomers of poly-carbenes Iwamura et al. were able to synthesize polycarbenes as given in figure 11 [41].

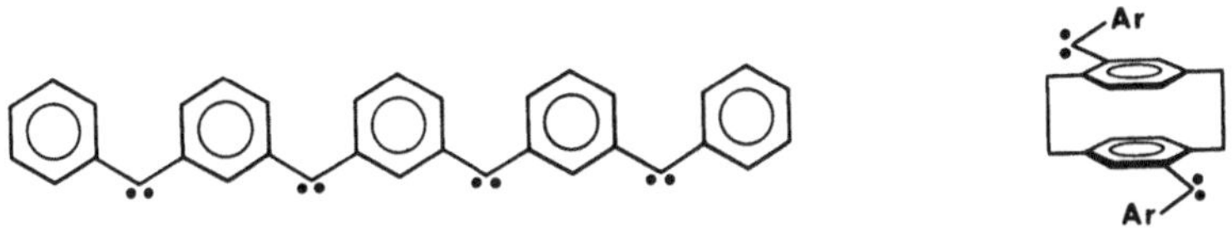

27

28

Figure 11

Although intramolecular alignment of the spins is observed in **27** and as a model for intermolecular coupling also in **28**, the bulk properties are those of an antiferromagnet.

Most successful up to now seems to be Ovchinnikov with the preparation of a polydiacetylene with dangling stable nitroxyl radicals **29** [35,42] . The structure of the polymer and its way of preparation are given in scheme 7. The solid state polymerization of **30** is claimed to be feasible by both thermal treatment and by irradiation. Selected samples of the polymer exhibit magnetization values above 1 Gauss and respond noticeably to the field of a permanent magnet. The observed Curie temperatures ranged from 150-190 K up to 310 K for the high magnetization samples. Despite the experimental evidence presented, some scepticism against the results seems justified. Several other laboratories have tried to reproduce Ovchinnikovs results, however, sofar unsuccessfully[43].

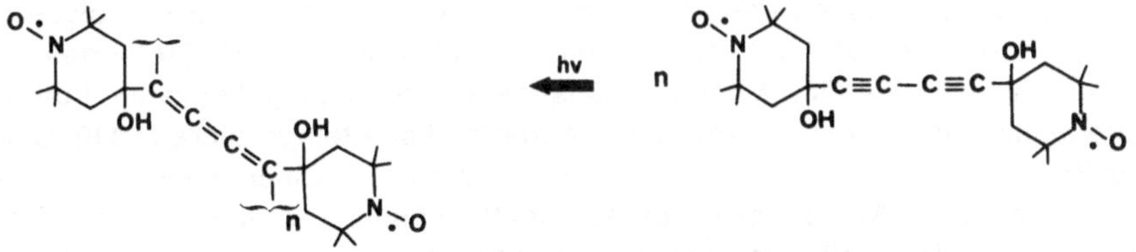

29

30

Scheme 7

A variety of other examples for possible high spin polymers possessing free nitroxyl radicals have been investigated. In most cases the spin density is low and hence polymers like **31** (figure 12) are antiferromagnetic[44].

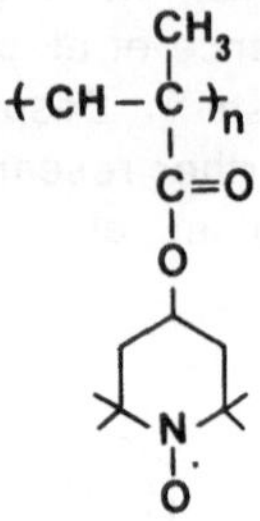

$$\left. \mathrm{+CH-\underset{\underset{\displaystyle C=O}{|}}{\overset{\displaystyle \overset{CH_3}{|}}{C}}} \right\}_n$$

(piperidine nitroxide structure with O at position 4 connected via C=O, and N–O• group)

Figure 12   **31**

3.1 APPROACH TO HIGH SPIN POLYNITROXIDE

One of the major drawbacks of the polynitroxides synthesized sofar is the limited concentration of stable radicals. In an approach to reduce the number of "useless" additional atoms, we have tried to synthesize polynitroxide **32**, one of the polymers with the highest concentration of radicals experimentally possible[45]. The synthetic plan is outlined in scheme 8.

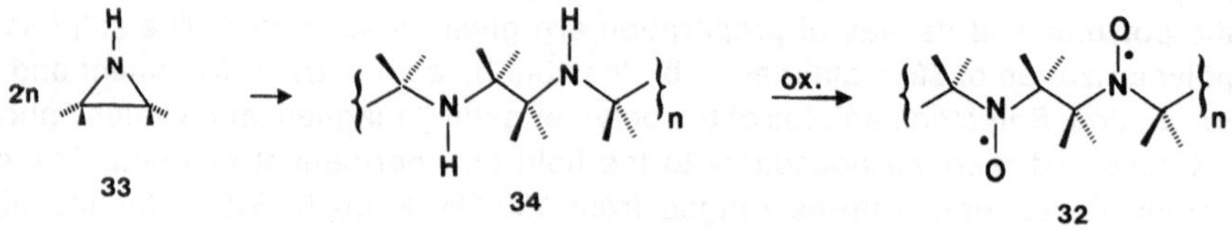

33   **34**   **32**

Scheme 8

The oxidation of a secondary amine to a nitroxyl radical is well-documented and known to proceed in high yield[46]. Furthermore, numerous papers concerning aziridine polymerization have appeared[47]. However, no reports are given on the polymerization of the most hindered derivative in the aziridine series, i.e. tetramethylaziridine (**33**). Monomer **33** is made using a slightly modified procedure published earlier[48]. Neat **33** is polymerized at a temperature of 110 ºC using BF_3 as the catalyst. After prolonged heating in an inert atmosphere, the polymer complex **34** / BF_3 precipitated from the solution. Although grafting is a well-known side-reaction in aziridine polymerization, structure elucidation of **34** (¹H- and ¹³C- NMR spectroscopy) showed that no grafting had occurred in the polymerization of **33**.

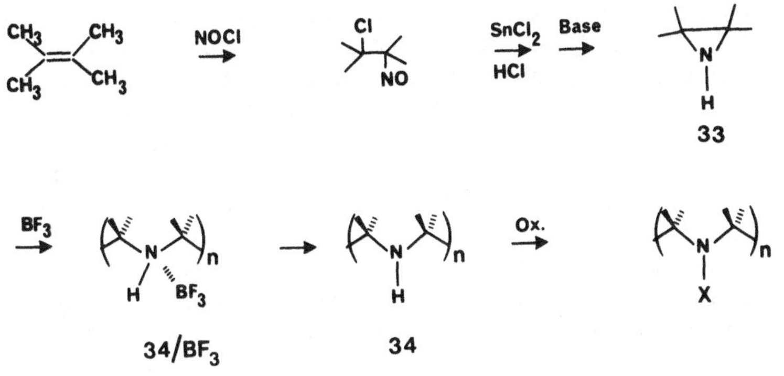

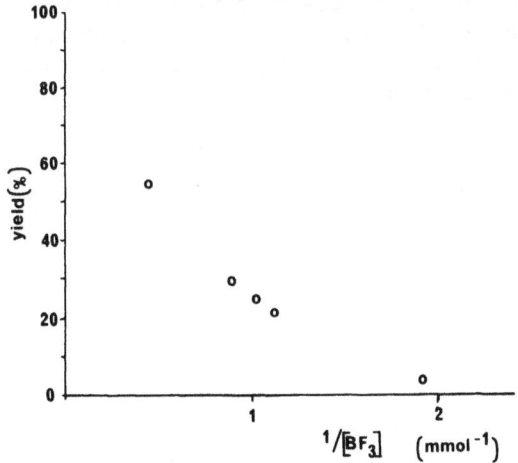

Scheme 9

The yield of the polymer proved to be strongly dependent on the ratio catalyst/monomer, as illustrated in figure 13. Furthermore, the estimated molecular weight on the basis of solubility decreased by increasing yield. The following picture emerges from the experiments: As in common aziridine polymerizations, the aziridine **33** is activated by BF_3 and polymerizes. The polyimine produced is a strongly hindered amine and has to be regarded as a very strong base, that upon formation complexes the acid catalyst BF_3, leading to an inhibition of the polymerization. At high catalyst concentrations many macromolecules can be formed, however, the complexing ratio BF_3 / (monomeric unit) will be high, and hence precipitation occurs at a low molecular weight. At a low catalyst concentration a limited number of higher molecular weight macromolecules is formed.

Figure 13 The yield of polymer 34/BF_3 as a function of catalyst concentration. The concentration of monomer 33 is in all cases 5 mmol.

Due to the strong complex formation , it proved to be difficult to obtain the free polyamine **34**. Ultimately **34** was isolated after decomplexation in hot DMSO. The oxidation of **34** to the polynitroxide **32** was performed under a variety of reaction conditions. The maximum conversion obtained up to now is low, being about 10 %. A possible explanation for this low conversion is based on the low solubility of **32** and **34** under the reaction conditions used. The characterization of the ultimate polymer is strongly limited by the presence of (para)magnetism and some inorganic impurities.

Some other approaches to synthesize polynitroxide **32**, to wit the direct polymerization of the monomeric unit tetramethylaziridin-yloxy or the polymerization of the α-nitroso radical were all unsuccessful (scheme 10)[45].

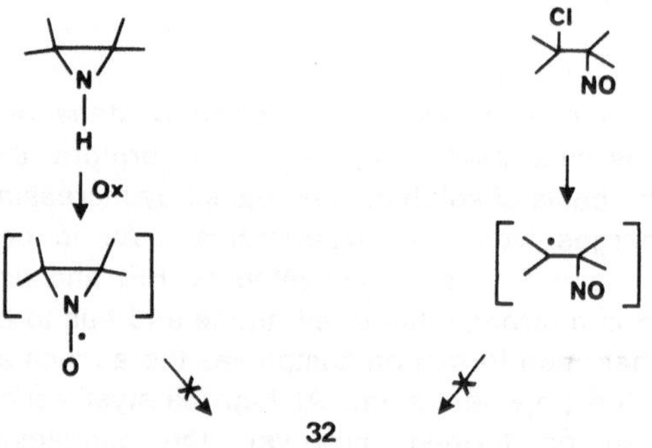

Scheme 10

Many problems are still underlying the research to polymeric ferromagnets. The work is strongly hampered by possible impurities. However, much progress has been made lately. For the time being it is still a scientific challenge to observe ferromagnetism in polymers and organics, and it is yet not worthwhile to speculate on possible applications.

4 POLYMERS WITH NONLINEAR OPTICAL PROPERTIES

Although nonlinear optical properties of materials are known for more than a century, the relevance of organics and polymers in this fascinating area of research originates from a more recent development[49,50,51]. To introduce the subject of this paragraph, the fundamental concepts of nonlinear optics and their relation to chemical structure are briefly summarized.

The dipole moment, p, of a molecule will change if the molecule is subjected to an electric field, E. Since the response is mainly linear, a power expansion can be used to describe the resulting dipole moment[52]:

$$p = p_0 + \alpha.E + \beta.EE + \gamma.EEE + ...$$

Here p, p_0, and E are vectors and α, β and γ tensors, normally referred to as polarizability, hyperpolarizability and second order hyperpolarizability, respectively. Similarly, the polarization in bulk or macroscopic media is given by:

$$P = P_0 + \chi^{(1)}.E + \chi^{(2)}.EE + \chi^{(3)}.EEE + ...,$$

where $\chi^{(1)}$, the linear susceptibility, $\chi^{(2)}$ and $\chi^{(3)}$, the second and third order susceptibilities, are the macroscopic counterparts of α, β and γ. The exact relation between molecular polarizabilities and macroscopic susceptibilities is rather complicated and depends on the positions and orientations of the molecules. As a consequence of symmetry, the second order susceptibility is zero in centrosymmetric media. For $\chi^{(3)}$- materials such a rule does not exist. The magnitudes of $\chi^{(3)}$ and $\chi^{(2)}$ depend critically on the directions of field(s) and polarization(s), and both susceptibilities can be greatly enhanced by anisotropy.

For most materials and reasonable fields, the effect of nonlinear response is only small. Therefore, only molecules with an exceptionally high hyperpolarizability β are of interest when applications of nonlinear optics are concerned. The recent increase in interest in second order nonlinearities is motivated by two major applications, to wit light modulators by electro-optical effects (Pockels-effect) and the frequency doubling of laser light. The electro-optical effect is based on the property of materials with a high value of $\chi^{(2)}$ whereby the refractive index is markedly dependent on the applied electric field. The frequency doubling of light by means of nonlinear optical materials (Second Harmonic Generation, SHG) is another manifestation of second-order nonlinearities. For example, light with low energy (e.g. 1.064 μm of a Nd:YAG laser) can be doubled in frequency to the second harmonic (532 nm).

Proposed applications of third order nonlinearities are very intriguing and include all-optical computing and communication[53]. However, the proposed materials and technologies are still far from adequate. Third order nonlinearities are only large in extended π-electron conjugated systems, and therefore the high $\chi^{(3)}$-polymers are of the same class as those in figure 1. The polymers of choice are polydiacetylenes, as these can be prepared well-oriented in a topochemical polymerization. The solid-state monomer can either be in the form of a crystal or of a Langmuir-Blodgett thin film[54]. Recently a variety of other polymers are the subject of investigation and oriented polyacetylene seems to be comparable to polydiacetylenes[55]. A major drawback of possible applications of $\chi^{(3)}$-polymers is stressed by Stegeman[56], viz. the low absorption demanded for application is hard to combine with the high nonlinearities found today. In this review we will concentrate on the $\chi^{(2)}$- materials.

As stated above, non-zero β- and $\chi^{(2)}$- values can only be found in non-centrosymmetric molecules and media. Even a molecule with an asymmetric charge distribution in a centrosymmetric crystal or in an isotropic liquid or amorphous polymer matrix will exhibit a vanishing small value of $\chi^{(2)}$. Only surface effects, due to the intrinsic anisotropy of surfaces will lead to a minor nonlinear response. The value of β is strongly dependent on the molecular structure. Typically, high values for β are found in molecules in which low-lying charge-transfer transitions are present[46]. In general such molecules possess an electron donor at one side of the molecule and an electron acceptor at the other side, separated by a π-conjugated system. Enormous enhancement of β is found by extending the π-conjugation. Examples are given in figure 14, whilst many others are subject of research.

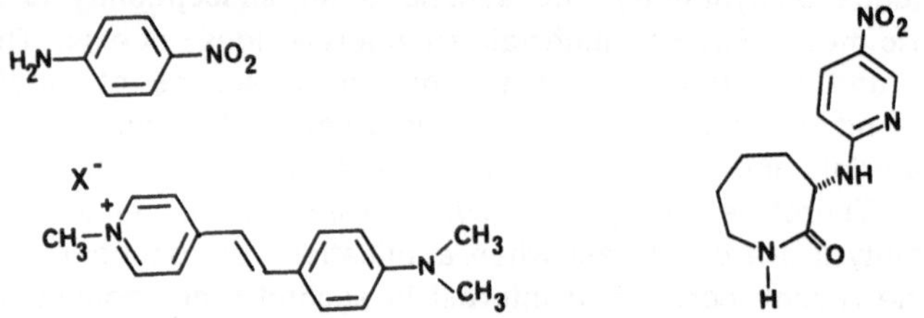

Figure 14

β -Values can be measured by several techniques, including electric field induced second harmonic generation and solvatochromic measurements of absorption and fluorescence[57]. Moreover, quantum mechanical calculations can be used to estimate these nonlinearities[58]. For a more detailed description of the physical backgrounds of nonlinear optics we like to refer to excellent reviews and books[49,50,51].

In order to transform a molecule with a high β value into a material without a centrosymmetric structure three main techniques of material engineering are available:

* Crystal engineering [59] .

The structures that exhibit high β-values possess in many cases a large ground-state dipole moment. As a result of the latter the electrostatic interaction between the molecules is large too, leading to a preference for antiparallel order and, hence, to $\chi^{(2)} = 0$. One way to achieve non-centrosymmetric crystals is obtained by introducing asymmetry in the molecule. Optically active molecules are of special interest since nature does not permit centrosymmetry in crystals of these molecules.

* Langmuir-Blodgett thin film technology [60] .

Amphiphilic molecules with a polar head and an apolar chain form highly ordered monomolecular films at the solvent-air interface. These so-called Langmuir-Blodgett films can be deposited on a substrate, retaining their high ordering. Stable polymer Langmuir-Blodgett films are formed upon topochemical polymerization of amphiphiles provided with polymerizable functionalities like double bonds or diacetylenes. High surface nonlinearities are observed for these thin films when they are built up in such a manner that the amphiphiles with the charge-transfer groups are deposited in a polar alignment.

* The poling of dipoles in a polymer

This subject will be discussed in the next paragraph.

4.1 POLING OF DIPOLES IN A POLYMER

Useful optical polymers with high second order susceptibilities should be both completely transparent. Hence semicrystalline and polycrystalline samples as well as incompatible blends have to be avoided, while anisotropy by a large number of oriented β -molecules or fragments is demanded. The latter can be achieved by applying an external electric field to the polymer. The sequence as used by many investigators and introduced some 10 years ago at our laboratories is outlined in figure 15 for a dipole dissolved in PMMA, the pre-eminent optical polymer[61,62].

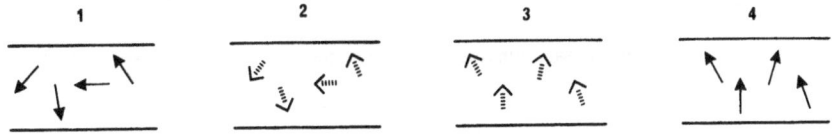

Figure 15 The poling sequence. 1) Glassy PMMA at room temperature with a frozen-in random orientation of the dipoles. 2) Heating above Tg yields a rubbery state with a random orientation and dipoles that are free to rotate. 3) An E-field applied on the PMMA at high T yields a rubbery state with aligned dipoles. 4) Cooling to room tempearture followed by removing the E-field yields glassy PMMA with aligned (frozen-in) dipoles.

A PMMA sheet with a thickness between 1-100 μm is heated just above its Tg to introduce molecular mobility. At that temperature, an external dc-field is applied to the polymer film and as a consequence the dipoles are oriented. The degree of orientation is dependent on the dipole moment of the dipole, the field-strength and the temperature of the experiment[63]. This non-equilibrium state is frozen-in by lowering the temperature, while the field remains. After reaching room temperature the field is removed. The electric field can be applied by using two electrodes sandwiching the polymer film. One of the electrodes can be replaced by a Corona discharge that is applied to the polymer film. With the latter higher fields can be obtained and hence a better alignment.

The degree of orientation obtained can be measured in several ways. Obviously, second harmonic generation is one of them, however interpretation of the data in a quantitative manner is subject to a variety of assumptions and measurement inaccuracies. In many cases, SHG is only given in arbitrary units or related to reference samples, and is strongly dependent on the wavelengths used due to resonance effects. Another method uses the changes in the absorption spectrum of the polymer sheet upon poling due to the alignment of the molecular dipoles and, hence, of the transition moments (electrochromism)[61].

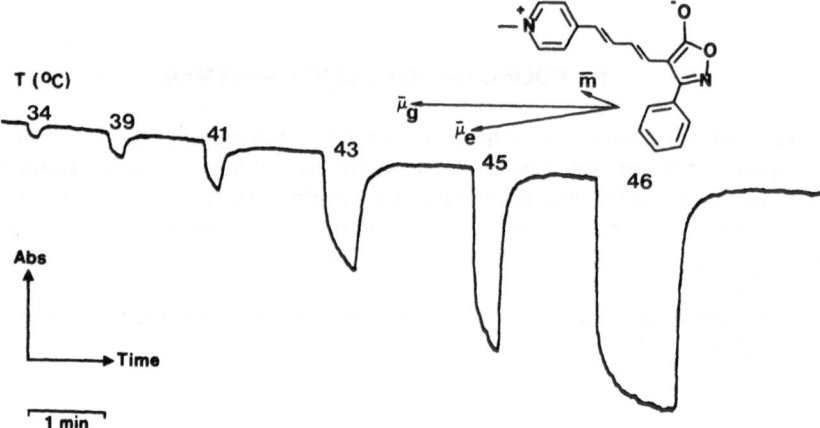

Figure 16   Electrochromic measurements at different temperatures of a dye in a Epikote polyepoxide.

In figure 16, the decrease in absorption of a dye with a high dipole moment dissolved in an epoxide polymer network is given. The alignment is strongly dependent on the temperature used. Only at temperatures reaching Tg a large decrease in absorption is found. The quadratic dependence of applied field and decrease in absorption is illustrated in figure 17 for another dye poled in PMMA at T = 125 °C.

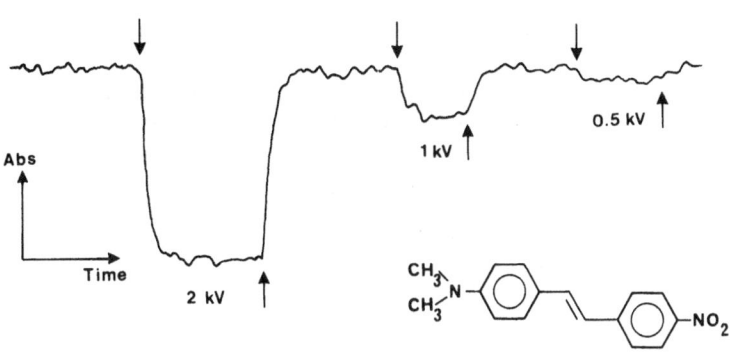

Figure 17 Electrochromic measurements of DANS (4-dimethylamino-4'-nitrostilbene) in PMMA at 125 °C. The change in absorption is measured at a wavelength of 435 nm.

Several issues are under investigation with this alignment strategy in mind. First, how can we increase the number of dipoles in the polymer matrix? In most polymers the maximum amount of dipoles that can be dissolved is in the range of 1-5 ww%, only. Polymers with pendent dipoles attached covalently to the polymer backbone allow much larger concentrations. In figure 18, a few examples of monomers from our laboratories are given[64]. Another approach currently under investigation is based on liquid-crystalline materials and their use in nonlinear optics[64].

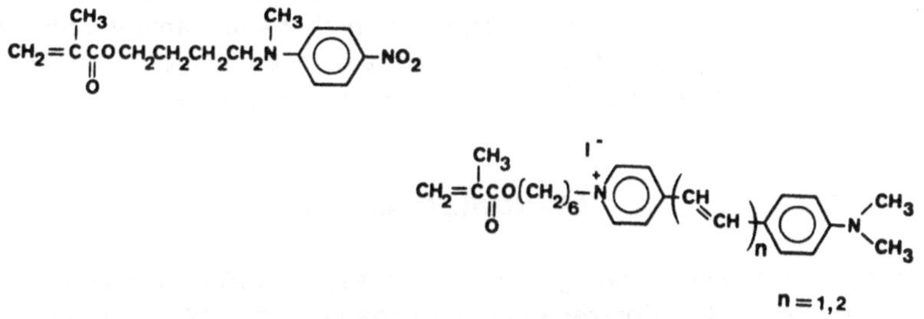

Figure 18

Second, what is the stability of the aligned, non-equilibrium state. As is known from a variety of studies of the molecular mobility of polymers and polymer networks, it is reasonable to assume, that even far below Tg most polymers have some motional freedom. This will lead to a relaxation of the dipoles. The general feeling is, however, that sufficiently stable, highly nonlinear ($\chi^2 > 50$ pm/V, compared to 6 pm/V for $LiNbO_3$) polymers can be made.

4.2 APPLICATIONS OF POLED POLYMERS AND INTEGRATED OPTICS

Being able to produce poled polymers with stable second order nonlinearities, it is possible to construct prototype devices in which the polymer is the active component[65]. As mentioned above major applications are foreseen in light modulation and frequency-doubled diode lasers. Despite the high nonlinearities obtained, fruitful use of these nonlinearities is only possible when an appropriate interactionlength is used, hence waveguide structures are necessary. An example is given in figure 19.

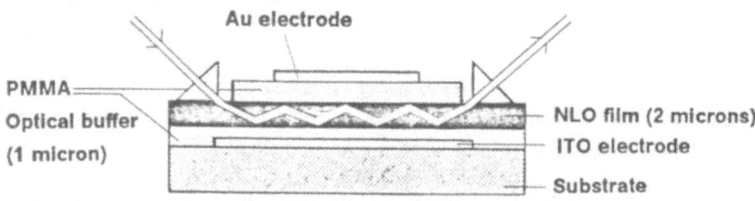

Figure 19 A substrate is covered with a thin (ITO) electrode and an optical buffer is spin-coated on top. PMMA, polyepoxides and silicon rubbers are useful buffers with low refractive index. The nonlinear optical polymer is followed by another optical buffer and a topelectrode. The light is coupled into the waveguide using prisms. The electrodes can be used for both the poling experiment as well as to modulate the refractive index of the polymer.

For frequency doubling, some additional demands should be fulfilled, viz. phase matching and a very low absorption at both the fundamental and second harmonic wavelength. Several proposals have been made to obtain phase matching in polymer waveguides and work along these lines is in progress.

ACKNOWLEDGEMENT

The authors like to acknowledge L.W. van Horssen, F.C.B.M. van Vroonhoven, I. Rotte, T.R. Vries, K.L. Pouwer (Philips Research Labs), W. ten Hoeve and H. Wynberg (University of Groningen) for fruitful discussions and experimental support.

REFERENCES

1. "Polymers in Electronics" Ed. T. Davidson, ACS Symp. Series 242, American Chemical Society, Wasingthon 1984.

2. J.G. Kloosterboer, Adv. Pol. Sci.,84, 1 (1988); F.A. Vollenbroek, E.J. Spiertz, Adv. Pol. Sci., 84, 85 (1988); "Polymers for High Technology" ACS Symposium Series 346, Eds M.J. Bowden, S.R. Turner, American Chemical Society, Washington DC, 1987.

3. A.F. Diaz, J.F. Rubinson, H.B. Mark, Adv. Pol. Sci., 84, 113 (1988); "Polymers for Electronics and Photonics" ACS Symposium Series

4. H. Shirakawa, E.J. Louis, A.G. MacDiarmid, C.K. Chiang, A.J. Heeger, J. Chem. Soc., Chem. Commun. 1977, 578; C.K. Chiang, C.R. Fincher, Y.W. Park, A.J. Heeger, H. Shirakawa, E.J. Louis, S.C. Gau, A.G. MacDiarmid, Phys. Rev. Lett 39, 1098 (1977).

5. A.J. Heeger in "Handbook of Conducting Polymers", Ed. T.A. Skotheim, J.C.W. Chien "Polyacetylene: Chemistry, Physics, and Material Science", Academic Press, New York, 1984.

6. H. Naarman, Synth. Met. 17, 2233 (1987); H. Naarman, N. Theophilou, Synth. Met. 22, 1 (1987).

7. R.R. Chance, D.S. Boudreaux, J.L. Bredas, R. Silbey in "Handbook of Conducting Polymers" Ed. T.A. Skotsheim, Marcel Dekkers, New York, 1986, p.825.

8. A.O. Patil, A.J. Heeger, F. Wudl, Chem. Rev. 88, 183 (1988).

9. "Handbook of Conducting Polymers" Ed. T.A. Skotheim, Marcel Dekker, New York, 1986, Vol 1 and 2.

10. F. Wudl, M. Kobayashi, A.J. Heeger, J. Org. Chem., 49, 3382 (1984); N. Colaneri, M. Kobayashi, A.J. Heeger, F. Wudl, Synth. Met., 14, 45 (1986); J.L Bredas, A.J. Heeger, F. Wudl, J. Chem. Phys., 85, 4673 (1986); S.A. Jenekhe, Nature (London), 345, 322 (1986).

11. W.J. Feast in "Handbook of Conducting Polymers" Ed. T.A. Skotheim, Marcel Dekker, New York, 1986, p1.

12. J.H. Edwards, W.J. Feast, Polymer 21, 595 (1980);

13. M. Kanbe, M. Okawara, J. Pol. Sci. A1, 6, 1058 (1968); D.R. Gagnon, J.O. Capistron, F.E. Karasz, R.W. Lenz, Polym. Bullet., 12, 293 (1984).

14. K.L. Pouwer, T.R. Vries, E.E. Havinga, E.W. Meijer and H. Wynberg, J. Chem. Soc. Chem. Commun., in press.

15. M. Sato, S. Tanaka and K. Kaeriyama, J. Chem. Soc., Chem. Commun, 1985, 713; A.F. Diaz and J. Bargon in 'Handbook of Conducting Polymers' ed. T.A. Skotheim, Marcel Dekker, New York 1986, p81.

16. V. Bocchi and G.P. Gardini, J. Chem Coc., Chem. Commun., 1986, 148.

17. K.Y. Jen, G.G. Miller and R.L. Elsenbaumer, J. Chem. Soc., Chem. Commun, 1986, 1349; M. Kobayashi, J. Chen, T.-C. Chung, F. Moraes, A.J. Heeger and F. Wudl, Synthetic Metals 1984, 9, 77.

18. E. Campaigne and W.O. Foye, J. Org. Chem., 1952, 17, 1405.

19. J.E. Frommer, Acc. Chem. Rev. 19, 2 (1986); J.E. Frommer, R.L. Elsenbaumer, R.R. Change, Org. Coat. Appl. Polym. Sci. Proc. 48, 552 (1983); S.A. Jenekhe, S.T. Wellinghoff, J.F. Reed, Mol. Cryst. Liq. Cryst. 105, 175 (1984).

20. M. Sato, S. Tanaka, K. Kaeriyama, J. Chem. Soc., Chem., Commun. 1986, 873.

21. K.Y. Jen, G.G. Miller, R.J. Elsenbaumer, J. Chem. Soc., Chem. Commun., 1986, 1346; R.L. Elsenbaumer, K.Y. Jen, R. Oboodi, Synth. Met. 15, 169 (1986).

22. E.E. Havinga, L.W. van Horssen, Makrom. Chemie, in press.

23. E.E. Havinga, L.W. van Horssen, W. ten Hoeve, H. Wynberg, E.W. Meijer, Polymer., Bull. 18, 277 (1987).

24. A.O. Patil, Y. Ikenoue, F. Wudl, A.J. Heeger, J. Am. Chem. Soc. 109, 1858 (1987); A.O. Patil, Y, Ikenoue, N. Basescu, N. Colaneri, J. Chen, F. Wudl, A.J. Heeger, Synth. Met. 20, 151 (1987); Y. Ikenoue, J. Chiang, A.O. Patil, F. Wudl, A.J. Heeger, J. Amer. Chem. Soc., 110, 2983 (1988).

25. E.E. Havinga, W. ten Hoeve, E.W. Meijer, H. Wynberg, manuscript in preparation

26. G.B. Street in "Handbook of Conducting Polymers" Ed. T.A. Skotheim, Marcel Dekker, New York, 1986, 278; M.J. Nowak, S.D.D.V. Rughooputh, S. Hotta, A.J. Heeger, Macromolecules, 20, 965 (1987).

27. E.T. Turner Jones, O. M. Chyan, M.S. Wrighton, J. Am. Chem. Soc., 109, 5526 (1987); S.Chao, M.S. Wrighton, J. Am. Chem. Soc., 109, 6627 (1987); G.P. Kittleson, M.S. Wrighton, J. Mol. Electr., 2, 23 (1986).

28. E.W. Meijer, S. Nijhuis, F.C.B.M. van Vroonhoven, J. Am. Chem. Soc., in press.

29. Azides and Nitrenes, Reactivity and Utility, Ed. E.F.V. Scriven, Academic Press, New. York, 1984.

30. L. Horner, A. Christmann, A. Cross, Chem. Ber. 96, 399 (1963); A. Reiser, L.J. Leyshon, J. Am. Chem. Soc. 93, 4051 (1971); R.A. Abramovitch, S.R. Challand, E.F.V. Scriven, J. Am. Chem. Soc., 94, 1374 (1972); E. Leyva, M.J.T. Young, M.S. Platz, J. Am. Chem. Soc., 108, 8307 (1986).

31. W.H. Waddell, C.L. Go, J. Am. Chem. Soc. 104, 5804 (1982); E. Leyva, M.S. Platz, G. Persy, J. Wirz, J. Am. Chem. Soc., 108, 3783 (1986); A.K. Schrock, G.B. Schuster, J. Am. Chem. Soc., 106, 5228 (1984); O.L. Chapman, R.S. Sheridan, J.P. Le Roux, Recl. Trav. Chim. Pays-Bas, 98, 334 (1979); C.J. Shields, D.R. Chrisope, G.B. Schuster, A.J. Dixon, M. Poliakoff, J.J. Turner, J. Am. Chem. Soc., 109, 4723 (1987).

32. K.L. Rinehart Jr., A.C. Buchholz, G.E. van Lear, J. Am. Chem. Soc., 90, 1073 (1968); M.J.S. Dewar, N. Trinajstic, Tetrahedron 26, 4269 (1970); E.G.J. Staring, E.W. Meijer, manuscript in preparation.

33. J.R. Ellis in "Handbook of Conducting Polymers" Ed. T.A. Skotheim, Marcel Dekker, New York, 1986, p.489.

34. A.G. MacDiarmid, R.B. Kaner in "Handbook of Conducting Polymers" Ed. T.A. Skotheim, Marcel Dekker, New York, 1986, p.689.

35. Yu. V. Korshak, T.V. Medvedeva. A.A. Ovchinnikov, V.N. Spector, Nature 326, 370 (1987); Yu. V. Korhak, A.A. Ovchinnikov, A.M. Shapiro, T.V. Medvedeva, V.N. Spector, J.E.T.P. Letters 43, 399 (1986).

36. J.B. Torrance, S. Oostra, A. Nazzal, Synthetic Metals, 19, 709 (1987).

37. J.S. Millar, J.C. Calabrese, A.J. Epstein, R.W. Bigelow, J.H. Zhiang, W.M. Reiff, J. Chem. Soc. Chem. Commun. 1986, 1026; J.S. Miller, A.J. Epstein, W.M. Reiff, Science, 240, 40 (1988).

38. H.M. McConnell, J. Chem. Phys. 39, 1910 (1963)

39. N. Mataga, Theoret. Chim. Acta (Berl.) 10, 372 (1968).

40. R.S. Breslow, P. Maslak, J.S. Thomaides, J. Am. Chem. Soc. 106, 6433 (1984); R. Breslow, Pure and Appl. Chem. 54, 927 (1982); R. Breslow, Mol. Cryst. Liq. Cryst. 125, 261 (1985); R. Breslow, B. Jaun, R.Q. Klutz, C.-Z. Xia, Tetrahedron, 38, 863 (1982); T.J. LePage, R. Breslow, J. Am. Chem. Soc. 109, 6412 (1987).

41. H. Iwamura, Pure & Appl. Chem. 58, 187 (1986); H. Iwamura, T. Sugawara, K. Itoh, T. Takeji, Mol. Cryst. Liq. Cryst. 125, 251 (1985); Y. Teki, T. Takui, K. Itoh, H. Iwamura, K. Kobayashi, J. Am. Chem. Soc. 105, 3722 (1983); T. Sugawara, S. Bandow, K. Kimura, H. Iwamura, K. Itoh, J. Am. Chem. Soc. 108, 368 (1986); Y. Teki, T. Takui, K.Itoh, H. Iwamura, K. Kobayashi, J. Am. Chem. Soc. 108, 2147 (1986).

42. A.A. Ovchinnikov, Theoret. Chim. Acta. (Berl.), 47, 247 (1978).

43. J.S. Millar, D.T. Glatzhofer, J.C. Calabrese, A.J. Epstein, J. Chem. Soc., Chem. Commun.

44. T. Kurosaki, K.W. Lee, M. Okawara, J. Pol. Sci. Pol. Chem. Ed. 10, 3295 (1972); T. Kurosaki, O. Takahashi, M. Okawara, J. Pol. Sci. Pol. Chem. Ed. 12, 1407 (1974); M. Kamachi, M. Tamaki, Y. Morishima, S. Nozakura, W. Mori, M. Kishita, Polymer Journal, 14, 363 (1982); M.Kamachi, H. Enomoto, M. Shibasaka, W. Mori, M. Kishita, Polymer Journal 18, 439 (1986); R.Seidemann, L. Dulog, Makromol. Chem. 1987, 2545 (1986); S. Nozakura, M. Kamachi, Makromol. Chem. Suppl. 12, 255 (1985);

45. E.W. Meijer, F.C.B.M. van Vroonhoven, unpublished results.

46. E.G. Rozantsev, V.D. Sholle, Synthesis 1970, 190; J.F. Keana, Chemical Rev. 78, 37 (1978).

47. E.J. Goethals in "Ring Opening Polymerization" ed K.J. Ivin and T. Saegusa , Elsevier, Barking 1984, p.715.

48. G.L. Closs, S.S. Brois, J. Am. Chem. Soc., 82, 6068 (1960).

49. D.J. Williams, Angew. Chem. Int. Ed. Engl. 23, 490 (1984).

50. "Nonlinear Optical Properties of Organic Molecules and Crystals" Eds. D.S. Chemla, J. Zyss, Academic Press, New York 1987, Vols 1 and 2.

51. "Nonlinear Optical Properties of Organics and Polymeric Materials" Ed. D.J. Williams, ACS Symp. Series Vol. 233, American Chemical Society, Washington, 1983.

52. Y.R. Shen "The Principles of Nonlinear Optics", Wiley, New York, 1984.

53. "Nonlinear Optical Properties of Polymers" MRS Symposium Proceedings Vol.109 Eds A.J. Heeger, J. Orenstein, D.R. Ulrich, Material Research Society, Pittsburgh, 1988.

54. M. Thakur, B. Verbeek, G.C. Chi, K.J. O'Brian in ref 53, p.41.

55. M. Sinclair, D. Moses, K. Akagi, A.J. Heeger in ref 53, p.205.

56. G.I. Stegeman, R. Zanoni, C.T. Sealon in ref 53, p.53.

57. C.G. Bethea, J. Chem. Phys. 68, 1312 (1978); J.L. Oudar, J. Chem. Phys., 67, 446 (1977); B.F. Levine, C.G. Bethea, J. Chem. Phys., 63, 2666 (1975); G.R. Meredith, Rev. Sci. Instrum., 53, 48 91982).

58. D. Li, M.A. Ratner, T.J. Marks, J. Am. Chem. Soc., 110, 1707 (1988); J. Zyss, G. Berthier, J. Chem. Phys., 77, 3635 (1982).

59. J.F. Nicoud, R.J. Twieg in ref 50, p.227.

60. I.R. Girling, N.A. Cade, P.V. Kolinsky, J.D. Earls, G.H. Cross, I.R. Peterson, Thin Solid Films, 132, 101 (1985); N. Carr, M.J. Goodwin, A.M. McRoberts, G.W. Gray, R. Marsden, R.M. Scrowston, Makromol. Chem., Rapid. Commun., 8, 487 (1987).

61. E.E. Havinga, P. van Pelt, Ber. Bunsenges. Phys. Chem. 83, 816 (1979); E.E. Havinga, P. van Pelt, Mol. Cryst. Liq. Cryst.,52, 145 (1979).

62. C. Ye, T.J. Marks, J. Young, G.K. Wong, Macromolecules, 20, 2322 (1987); K.D. Singer, J.E. Sohn, S.J. Lalama, Appl. Phys. Lett., 49, 248 91986); G.R. Meredith, J.G. van Dusen, D.J. Williams, Macromolecules, 15, 1385 (1982); H.L. Hampsch, J. Yang, G.K. Wong, J.M. Torkelson, Macromolecules, 21, 526 (1988).

63. E.W. Meijer, S.Nijhuis, unpublished results

64. J.I. Thackara, G.F. Lipscomb, M.A. Stiller, A.J. Ticknor, R. Lytel, Appl. Phys. Lett., 52, 1031 (1988); J. Zyss, J. Molec. Electr., 1, 25 (1985).

ENGINEERING PLASTICS

**An analysis of the status and trends on the
commercial and technological aspects.
J. BUSSINK, GE Plastics Europe / TU Eindhoven.**

Since there is no clear definition of what engineering plastics really are, it is necessary to define what in this analysis is understood by this term. Generally engineering plastics constitute a group of plastics which distinguish themselves from the so called commodity plastics by primarily demonstrating a greatly improved mechanical behaviour over a wide range of temperatures, load and time. Especially the long term applicability at temperatures over 90°C up to at least 150°C is considered an important criteria. As such all the "high temperature resistant" plastics are being regarded as engineering plastics.

The total group of ABS plastics, having a just about 90°C capability and which are sometimes being mentioned as "transition plastics" will also be regarded as belonging to the engineering plastics group. Special glassfibre reinforced PP grades which certainly do possess a great deal of the required mechanical properties are not being included in this survey. See the following scheme.

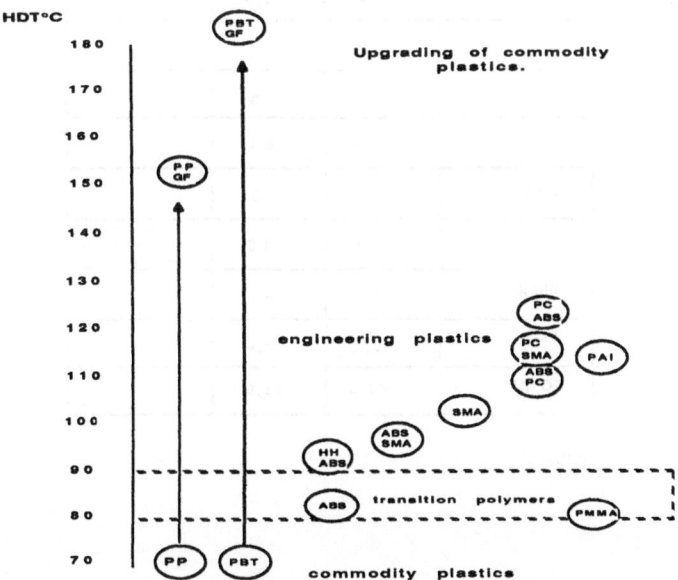

WORLD DEMAND ENGINEERING PLASTICS IN 1000 TONS

	1987	1992 estimate	
ABS	1.600	1.800	
PA	650	750	
PC	400	570	
POM	240	300	
PPE (mod.)	190	250	
PBT	100	200	
mould. PET	50	80	* extreme growth packaging expect·
REST	30	50	
TOTAL	3.260	4.000	

PMMA	550	600	not regarded as an engineering plastic.

table 1a.

'REST' ENGINEERING PLASTICS WORLD DEMAND In 1000 tons

	1987	1992 estimated	
amorph PA	1,5	3	new growth expected or its barrier prop.
PSU	9	1,5	
PES	3,5	5	
PEI	1,5	3	
PAI	0,5	0,6	
PAR	2	3	
PPS	11	15	new growth expected improved synthesis improved properties
PEEK	0,4	0,8	
PEK	0,1	0,2	
TOTAL	29,5	45,6	

table 1b.

By listing the actual sales data and the expected 1992 data in table 1a, we can draw the following conclusions:

- Estimating the annual total world sales of thermoplastics today at 62 MM ton, engineering plastics only have a 5,4% share of the total volume.
- Thus in spite of a much stronger average annual total growth (8%) over the commodity plastics (3%), the relative importance in volume will increase very slowly.
- Especially polycarbonate (PC) is demonstrating an excessive annual growth rate of 15%.
- The developments of blends having engineering plastics property- profiles is very important.
- 95% of the total volume of the engineering plastics market is being covered by only a few basic polymers.
- Volume wise therefore, engineering plastics do not form an impressive group. Value- and profitwise however they are much more important. Their average selling price can be estimated as three times that of commodity plastics and especially their profit margins have, particularly over the difficult years 1974-1984, been much better.

In spite of the intrinsically higher prices, a very strong volume growth has been realised. This due to a number of technological reasons, being:
- The evolution of new markets, like the automotive industry replacing metal parts for all kind of reasons. The developments of the compact disc, twinwall-sheet, computer and data-processing equipment etc.
- The ability to serve these markets with specifically application-dedicated products, frequently realised by applying impact modification and blend technology.

Especially the automotive industry plays a very important role. The actual importance of this industry for the engineering plastics industry can be illustrated by a few numbers:
- Average plastic utilisation in the average European car amounts to 8 weight percent.
- Every percent increase accounts for over a 100 M tons market increase.
- To day in Europe at least 30% of all engineering plastic sales are coming from automotive applications.

Since it is to be expected that, especially for the more mechanically demanding applications of plastics in cars, like side panels, doors, roof and bonnet, even more engineering plastics will be used; many developments are aimed at this market.
Based upon these observations, over the past years a number of mainly commodity plastic producers decided to participate in this very interesting engineering plastics market.
This proved to be possible, because:
- Most basic patent positions have expired.
- The market size was developed to an extent that would allow to build an economic right size facility.
- The required know-how can be purchased (Japan), or is, with todays general available knowledge, relatively easy to acquire.

The two most important basic polymers, PC and PPE (poly 2,6 dimethylpara-phenylene ether) were a natural first choice. A survey of the resulting activities is given in table 2.

	PC-PRODUCERS		modified PPE-PRODUCERS	
	1984	1992	1984	1992
W.Europe	B a y e r GE-P l a s t i c s E n i c h e m	Dow DSM/Idemitsu	G E - P l a s t i c s	BASF HULS Enichem
U.S.A.	GE - P l a s t i c s M o b a y	Dow	G E - P l a s t i c s	Borg-Warner
Japan	T e j i n Mitsubishi Gas Chem Mitsubishi Chemical	Idemitsu GE-Plastics	G E - P l a s t i c s A s a h i - D o w	
Total capacity	320.000	820.000	200.000	300.000

table 2

A special case is **PPS**.

This polymer, originally only produced by Phillips suffered from extreme brittleness and the total sales volume amounts only to 11 M tons/year. Major improvements in the synthesis now seem to be possible, providing better mechanical and processing properties. This created a sudden interest of many companies to participate in the PPS development, as can be seen from table 3. The product is especially of todays interest for its high continuous use temperature (200°C), flame resistance, chemical resistance, dimensional stability, low melt viscosity (composites) and low cost monomer.

PPS Case	
1984	1992
PHILLIPS	PHILLIPS BAYER GE. PLASTICS RHONE POULENC CIBA-GEIGY TOYA SODA etc.

Impact modification and **blend technology** are very important tools for designing of plastics with specific end-uses. Each polymer provides a certain set of properties. The actual application requirements can frequently not be fulfilled by a single polymer-based product. In those cases a very special new, probably difficult to produce (co, block) polymer should be designed or we combine known properties of different polymers in a blend system. We distinguish 4 principal different possibilities, and examples of these are very well known.
See chart of blend possibilities.

BLEND POSSIBILITIES

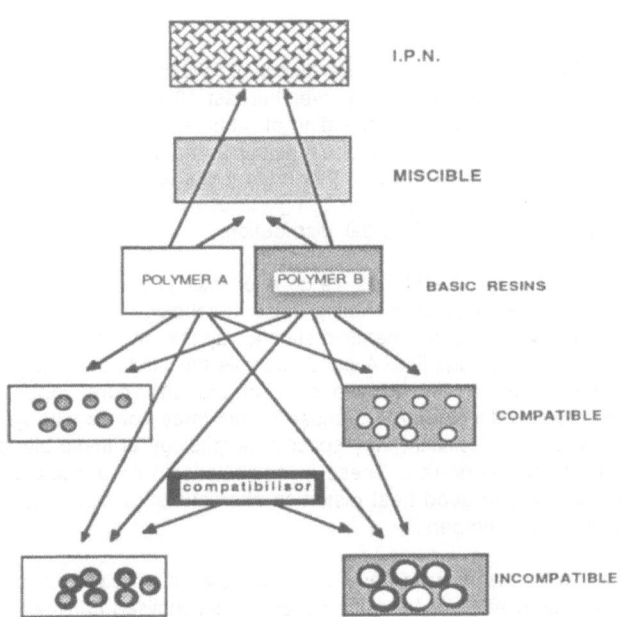

To illustrate the obtained properties combinations for a few well known blends table 4 is presented.

Polymer 1	Blend properties		Polymer 2
PPE	heat distortion toughness flame retardancy	processibility impact cost	HIPS
PPE	heat distotrtion toughness flame retardancy dimensional stabilty	processibility chemical resistance crystallinity	PA
PC	heat distortion toughness flame retardancy	impact cost chemical resistance	ABS
PC	heat distortion toughness	chemical resistance crystallinity	PBT

table 4.

The sophisticated application of the mechanistic and thermodynamic principles which have been worked out extensively over the last 10 years is really the basis of these technologies. From simple compounding of rubbers in hard polymers we have come to actual control over the mechanical behaviour as required by the foreseen application. For single polymer based systems this often leads to a specific combination of impact modifying additions. A simple case being the use of a combination of small and large rubber particles, socalled Binodal distribution.

For multiphase compatible blends we do encounter phase specific impact and modulus modification. A new way of controling those morphology features is found in tuning the surface tensions of the components in such a way that only the phase to be modified can wet the modifier to be applied. A good example for this is the impact modified PBT/PC blend called XENOY. The acrylic shell of the shell/core impact modifier will be completely surrounded by the PC-phase in this three component system by interfacial tension forces. The resulting, very specific morphology of these blends is the basic for its unique impact properties. These in combination with a good chemical resistance against gasoline and good heat distortion characteristics make the material suited for application in car bumpers.

The other well-known way to have phase selected modification exists in using grafting techniques like in ABS, or by using specific functionalised polymeric additives.

The possibilities for making useful blends based upon incompatible polymers are so far limited and exist first of all of using a compatibilising third block-copolymer, which preferrably is formed in situ during the compounding phase. In order to do this, at least one of the base polymers has to be functionalised. Functionalised polymers completely compatible with one of the base polymers can also be used for this technology.

An other way can be found in using a third polymer which is completely miscible or very compatible with both base polymers. In this way we can obtain a, in some instances three component single phase matrix, as well as a two phase compatible one. This approach could be called solubilisation. Examples of all these possibilities can be found in recent literature.

NEW POLYMERS.

In spite of the fact that for years it has been stated in many publications that the development and production of new polymers would be very rare, we can see from the following historical table that especially over the last ten years a great number of new engineering polymers has been introduced. Their molecular structure has been designed to produce the required high heat resistance, oxydative stability and low flammability. Consequently they contain mainly aromatic building units. In those cases where crystallinity is desired to increase chemical resistance the most simple and regular chain elements had to be used.

The self reinforcing liquid crystal polymers basically form a special case of the arguments used. The molecular design has to possess an additional chain element being a two dimensional regularity over the usual one dimensional one. Even above the actual melting point a highly ordered molten phase structure exists creating the special features of these polymers, being their rheological and self-reinforcing properties.

ENGINEERING POLYMERS.

Year of introduction		Tg°C	Tm°C
1948	PA66	50	260
1950	PA6	50	215
1950	ABS	95-110	
1958	PA12	40	180
1958	PC	145	
1958	POM	0	180
1962	PPE	215	
1966	PSU	175	
1970	PAS	230	
1970	PET	70	256
1970	PBTP	55	220
1970	PPS	82	280
1975	TMBPC	190	
1975	PES	215	
1976	P.AR I	165-190	
1978	PAI	120	140
1978	LCP I	120	300
1979	PEEK	142	320
1981	PPC	155	175
1982	PEI LCP II (VECTRA)	95	280
1986	PEK	175	340
	LCP III		
	PISA	± 250	
	PESA	± 230	
	P.AR II	160	175
1988	Polyamide 4,6	55	298

NEW BLENDS

With the availability of new polymers the possibilities for new blends increase again and can be expected. Actually they are already emerging and a great many publications were produced on this topic over the last years.

It may be interesting to observe that the mutual compatability and even complete miscibility of these aromatic polymers is surprisingly good, allowing just like in the PPE/PS blend case to produce products over a wide compositional range.

DEVELOPMENTS IN REINFORCED MATERIALS

Reinforcing thermoplastic matrices with primarily glass fibres and reinforcing fillers has long been recognised as a means of improving a number of very important mechanical parameters. This approach is very useful to boost engineering properties like dimensional stability, long term loading properties etc. A generally occuring negative feature of these products has been the often significant reduction in impact strength. The importance of the fibre length for the overall impact behaviour has also been known for a long time.

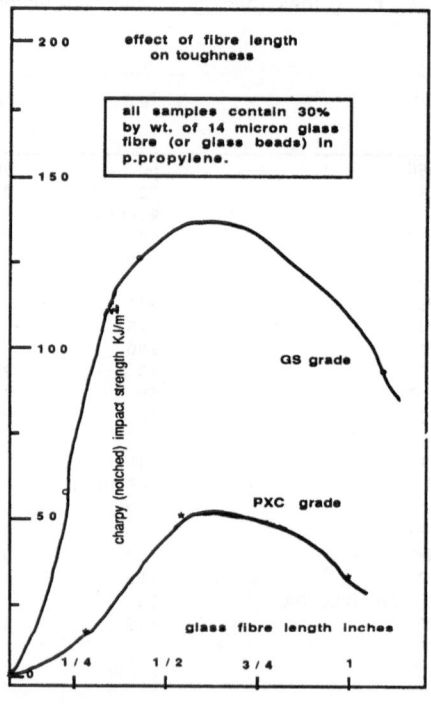

graph 1.

From graph 1, it can be seen that a maximum in impact resistance in long fibre reinforced matrices is reached at a length of plm. 1 cm. The actual breakdown of the curve has to be attributed to other facts associated with the production of these composite structures.

The development of long fibre reinforced thermoplastic and especially those of the engineering plastics has however started only recently. The reason for this is the fact that the utilisation of long fibres is difficult, due to the high l/d ratios providing a very straight forward relation between their ordering and the volume fraction. High volume fraction requiring high ordering. High ordering making it very difficult for taken up by the reinforcing fibres high viscous polymer melt impregnation. Special techniques had to be developed here. The easy but very costly way has been long fibre (fabrics) impregnation with polymer solutions, followed by evaporation of the solvent or by precipitating the polymer.

Another simple but not optimal way was melt-impregnation of filament bundels which were bound by matrix binding resins. The most elegant methods consist of utilising paper technology. Here the required volume fraction for long fibres dispersed in a random way can be easily provided by the dispersing medium which can be water or a foam. By dispersing finely milled polymer in the paper structure formed by depositing the long reinforcing fibres on the sieve belt we obtain a plastic filled fibre configuration which on hot pressing is shaped in a long fibre reinforced plastic sheet.

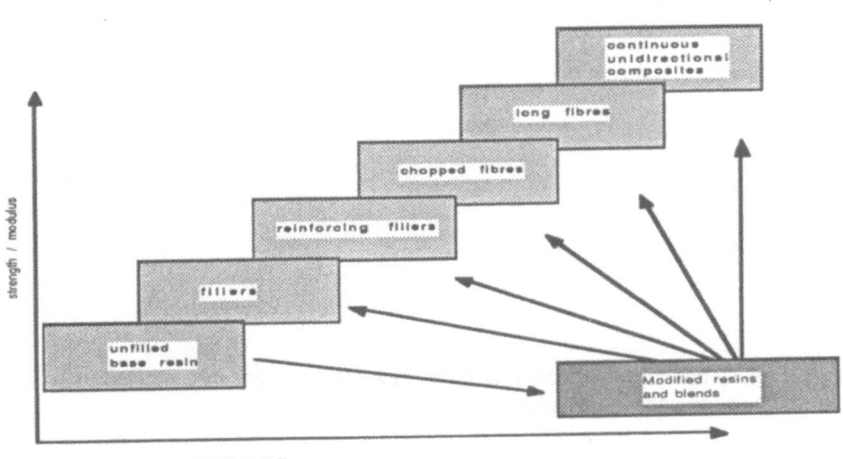

These types of composite sheet materials based upon engineering plastics like Poly-amides, PBT, XENOY etc. are now becoming available. By hot pressing, stamping
or flow moulding they can be formed for the application. Their mechanical property profile is excellent with respect to modulus, tensile strength, creep etc., but outstanding in impact. Impact values can be one order of magnitude better than comparable short fibre reinforced plastics.

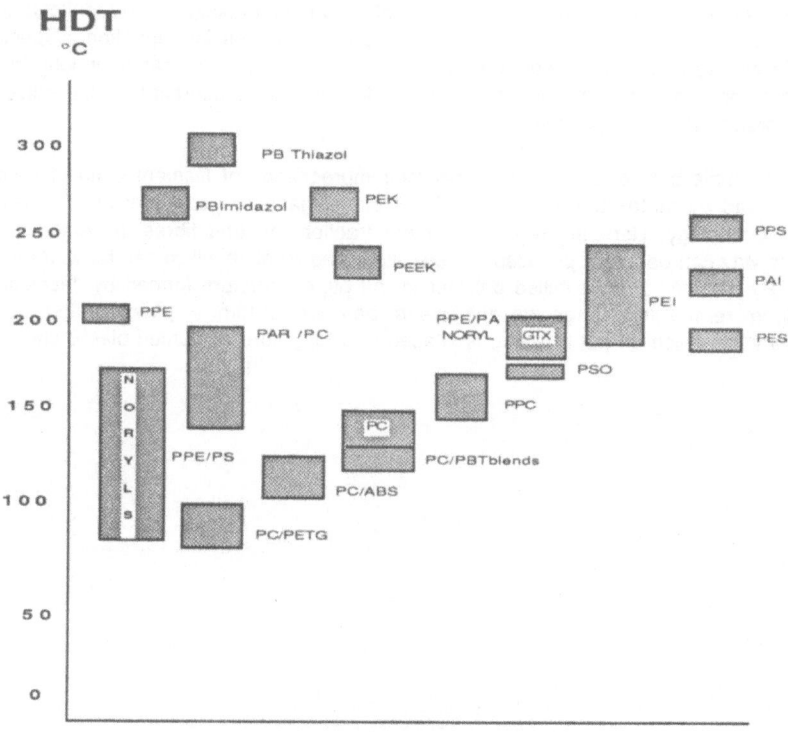

A great future has to be expected for these new material concepts, applied as well to "old" as "new" engineering plastics and blends.

Part 2

THERMODYNAMICS/BLENDS

Part 2
THERMODYNAMICS OF BULK

THERMODYNAMIC STABILITY OF COPOLYMER BLENDS

WILLIAM J. MACKNIGHT, HEUNG S. KANG, FRANK E. KARASZ
Polymer Science & Engineering Department,
University of Massachusetts,
Amherst, MA 01003, USA
and
RONALD KONINGSVELD
Polymer Research Institute ΣΠ,
University of Maastricht,
P.O. Box 5551, 6202 XA Maastricht, Netherlands

ABSTRACT

Liquid-liquid phase diagrams of polymer blends show a large variety in shape and depend very sensitively on average molar mass and molar-mass distribution. If one or both constituents are statistical copolymers phase behavior is still more complex and of surprising subtlety. A mixture of two statistical copolymers based on the same pair of monomers but differing in chemical composition is usually characterized by a maximum tolerable composition difference for the system to remain homogeneous. Mixtures of homopolymers and copolymers, whether sharing a common repeat unit or not, may show unexpected composition ranges of miscibility within vast areas of virtual immiscibility. Molecular modelling of such phenomena is possible on different levels of sophistication. We restrict the discussion to the rigid lattice model and explore its capability of predicting phase relations with a minimum of *a priori* knowledge. The Flory-Huggins-Staverman-Scott model represents the simplest version and has a considerable predictive power in locating regions of (im)miscibility in a semiquantitative fashion. For more precise descriptions the model must be amended, either accounting for minute differences in molecular volume or chain length, or using Staverman's contact statistical treatment.

INTRODUCTION

The thermodynamic stability of a binary liquid mixture can be characterized by the condition [1-3]

$$(\partial^2 \Delta G / \partial \phi_2^2)_{p,T} \begin{array}{l} > 0 \; stable \\ = 0 \; spinodal \\ < 0 \; unstable \end{array} \tag{1}$$

where ΔG is the free enthalpy (Gibbs free energy) of mixing and ϕ_2 is a measure of concentration, e.g.,the volume fraction of the second component. The equality sign represents the limit of stability, or spinodal, the locus of which in a temperature/ composition plane (phase diagram) is encompassed by the binodal curve. The latter is the locus of coexisting phase compositions (e.g., a and b in fig.1b), touching the spinodal curve at its extreme in the critical point where

$$(\partial^3 \Delta G / \partial \phi_2^3)_{p,T} = 0 \tag{2}$$

Given an expression for ΔG as a function of molecular parameters and experimental variables we can apply the operations (1) and (2) and obtain parameter values from experimental data. Such data are easiest to obtain by changing the temperature of a homogeneous mixture and observing the cloud-point temperature at which phase separation sets in. If the miscibility gap is symmetrical, as in fig.1a, the cloud point of a 50/50 mixture is described by either eq. (1) or (2). If the miscibility gap is asymmetrical, however, neither spinodal nor critical condition applies. We measure the temperature at which the 50/50 composition represents one of the two equilibrium phase compositions (b in fig.1b) and should then use the equilibrium conditions

$$\Delta \mu_{1_a} = \Delta \mu_{1_b}$$
$$\Delta \mu_{2_a} = \Delta \mu_{2_b} \tag{3}$$

to calculate the molecular parameters. The quantity $\Delta \mu_i$ is defined by

$$\Delta \mu_i = (\partial \Delta G / \partial n_i)_{p,T,n_j \neq n_i} \tag{4}$$

where n_i is the amount of component i in moles, and p and T are pressure and absolute temperature, respectively.

Since eqs (1) and (2) are more convenient to use than eqs (3) and (4), the distinction between critical, spinodal and coexisting phase points is sometimes ignored [4-7]. Though incorrect, such an approximation may still be useful provided it allows predictions of phase behavior that can be proven reasonably accurate. In this paper we discuss examples of that approach and also present a case in point where the approximation mentioned can be analyzed quantitatively and shown to involve negligible errors only.

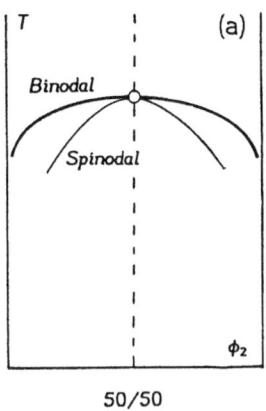 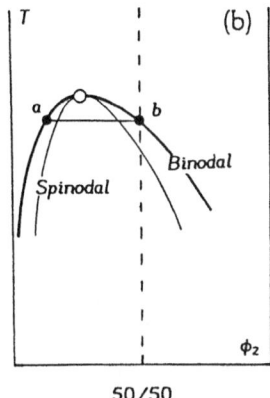

Figure 1. Binodal and spinodal for a symmetrical (a) and an asymmetrical (b) liquid system: (a) spinodal analysis permitted if $m_1 \equiv m_2$, $\phi_{2_c} = 0.5$; (b) binodal analysis required if $m_1 \neq m_2$, $\phi_{2_c} \neq 0.5$ (o: critical point, ϕ_{2_c} = critical concentration).

MISCIBILITY OF POLYMER BLENDS

Blends of homopolymers are usually either completely miscible or virtually immiscible. The former behavior is not abundant and is mostly limited to systems in which specific interactions occur. Non-polar polymers are generally found not to mix at all unless chain lengths representative of oligomers are involved.

Diagrams like fig.1 characterize upper critical miscibility behavior (UCM), mutual solubility increasing when the temperature is raised. The reverse behavior, lower critical miscibility (LCM) is also known and appears to be a general phenomenon in polymer mixtures. Fig.2 shows an example in which we see that the location of the miscibility gap is very sensitive to polymer chain length, an observation consistent with theoretical prediction [9]. Fig.3 illustrates the sensitivity to chain length, not only with respect to location but also with respect to the shape of the miscibility gap. In fig.4 we see that the molar-mass distribution (mmd) in one of the constituents may also influence the shape of the cloud-point curve.

The mmd leads to additional complications in that the cloud-point curve cannot be identified with a binodal [12] and the approximate data analysis mentioned in the introduction is then subject to further ambiguity. The situation sketched in fig 1a will hardly ever be met in practice.

Further complications arise if the system contains statistical copolymers that must be expected to have both a mmd and a distribution in chemical composition. Problems in connection with the latter among the various macromolecules may again be ignored as long as predicted behavior corresponds to experience.

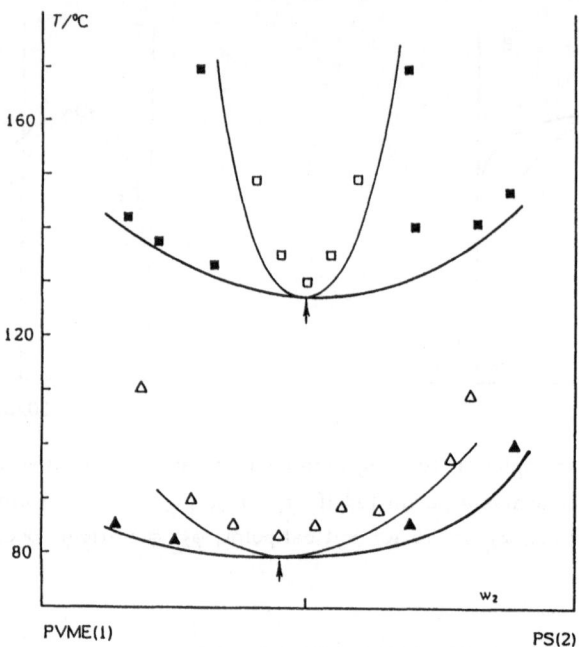

Figure 2. Lower critical miscibility in the system poly(vinylmethylether)/poly(styrene). Open symbols: spinodal decomposition, filled symbols: nucleation and growth. Weight-average molar mass values: PVME 75 kg/mol; PS (□) 50 kg/mol; PS (Δ) 17 kg/mol. [8].

MOLECULAR MODEL

The rigid-lattice expression for ΔG, the Gibbs free energy of mixing of monodisperse polymers 1 and 2 on N sites, reads [13-19]

$$\Delta G/NRT = (\phi_1/m_1)\ln\phi_1 + (\phi_2/m_2)\ln\phi_2 + \chi_{blend}\phi_1\phi_2 \qquad (5)$$

where ϕ_i is the volume fraction of polymer i, m_i the number of basic volume units in chain i, χ_{blend} the appropriate combination of binary interaction parameters for the case at hand and RT has its usual meaning.

We shall first apply eq. (5) to blends of statistical copolymers differing only in composition, then to blends of a homopolymer with a statistical copolymer, and finally to blends of two statistical copolymers containing different repeat units. More complicated situations will also be considered in which the rigid-lattice expressions for χ_{blend} must be modified to describe the data adequately.

Spinodal and critical conditions (1) and (2) can be derived from eq (5) to yield

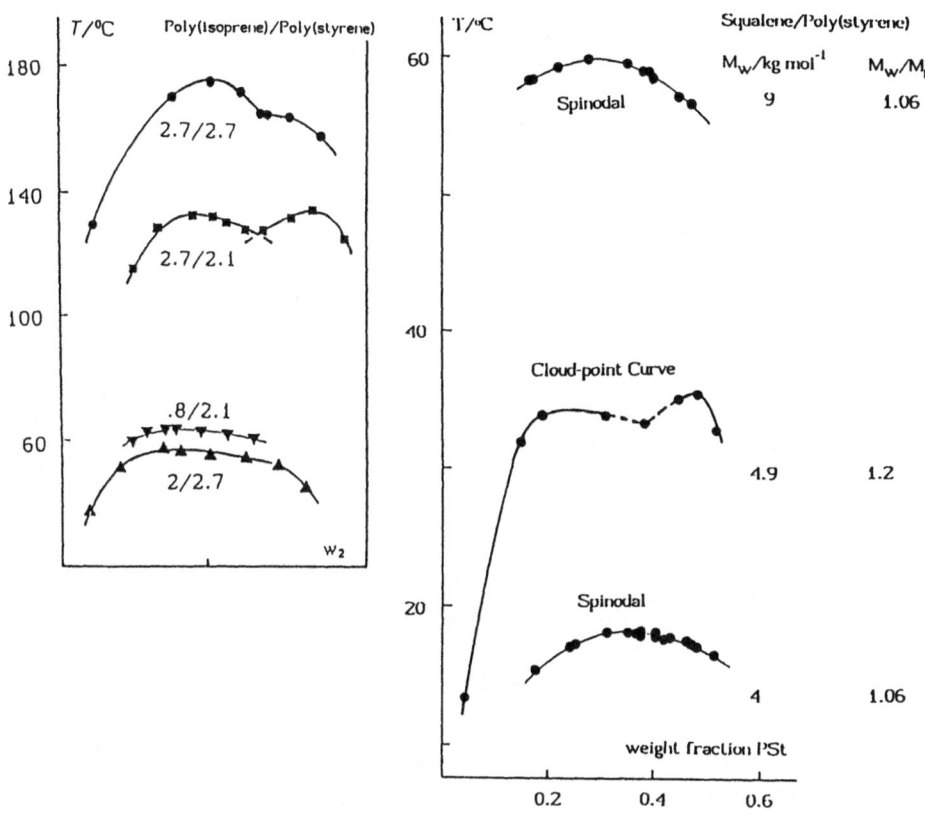

Figure 3. Upper critical miscibility gaps in the system poly(isoprene)/poly(styrene) for indicated M_n values in kg/mol. Heavy curves: cloud-points; light curve: spinodal by light scattering [10]; w_2 = weight fraction poly(styrene).

Figure 4. Spinodals and cloud-point curve in the system squalene/poly(styrene) for indicated values of M_w and M_w/M_n [11].

$$\underline{Spinodal}: \quad 1/m_1\phi_1 + 1/m_2\phi_2 - 2\chi_{blend} = 0 \qquad (6)$$

$$\underline{Critical}\ \underline{Point}: \quad 1/m_1\phi_1^2 - 1/m_2\phi_2^2 = 0 \qquad (7)$$

Blends of Statistical Copolymers differing only in composition $(P_{\alpha\beta_1}/P_{\alpha\beta_2})$
Blends of sulfonylated poly(2,6 dimethyl-1,4 phenylene oxide) copolymers were examined. The modified polymers were considered to represent statistical copolymers of the structure

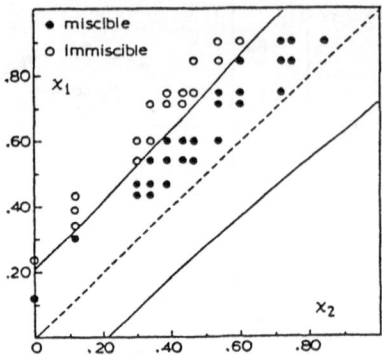

Figure 5. Isothermal miscibility diagram of 50/50 wt % blends of partly sulfonyl-ated poly(2,6 dimethyl-1,4 phenyleneoxide)(SPPO) at 290°C. Degrees of sulfonylat-ion indicated by mole fractions x_1 and x_2 of phenyleneoxide.

$$-\left\{-\left[\underset{CH_3}{\overset{CH_3}{\langle O \rangle}}-O-\right]_x-\left[\underset{\underset{\langle O \rangle}{SO_2}}{\overset{CH_3}{\langle O \rangle}}-O-\right]_{1-x}---\right\}-$$

It was found that the maximum tolerable composition difference, $x_1 - x_2 = \Delta$, ranged between 20 and 26 wt % of sulfonylated units (fig.5). In this case,

$$\dot{X}_{blend} = \Delta^2 g_{\alpha\beta} \tag{8}$$

and therefore we derive from eqs (5), (6) and (7)

$$g_{\alpha\beta} = 0.15$$

where $g_{\alpha\beta}$ is the pair interaction parameter for α-β contacts.

Blends of a Homopolymer and a Statistical Copolymer ($P_\gamma/P_{\alpha\beta}$)

Blends of poly(styrene) with partly sulfonylated poly(2,6 dimethyl-1,4 phenylene-oxide) were studied. In this case X_{blend} takes the form:

$$X_{blend} = x g_{\alpha\gamma} + (1-x)g_{\beta\gamma} - x(1-x)g_{\alpha\beta} \tag{9}$$

where the g's are the appropriate binary segmental interaction parameters and x is the mole fraction of phenylene units in the copolymer. The maximum degree of sulf-onylation tolerable before phase separation was found to be 31 % by weight. Using this fact and the values for $g_{\alpha\beta}$ and $g_{\alpha\gamma}$, previously determined ($g_{\alpha\gamma} = -0.043$, ref. 8), it is found that

$$g_{\beta\gamma} = 0.20$$

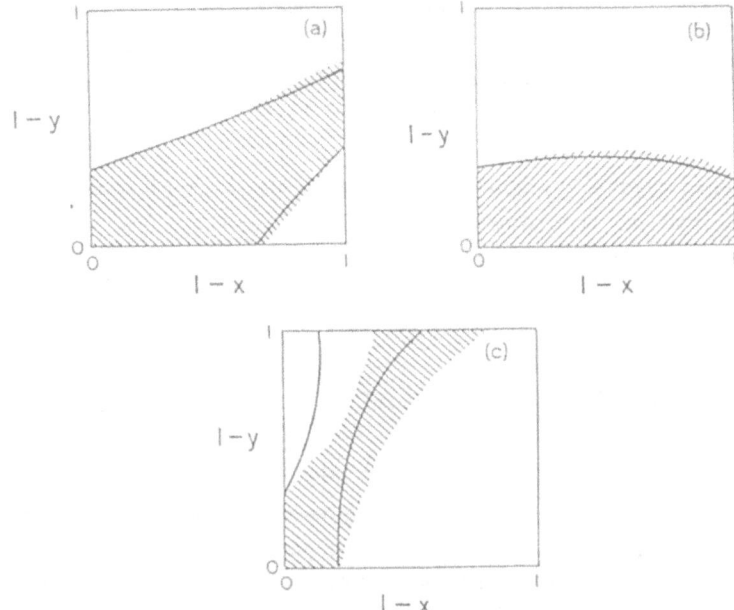

Figure 6. Domains of miscibility (shaded areas) at 290°C compared with calculated phase boundaries (curves): $g_{\alpha\gamma} = -0.043$, $g_{\alpha\delta} = 0.200$, $g_{\gamma\delta} = 0.150$.
(a) poly(styrene-co-p-fluorostyrene)/SPPO; $g_{\alpha\beta} = 0.080$, $g_{\beta\gamma} = 0.053$, $g_{\beta\delta} = 0.022$
(b) poly(styrene-co-o-fluorostyrene)/SPPO; $g_{\alpha\beta} = 0.040$, $g_{\beta\gamma} = 0.004$, $g_{\beta\delta} = 0.113$
(c) poly(styrene-co-acrylonitrile)/SPPO; $g_{\alpha\beta} = 0.700$, $g_{\beta\gamma} = 0.745$, $g_{\beta\delta} = 0.097$.

Blends of Statistical Copolymers with different Repeat Units ($P_{\alpha\beta}/P_{\gamma\delta}$)
Here χ_{blend} takes the form

$$\chi_{blend} = x^2 g_{\gamma\delta} + (g_{\alpha\gamma} - g_{\alpha\delta} - g_{\beta\gamma} + g_{\beta\delta})xy + y^2 g_{\alpha\beta} + (g_{\alpha\delta} - g_{\alpha\gamma} - g_{\gamma\delta})x$$
$$+ (g_{\beta\gamma} - g_{\alpha\gamma} - g_{\alpha\beta})y + g_{\alpha\gamma} \tag{10}$$

where x is the mole fraction of δ units in statistical copolymer $P_{\gamma\delta}$ and y is the mole fraction of β units in $P_{\alpha\beta}$.

Several styrene copolymer blends with sulfonylated PPO were studied and the results are summarized in the form of miscibility/immiscibility boundaries constructed in a Cartesian coordinate system as shown in fig.6(abc). The calculated miscibility regions are the shaded areas in the figure and these were obtained using g values from eq. (10). All of these blends have $g_{\alpha\gamma}$, $g_{\beta\gamma}$ and $g_{\alpha\beta}$ in common, and all exhibit very wide miscibility domains even though all the g's are positive except $g_{\alpha\gamma}$. The $g_{\beta\delta}$ values for the *ortho* - and *para* fluorostyrene containing blends are quite different from one another and produce very different miscibility diagrams (a,b). This isomer effect cannot be explained on the basis of a solubility parameter approach [20].

The calculated miscibility region reproduces the experimental boundary lines quite well, showing that the approximate and principally incorrect treatment based on eqs (8)-(10) is adequate in this instance.

PARTIALLY BROMINATED POLY(STYRENE)

Recently, Kambour *et al.* [6,7] reported on the miscibility of poly(styrene) and poly-p-bromostyrene) and its dependence on chain length and bromine content. The material studied, anionically prepared poly(styrene), was partially brominated so that the chain lengths of original sample and its substitution products could be assumed to be identical, but for a contribution of the bromine atoms. Kambour *et al* neglected the latter effect in their theoretical interpretation if the data.

In a first set of measurements poly(styrene) samples differing in chain length were mixed with their brominated derivatives and the authors determined the maximum tolerable bromine content, Δ, for a 50/50 (w/w) blend to remain homogeneous, *i.e.*, transparent. Δ was found to increase with decreasing chain length of the original poly(styrene). A second set concerned 50/50 mixtures of substituted poly(styrenes) of different bromine content. Kambour *et al.* found that the maximum tolerable difference in bromine content increased significantly with the average bromine content of the two constituents, at constant chain length.

Kambour *et al.* interpreted the data in terms of Scott's theory [13] which includes eqs (6), (7) and (8). For a system in which $m_1 = m_2 = m$, these equations can be transformed into

$$\Delta = (2/mg_{\alpha\beta})^{\frac{1}{2}} \tag{11}$$

which relates Δ to the inverse square root of chain length m, in agreement with Kambour *et al*'s findings for molar-mass values above 10^4 kg/mol. Lower chain lengths do not obey Scott's equation as is seen in fig.7. Eq. (11) also predicts the maximum tolerable composition difference in a mixture of two copolymers $P_{\alpha\beta_1}/P_{\alpha\beta_2}$ to be independent of the composition of either of the two constituents. Again, the experimental evidence does not confirm the prediction (fig.8).

A simple extension of Scott's model already suffices to deal with the situation. The above treatment ignores the small change in molecular volume brought about by the bromination. Accounting for it proves to be an adequate remedy [14].

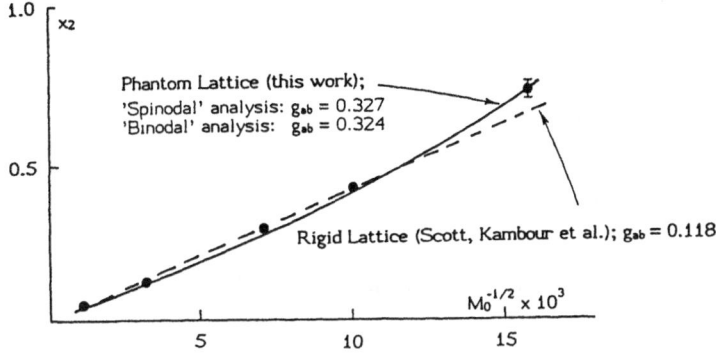

Figure 7. Maximum tolerable bromine content (x_2) for miscibility of 50/50 (w/w) mixtures of poly(styrene) and partially brominated poly(styrene) as a function of the molar mass of the primary homopolymer. Dashed curve: description by Scott's eq. (11) with $g_{\alpha\beta} = 0.118$; drawn curve: description with either eq. (15) (spinodal analysis, $g_{\alpha\beta} = 0.327$) or eqs (3) (binodal analysis, $g_{\alpha\beta} = 0.324$). Data by Kambour et al.[6,7].

We represent the partially brominated poly(styrene) (copolymer) molecules by chains, the beads of which are styrene units α, randomly carrying 'dangling' β groups, bromine atoms in the present case. The mole fraction of modified α units is x, the numbers of sites occupied by α units and β groups are 1 and b, respectively. If the primary chain contains m_0 units, the number of sites m, occupied by a modified chain is

$$m = m_0(1 + xb) = m_0p \tag{12}$$

Calculation of ΔH, the heat of mixing two samples differing in degree of modification (x_1 and x_2, respectively), along the path defined by regular-solution rules yields

$$\Delta H/NRT = [b\Delta/p_1p_2]^2 g_{\alpha\beta}\phi_1\phi_2 = \Lambda g_{\alpha\beta}\phi_1\phi_2 \tag{13}$$

where $p_i = 1 + x_ib$. The free enthalpy of mixing now reads

$$\Delta G/NRT = (\phi_1/m_0p_1)\ln\phi_1 + (\phi_2/m_0p_2)\ln\phi_2 + \Lambda g_{\alpha\beta}\phi_1\phi_2 \tag{14}$$

In this first approximation we have introduced the factors p_i which deviate from unity because the bromination changes the numbers of occupied sites. Consequently, the 50/50 (v/v) composition is no longer a critical concentration, and the measured Δ value does not refer to a spinodal situation. Ignoring this source of error we may use the spinodal equation to derive values for $g_{\alpha\beta}$. In the present approximation it reads

$$\Delta^2 = \{1 + b(x_1\phi_1 + x_2\phi_2)\}p_1p_2/2\phi_1\phi_2b^2 g_{\alpha\beta}m_0 \tag{15}$$

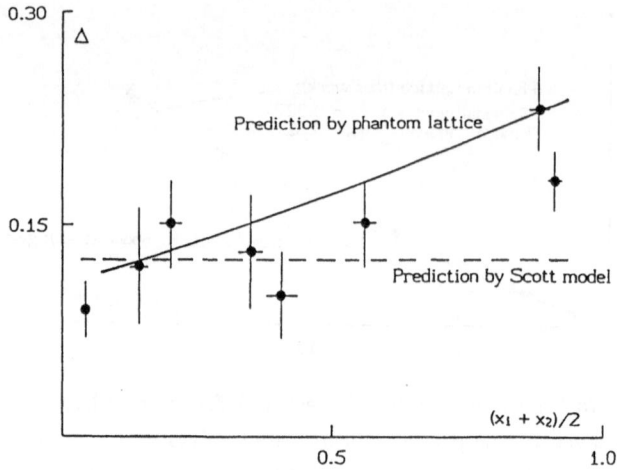

Figure 8. Maximum tolerable composition difference, Δ, for miscibility of 50/50 (w/w) mixtures of partially brominated poly(styrenes). Dashed curve: prediction with Scott's eq. (11) with $g_{\alpha\beta} = 0.118$; drawn curve: prediction with eq.(15) ($g_{\alpha\beta} = 0.327$) or eqs (3) ($g_{\alpha\beta} = 0.325$). Data by Kambour et al.[6,7].

Allowing for the fact that the 50/50 (w/w) composition is usually not a critical concentration complicates the analysis. The 50/50 cloud point may be considered to represent a binodal point (provided the samples have very narrow molar-mass distributions). The composition of the phase coexisting with that of the measured binodal point is unknown but we have an extra equation as well, two-phase equilibria being characterized by conditions (3).

The evaluation of the g values requires a convention with respect to the calculation of ϕ_i and m_i. The most convenient basic unit for the system is a mole of repeat units of one of the constituents. Then, ϕ_i equals the weight fraction and m_0 is the degree of polymerization of the primary chains. Further, $b = M_\beta/M_\alpha$, where M_α and M_β are the molar masses of α units and β groups, respectively.

Applying the, strictly speaking, irrelevant eq. (15) to the measured maximum tolerable composition differences Δ we may use the measured point for a particular primary molar mass to 'calibrate' $g_{\alpha\beta}$, and then use eq. (15) to calculate Δ values for the other molar masses measured. If the data for the shortest primary chains among Kambour et al.'s data are used for the calibration, one finds $g_{\alpha\beta} = 0.327$. Since the other data were presumably taken at the same temperature, the same value of $g_{\alpha\beta}$ should apply. Fig.7 shows that the curve so calculated fits the data quite well, and includes the upturn at small m_0, an effect not covered by Scott's equation (11). Next, eq. (15) is used to predict the dependence of Δ on the average bromine content of blends of modified primary chains. Fig.8 demonstrates that the prediction follows the experimental trend within the accuracy of the measurements, again in contrast to Scott's equation.

Though this result seems to present one more example of the extreme sensitivity of cloud points in polymer blends to variations in chain length (or sites occupied), it is preferable to treat the data as binodal points, even if this involves the assumption that the system may be described as a strictly-binary mixture. We accept this uncertainty because Kambour et al.'s primary samples were narrow-distribution poly(styrenes). Application of this binodal treatment [14] leads to a value for $g_{\alpha\beta}$ of 0.324, again based on the sample with the shortest chains. The other Δ values can now be calculated and the curve so obtained is indistinguishable from that based on the spinodal analysis (fig.7). The predicted dependence of Δ on average composition shows the same feature (fig.8). In this particular example the two modes of evaluation are consistent, and include a high degree of agreement of the value of the interaction parameter $g_{\alpha\beta}$. This far from obvious result is obtained because the p_i values involved are small. Moreover, when samples with wide molar-mass distributions are used, such a fortuitous agreement will probably not be encountered [15].

ACRYLIC COPOLYMERS

The simple model and its equally simple extension discussed above do not suffice in the description of maximum tolerable composition differences found in mixtures of acrylic copolymers. The extensive data by Kollinsky and Markert [21,22] refer to systems of the type $P_{\alpha\beta_1}/P_{\alpha\beta_2}$ where α is methylmethacrylate throughout and β-groups of various sizes were studied. One might therefore expect the procedure of the preceding section to apply but this has been found not to be the case [23].

If the two repeat units differ in size, they will also show a disparity in number of nearest neighbor contacts, or coordination number. We follow Staverman [24] and account for such differences assuming that ratios of coordination numbers can be identified with ratios of molecular surface areas. The latter can be estimated with Bondi's group increment method [25].

Using the rules of regular solution theory [26] we derive the free enthalpy of mixing two monodisperse copolymers $P_{\alpha\beta}$ and find [27]

$$\Delta G/NRT = (\phi_1/m_1)\ln\phi_1 + (\phi_2/m_2)\ln\phi_2 + \Gamma\phi_1\phi_2 \qquad (16)$$

$$\Gamma = \Delta^2 s_{\beta\alpha}g_{\alpha\beta}/\tau_1\tau_2Q$$

$$s_{\beta\alpha} = z_{\beta}/z_{\alpha}$$

$$\tau_i = \phi_{\alpha i} + s_{\beta\alpha}\phi_{\beta i}$$

$$Q = \tau_1\phi_1 + \tau_2\phi_2$$

Applying the spinodal condition (1) to 50/50 (v/v) mixtures in which $m_1 = m_2 = m$ we obtain

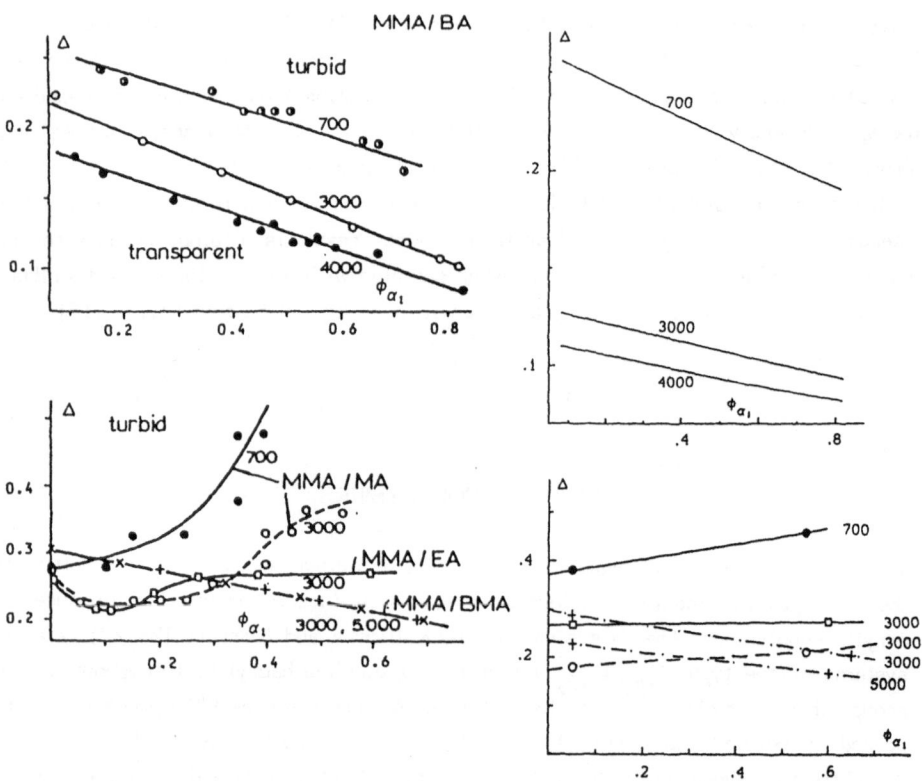

Figure 9. Maximum tolerable composition difference Δ for 50/50 (w/w) blends of acrylic copolymers $P_{\alpha\beta}$ as a function of ϕ_{α_1}, the α content of the first of the two copolymers. Left: data by Kollinsky and Markert [21,22]: α = MMA (methyl methacrylate); β = BA (butylacrylate), MA (methyl acrylate), EA (ethyl acrylate) or BMA (butyl methacrylate). The weight-average degree of polymerization of the two polymers is identical, values indicated. Right: Δ (ϕ_{α_1}) calculated with eq. (17) for indicated m values. Top: MMA/BA, $g_{\alpha\beta} = 0.065$, $s_{\beta\alpha} = 1.32$. Bottom: ●,o: MMA/MA, $g_{\alpha\beta} = 0.015$, $s_{\beta\alpha} = 0.80$; ▢: MMA/EA, $g_{\alpha\beta} = 0.009$, $s_{\beta\alpha} = 0.97$; +: MMA/BMA, $g_{\alpha\beta} = 0.015$, $s_{\beta\alpha} = 1.52$.

$$\Delta^2 = (\tau_1 + \tau_2)/4g_{\alpha\beta}ms_{\beta\alpha} \qquad (17)$$

and use the spinodal analysis to try to simulate Kollinsky and Markert's data. The ratio $s_{\beta\alpha}$ of molecular surface areas can be found in Bondi's paper and we adjust the pair-interaction parameter $g_{\alpha\beta}$ to let the calculated $\Delta(\phi_{\alpha_1})$ curves fall in the range of the measured data. We see in fig.9 that the introduction of disparity of molecular surface areas reproduces the trend of the data in correctly predicting the sign of the slope of the $\Delta(\phi_{\alpha_1})$ curve but neither its magnitude, nor the curvature, and that at the cost of $g_{\alpha\beta}$ values that need to differ more among the various acrylic groups than is probably reasonable.

These shortcomings can at least be partly removed when non-randomness of mixing [5], or higher-order contact statistics are taken into account [28-30].

DISCUSSION

The classic rigid lattice model still proves to be a valuable tool for predictive calculations of phase stability in polymer blends. It depends on the required accuracy whether the simple model needs improvement. Miscibility behaviour of blends containing statistical copolymers do not seem to require such improvements since values of pair-interaction parameters obtained on one system appear to be transferable to another, fig.6 bears witness of this welcome feature. A similar transferability was observed previously in an analysis of critically demixing poly(styrene) solutions in n-alkanes and n-alcohols varying in number of carbon atoms [31]. The data allowed extraction of the end/middle group interaction function for n-alkanes that correctly predicted the lower critical miscibility behavior of n-alkane /poly(ethylene) systems.

If more precise predictions are required the first feature to be aware of is the considerable sensitivity of blend miscibility to chain length or molecular volume. In particular, when the chains are modified and carry 'dangling' groups of considerable size, an influence on miscibility relations must be expected to be noticable. Values for the interaction parameter are very sensitive even to the minute changes in molecular volume such as those caused by partial bromination. Literature values of the interaction parameter should therefore always be accompanied by a precise description of the model on which they are based.

Another source of deviations from rigid-lattice behavior lies in disparity of size between the various repeat units or molecules in the system. A first approximation is then supplied by Staverman's contact number statistics [24] which, if need be, can be extended to more sophisticated levels of development [28-30,32]. Here the emphasis is on ratios of coordination numbers, assumed identical to ratios of molecular surface areas. The latter can be estimated with Bondi's method [25] and, hence, do not present extra parameters.

REFERENCES

1. Gibbs, J.W., The Scientific Papers, Dover Reprint, New York, Vol.I ,1961.
2. Van der Waals, J.D.and Kohnstamm, Ph., Lehrbuch der Thermodynamik, Barth, Leipzig, Vol II, 1912.
3. Koningsveld, R., Stockmayer, W.H. and Nies, E., Polymer Phase Diagrams, Oxford University Press, 1988.
4. Ten Brinke, G., Karasz, F.E. and MacKnight, W.J., Macromolecules, 1983, 16, 1827.
5. Balazs, A.C., Karasz, F.E., MacKnight, W.J., Ueda, H. and Sanchez, I.C., Macro-molecules, 1985, 18, 2784.
6. Kambour, R.P.and Bendler, J.T., Macromolecules, 1986, 19, 2679.
7. Strobl, G.R., Bendler, J.T., Kambour, R.P. and Shultz, A.R., Macromolecules, 1986, 19, 2683.
8. Voigt-Martin, I.G., Leister, K.-H., Rosenau, R. and Koningsveld, R., J. Polym. Sci., Part B: Polym. Phys., 1986, 24, 723.
9. McMaster, L.P., Macromolecules, 1973, 6, 760.
10. Koningsveld, R.; Kleintjens, L.A. in: Polymer Blends and Mixtures, eds., D.J. Walsh, J.S. Higgins and A. Maconnachie, Nijhoff, Dordrecht, 1985, p.89.
11. Onclin, M.H., Kleintjens, L.A. and Koningsveld, R., Brit. Polym. J., 1980, 12, 221.
12. Koningsveld, R., Adv. Coll. Interf. Sci., 1968, 2, 151.
13. Staverman, A.J. and Van Santen, J.H., Recl. Trav. Chim., 1941, 60, 76.
14. Staverman, A.J., Recl. Trav. Chim., 1941, 60, 640.
15. Huggins, M.L., J. Chem. Phys., 1941, 9, 440.
16. Huggins, M.L., Ann. N.Y. Acad. Sci., 1942, 43, 1.
17. Flory, P.J., J. Chem. Phys., 1941, 9, 660.
18. Flory, P.J., J. Chem. Phys., 1942, 10, 51.
19. Scott, R.L., J. Polym. Sci., 1952, 9, 423.
20. Kang, H.S., MacKnight, W.J. and Karasz, F.E., Polym. Prepr. ACS, in print.
21. Kollinsky, F.and Markert, G., Makromol. Chem., 1969, 121, 117.
22. Kollinsky, F.; Markert, G., Adv. Chem. Ser., 1971, No. 99, 325.
23. Koningsveld, R.and MacKnight W.J., Makromol. Chem., in print.
24. Staverman, A.J., Recl. Trav. Chim., 1937, 56, 885.
25. Bondi, A., J. Phys. Chem., 1964, 68, 441.
26. Guggenheim E.A., Mixtures, Clarendon Press, Oxford, 1952.
27. Koningsveld R.and Kleintjens, L.A. Macromolecules, 1985, 18, 243.
28. Koningsveld R., Kleintjens, L.A. and Leblans-Vinck, A.M., J. Phys. Chem., 1987, 91, 6423.
29. v.d.Haegen, R., work in progress.
30. v.Opstal, L., work in progress.
31. Koningsveld, R.and Kleintjens, L.A., J. Polym. Sci., Polym. Symp., 1977, 61, 225.
32. Staverman, A.J. in Integration of Fundamental Polymer Science and Technology, eds., L.A. Kleintjens and P.J. Lemstra, Elsevier, London, 1986, p.19.

INFLUENCE OF MOLAR MASS DISTRIBUTION ON THERMODYNAMICS OF POLYMER SYSTEMS

ERIK NIES and ALEXANDER STROEKS

Laboratory of Polymer Technology

Eindhoven University of Technology

P.O. Box 513, 5600 MB Eindhoven, The Netherlands

ABSTRACT

For the description and prediction of thermodynamic data, e.g. volumetric and compositional derivatives of thermodynamic functions of state, many theoretical models are available. The Simha-Somcynsky theory can be considered to be very succesful if one is interested in the quantitative description of thermodynamic properties. Especially, for the equation of state properties this has been shown on many occasions. For the phase behavior of polymer systems, the theory hasn't been evaluated yet in great detail. In this contribution the influence of composition, temperature and molar mass distribution of the polymer is studied for the system polystyrene/cyclohexane.

INTRODUCTION

Recently, the S-S theory has been adapted to deal with the phase behavior of polymer systems. So far, a detailed evaluation of miscibility behavior according to the S-S theory and comparison with experimental data has not been presented. For polymer solutions, a wealth of accurate and detailed information is available allowing for such a detailed study of the phase behavior. The influence of some molecular variables, e.g. molar mass, molar mass distribution and flexibility will be discussed.

THEORY

The Simha-Somcynsky theory: Some thermodynamic equations

In the S-S theory, a molecular liquid is modelled on a lattice with cells
either vacant or occupied by a segment of a molecule. A real polymer is
represented as an s-mer on the lattice. Segments interact according to a
Lennard-Jones pair potential characterised by a maximum attraction energy
ε^* and accessory segmental volume v^*. Furthermore, a molecule is
postulated to possess 3c external degrees of freedom contributing to the
configurational free energy. Equations for thermodynamic properties of
pure components and multicomponent systems have been presented on several
occasions [1-8]. The extension of the theory to mixtures of components A
and B gives rise to cross interaction parameters ε^*_{AB} and v^*_{AB}
characterising the interactions between segments of different components.
For mixtures, it is possible that the heterogeneous state becomes the
equilibrium state in a certain temperature and pressure region of the
phase diagram. The spinodal condition, defining the boundary between the
stable and unstable regions of the phase diagram, is given by the first
Gibbs determinant, J_{sp}, vis [9]

$$J_{sp} = \left| \frac{\partial^2 G}{\partial x_i \partial x_j} \right|_{P,T} \tag{1}$$

where G is the Gibbs free energy.

The critical conditions satisfy eq. (8) simultaneously with the second
Gibbs determinant J_C, obtained by substituting any row from eq. (8) by
the row vector [9]

$$[\partial J_{sp} / \partial x_i] \tag{2}$$

RESULTS AND DISCUSSION

a. Monodisperse polymer solutions

For three polystyrene fractions with a narrow molar mass distribution
spinodal [11,12] and critical [10] conditions in cyclohexane are shown in
Figure 1. The molar mass distribution characteristics for the PS-samples
are summarized in Table 1.

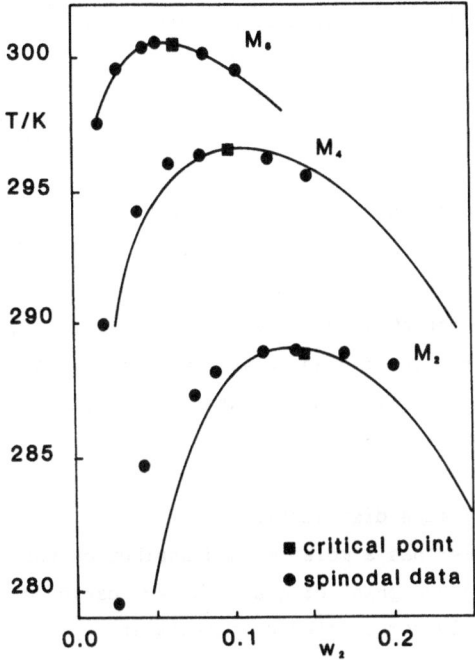

Figure 1. Experimental spinodal (●) and critical (■) conditions for three
different PS fractions in CH [10–12]. Predicted spinodals and
critical conditions for molecular according to the S–S theory
(——,■).

TABLE 1.

Molecular distribution characteristics of PS fractions.

Sample	M_n [kg/mole]	M_w [kg/mole]	M_z [kg/mole]
PS2	49.0	51.0	55.0
PS4	154.0	166.0	181.0
PS6	436.0	520.0	593.0
PS7	450.0	498.0	590.0
PS8	211.0	522.0	790.0
PS9	154.0	672.0	1800.

The molecular parameters characterizing the pure components and the
mixtures in the S-S theory, are taken from reference [6]. The pure
component parameters were estimated from equation of state data [13,14].
Values for the mixing parameters ε^*_{12} and v^*_{12} were adjusted to give
quantitative agreement between the computed and experimental critical
conditions. Since all the model parameters are available, we are in a
position to predict other thermodynamic properties. As an example,
spinodal conditions are considered. Details concern.ng the computational
methods have been presented elsewhere [5]. It can be observed in Figure 1
that, in comparison to the experimental spinodals, the predicted
spinodals become too narrow with decreasing molar mass. If the
flexibility parameter c is allowed to vary with molar mass in a manner
dictated by the experimental spinodal data, a quantitative description of
these data can be obtained [6].

a. Influence of molar mass distribution

Every synthetic polymer has a molar mass distribution which may have a
profound influence on the phase diagram. The evaluation of thermodynamic
properties is certainly complicated by polydispersity. To exemplify the
influence of polydispersity, spinodal data for three polydisperse PS
samples having comparable mass average molar masses M_w summarised in
Table 1. In the computations, the polymer samples were presented by
r-equivalent distributions [15] with the molar mass averages M_n, M_w and
M_z equal to the experimental ones [6].
In the hole theory the spinodal is not solely determined by the mass
average molar mass. This statement is in qualitative agreement with
experimental facts. Especially, the observed temperature shift is
reproduced quite well. However, quantitative agreement with experimental
spinodal data has not been achieved yet. It still remains to be evaluated
which moments of the molar mass distribution determine the spinodal
conditions in the hole theory. For Flory-Huggins like models, it can be
proven that spinodal conditions are determined only by the mass average
molar mass M_w [16-18]. The hole theory behaves quite differently and new
insight in the influence of the molar mass distribution may result.

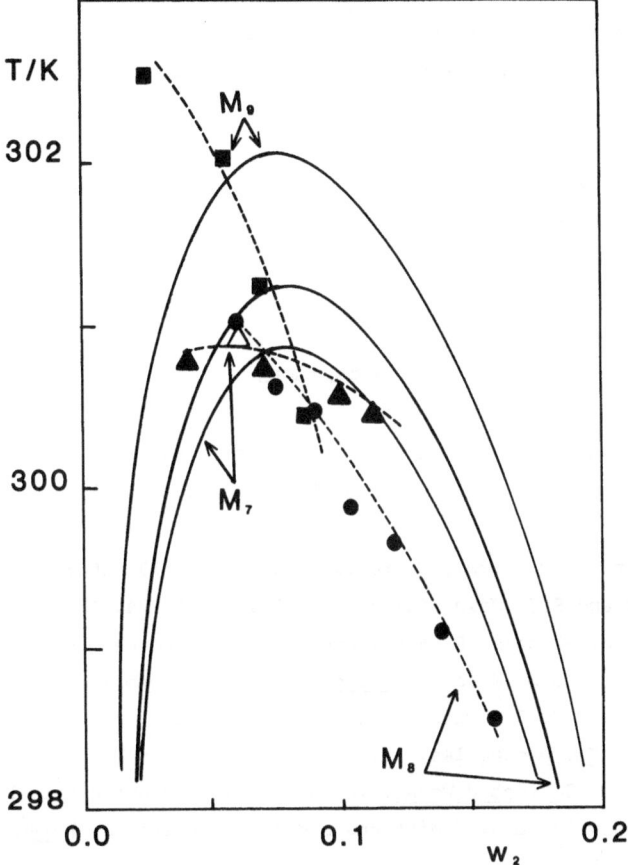

Figure 2. Experimental spinodal data for the polydisperse PS fractions
with similar mass average molar masses (■,●,▲, ---) [10,12].
Computed spinodal conditions according to the S-S theory (——).

CONCLUSIONS

The hole theory offers an excellent basis to evaluate the phase behavior
of polymer systems. The description of the spinodal conditions are almost
quantitative without the introduction of empirical parameters. The cell
free volume is very important for this quantitative success. The
influence of polydispersity on the spinodal conditions in the
Simha-Somcynsky theory is not restricted to the mass average molar mass.

For Flory Huggins like models it can be shown that the mass average molar mass determines the shape and position of the spinodal. For the hole theory other moments of the molar mass distribution become important also. The exact influence of the molar mass distribution on the complete phase diagram of polymer solutions is a topic of current research. The description of the spinodal conditions are almost quantitative without the introduction of empirical parameters. The cell free volume is very important for this quantitative success.

REFERENCES

1. R.Simha and R.K. Jain, Colloid Polym. Sci. 1985, **263**, 9052.
2 R.K. Jain, R. Simha and C.M. Balik, Indian J. Pure Appl. Phys., 1984, **22**, 651.
3. R.K. Jain and R. Simha, Macromolecules, 1984, **17**, 2663.
4. R. Simha and R.K. Jain, Polym. Eng. Sci, 1984, **24**, 1284.
5. E. Nies, A. Stroeks, R. Simha and R.K. Jain, to be published.
6. A. Stroeks and E. Nies, Pol. Eng. Sci., to be published.
7. R. Simha and T. Somcynsky, Macromolecules, 1969, **2**, 342.
8. R. Simha, Macromolecules, 1977, **10**, 905.
9. J.W. Gibbs, Collected Works, Vol 1, 1984, Yale University Press
10. R. Koningsveld, L.A. Kleintjens and A.R. Schultz, J. Polym. Sci., A2, 1970, **8**, 1261.
11. J. Goldsbrough, Sci. Progress (Oxford), 1970, **60**, 281
12. K.W. Derham, J. Goldsbrough and M. Gordon, Pure Appl. Chem., 1974, **38**, 97.
13. J. Jonas, D. Hasha and S.G. Huang, J. Phys. Chem.,1974, **84**, 109.
14. A. Quach and R. Simha, J. Appl. Phys.,1971, **42**, 4592.
15 P. Irvine and J.W. Kennedy, Macromolecules, 1978, **11**, 1145.
16. W.H. Stockmayer, J. Chem. Phys., 1949, **17**, 588.
17. R. Koningsveld, H.A.G. Chermin and M. Gordon, Proc. Roy. Soc., 1970, **A319**, 331.
18. P. Irvine, Ph.D. Thesis, University of Essex, 1979.

APPLICATION OF CRITICAL CONDITIONS
IN THE MEAN-FIELD LATTICE GAS MODEL

R. Van der Haegen*
Universitaire Instelling Antwerpen,
Dept. of Chemistry,
B-2160 Wilrijk,
Belgium

* presently at DSM—Research, Geleen, NL

SYNOPSIS

The mean-field lattice gas model is a molecular model for small-molecular and macromolecular systems that can reliable predict thermodynamic equilibrium properties. The model contains several adaptable empirical parameters for pure substances and for the mixtures, whereas no mixing rules are involved. The use of critical conditions in the adaption of parameters for binary mixtures to experimental data is discussed here, taking as an example the vapour-liquid critical behaviour of the system ethylene-naphtalene.

INTRODUCTION

In the mean-field lattice gas (MFLG) model, as applied and developed

by Kleintjens and Koningsveld[1], a pure substance, liquid or gas, is

represented by a lattice with occupied and vacant sites (holes). It is

assumed that only first nearest-neighbour contacts between molecules

contribute to the internal energy. Following Stavermans' idea[2], these

contacts are related to molecular surface areas. The entropy contribution

to the free energy is calculated using strictly regular solution assump-

tions. Molecules are allowed to occupy more than one lattice site and the

molar site volume is kept constant for convenience.

Introduction of free volume allows one to deal with .volume- and pressure variations of the system. Expressions for the Helmholtz free energy, the equation of state, spinodal (i.e. the limit of thermodynamic stability) and critical condition were derived before and do not contain mixing rules. The mean-field lattice equation of state of pure compounds contains six empirical parameters for which a possible molecular basis has been put forward by testing thermodynamic data on a wide variety of systems and by extension of the model[3,4].

The method has been successfully applied to fluid phase behaviour of pure non-polar systems[5], polar pure compounds and mixtures over a large temperature and pressure range[6,7], including supercritical behaviour, solid-liquid equilibria[8] and gas-gas demixing phenomena. The MFLG model can deal with lower- and upper critical demixing in polymer solutions[1,5,9] and can be extended to blockcopolymer systems[10].

MFLG DESCRIPTION OF BINARY SYSTEMS

A two component system is in the MFLG approximation treated as a pseudo-ternary mixture of constituents 1, 2 and holes (index 0). The appropriate thermodynamic function for the description of fluid phase equilibria is the Helmholtz free energy of 'mixing' vacant and occupied sites, and reads in the simplest version of the model[11].

$$\frac{\Delta F}{N_\phi RT} = \phi_0 \ln \phi_0 + \frac{\phi_1}{m_1} \ln \phi_1 + \frac{\phi_2}{m_2} \ln \phi_2 + g_{01}\phi_0\phi_1 + g_{02}\phi_0\phi_2 + g_{12}\phi_1\phi_2 \quad (1)$$

Where $\phi_i = \frac{m_i v_0}{V}$ is the volume fraction of component i. V is the total volume of the lattice and v_0 is the molar volume of a vacant site (taken as a constant, independant of pressure p and temperature T), m_i is the number of sites per molecule i, N_ϕ being the total number of lattice sites.

Obviously: $\phi_0 = 1 - \phi_1 - \phi_2$

Relating the intermolecular contacts to molecular surface areas and introduction of empirical correction parameters α_i and α_m leads to the following expressions for the interaction functions:

$$g_{oi} = \alpha_i + \frac{\beta_{oi} + \frac{\beta_{ii}}{T}}{1 - \gamma_1\phi_1 - \gamma_2\phi_2} \qquad i = 1,2$$

Where $\gamma_i = 1 - \frac{\sigma_i}{\sigma_o}$ are parameters related to contact surface areas of occupied and vacant sites (σ_i resp. σ_o). Furhter, β_{ii} can be related to the molecular interaction energy per unit contact surface area:

$$\beta_{ii} (T) = - \frac{1}{2} \frac{W_{ii}}{RT} \sigma_i$$

The interaction function for the mixture can be written as:

$$g_{12} = \alpha_m + \frac{g_m (1 - \gamma_1) (1 - \gamma_2)}{1 - \gamma_1\phi_1 - \gamma_2\phi_2}$$

with $g_m (T) = \frac{\Delta W_{12} \sigma_o}{RT}$

and $\Delta W_{12} = W_{12} - \frac{1}{2} (W_{11} + W_{22})$

Having properly described the pure component systems, the adaptable parameters α_m and $g_m (T)$ remain to be determined. It was already showed by Van der Waals[12] that such can be very well done from accurate experimental data; liquid-vapour critical points in small-molecule mixtures or liquid-liquid critical points in polymer systems showing partial miscibility.

The equation of state, which gives the pressure of the system, can be derived from:

$$p = - \left(\frac{\partial F}{\partial V}\right)_T$$

which results in short notation in:

$$P \equiv - \frac{p v_0}{RT} = A - A_1 \phi_1 - A_2 \phi_2 \tag{2}$$

where: $A = \frac{\Delta F}{N_\phi RT}$ and $A_i = (\frac{\partial A}{\partial \phi_i})_T$

For critical points we can derive two equations, viz., the spinodal expression and the critical condition. The spinodal is defined by:

$$J_s \equiv A_{11} \cdot A_{22} - (A_{12})^2 = 0 \tag{3}$$

where: $A_{11} = (\frac{\partial^2 A}{\partial \phi_1^2})_T$, $A_{22} = (\frac{\partial^2 A}{\partial \phi_2^2})_T$, $A_{12} = (\frac{\partial^2 A}{\partial \phi_1 \partial \phi_2})_T = A_{21}$

For the critical condition, one usually writes:

$$J_c \equiv \frac{\partial J_s}{\partial \phi_1} \cdot A_{22} - \frac{\partial J_s}{\partial \phi_2} \cdot A_{12} = 0 \tag{4}$$

The volume fraction ϕ_2 in the pseudo-ternary mixture is related to the molefraction by:

$$x_2 = \left(1 + (\phi_2^{0^{-1}} - 1) \; m_2 m_1^{-1} \right)^{-1}$$

where: $\phi_2^0 = \frac{\phi_2}{1 - \phi_0}$ is the volume fraction in the 'hole free' mixture.

The expressions for A_1, A_2, A_{12}, $\frac{\partial J_s}{\partial \phi_1}$ etc., written before[1]), are somewhat involved and require the aid of a computer.

THE SYSTEM ETHYLENE-NAPHTALENE

The binary mixture ethylene-naphtalene shows sudden changes in solubility of naphtalene in supercritical ethylene upon slight variations in pressure or temperature, because of the appearance of binary critical endpoints. The phenomenon of solubility of solids in supercritical solvents has been reviewed and discussed before[13,14]. The $p(T, x_b)$ diagram in figure 1 schematically shows the various equilibria in such binary systems. From figure 2, it can be seen from the interference of the vapour- and liquid branches with the critical curve at the second critical endpoint what kind of effect has to be expected.

<u>Figure 1</u>: $p(T, x_b)$ Bakhuis-Roozeboom phase diagram for a binary system with two critical endpoints $(V = L)_1$ and $(V = L)_2$ (C_1 and C_2 in figure 2).

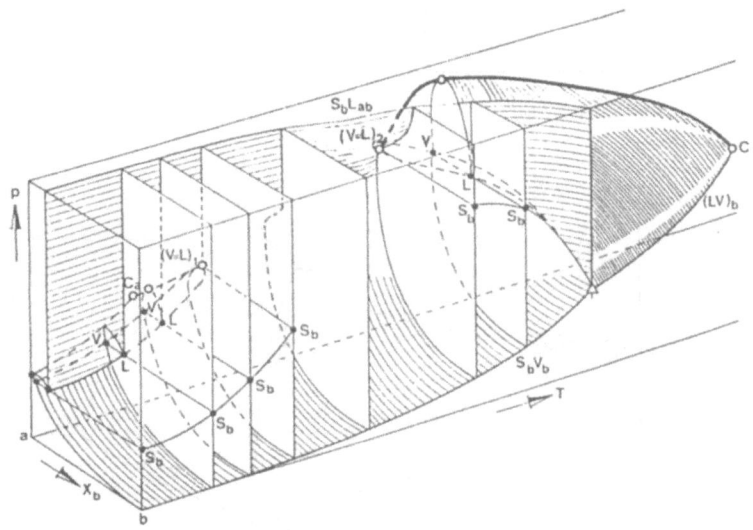

MFLG calculation of the (V = L) critical curve in the system ethylene-naphtalene

Having obtained the MFLG parameters for pure naphtalene and ethylene by standard procedures, the parameters for the mixture can be adjusted to reliable vapour-liquid critical data like those mentioned above[8,14]. The values of α_m and g_m (T) have been determined for five temperatures above that of the second critical endpoint in such a way, so that the calculated pressure (eq. 2) shows a maximum on the spinodal (eq. 3) that coincides with the experimental critical pressure. It is then verified whether the critical condition (eq. 4) is fullfilled. This leads to: $\alpha_m = 0.75$, and g_m

(T) is taken

$$g_m (T) = g_{m0} + g_{m1} T + g_{m2} T^2$$

with:

$g_{m0} = -0.6776;$

$g_{m1} = 51.516 \cdot 10^{-4};$

$g_{m2} = -8.105 \cdot 10^{-6}.$

Figure 2: p(T) en T(x_b) projections for figure 1. Metastable part of critical curve : _ _ _ _ _ _ _ _ , ⋯⋯⋯⋯⋯⋯⋯⋯⋯

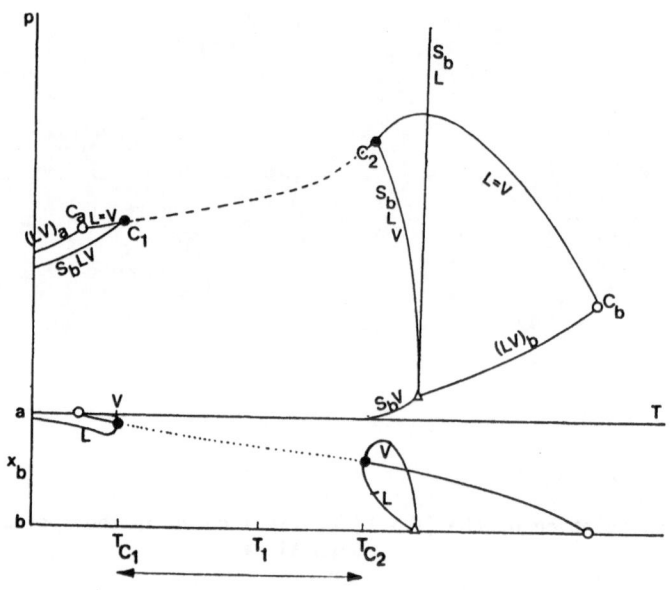

Critical points between C_2 and C_b have been accurately measured by Van
Welie and Diepen in the system ethylene-naphtalene[15]).

Using these well-determined values for the parameters of the mixture as
starting values, a simultanious fit of α_m and g_m (T) to eight experimental
critical data was set up using equations 2, 3 and 4 althogether. The use
of the critical condition proved to be troublesome and ambigious. Further,

calculation of the critical line using the above values for α_m and g_m (T) leads to unreal solutions as can be seen from figures 3 and 4.

Figure 3: MFLG calculated p (T) and T (x) projections of the critical curve(s) (L = V) for the system ethylene/naphtalene, using the critical determinant J_c (x = mole % $C_{10}H_8$).

0: Data by Van Welie and Diepen.

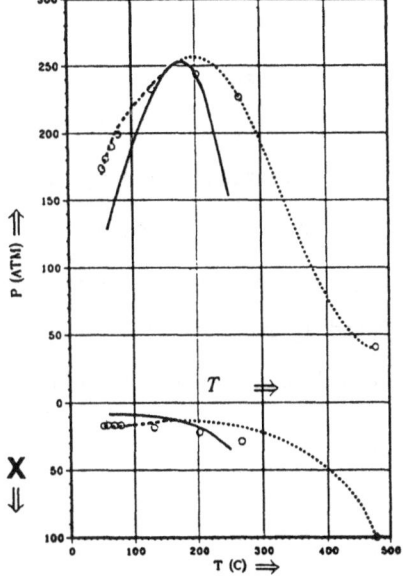

Figure 4: Representation of the calculated critical line(s) of figure 3 in the composition triangle ethylene (1)/naphtalene (2)/holes (0)

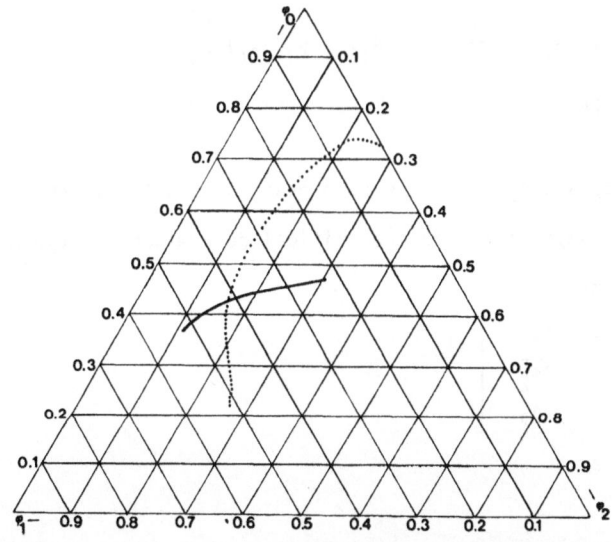

RELEVANT EXPRESSION FOR THE CRITICAL DETERMINANT IN THE MFLG MODEL
FOR BINARY MIXTURES

The procedure, used so far to adjust parameters for the mixture to
(V = L) critical data, outlined in par. 4 can be replaced by the
following. We are dealing here with the problem to find the (absolute)
extreme of the pressure (eq. 2) on the spinodal curve at constant temperature. This leads to the so called constrained extremum problem, which
can be solved by the method of Lagrange multipliers. Therefore it is
required to construct a helping function Z of the independant concentration variables ϕ_1 and ϕ_2:

$$Z\,(\phi_1,\ \phi_2) = P\,(\phi_1,\ \phi_2) + \lambda\,J_s\,(\phi_1,\ \phi_2)$$

where P and J_s are defined by equations 2 and 3 and λ is a constant, the
undetermined coefficient. The following conditions are to be fullfilled:

$$\begin{cases} \dfrac{\partial Z}{\partial \phi_1} = 0 & (5) \\[2mm] \dfrac{\partial Z}{\partial \phi_2} = 0 & (6) \\[2mm] J_s = 0 \end{cases}$$

Straight forward algebra and elimination of λ from (5) and (6) leads to
the solution of the constrained extremum problem:

$$\begin{cases} (\phi_1\,A_{12} + \phi_2\,A_{22})\,\dfrac{\partial J_s}{\partial \phi_1} - (\phi_1\,A_{11} + \phi_2\,A_{21})\,\dfrac{\partial J_s}{\partial \phi_2} = 0 & (7) \\[3mm] A_{11}\,A_{22} - (A_{12})^2 = 0 \end{cases}$$

Here A_{11}, A_{22} and A_{12} have the same meaning as before. From eq. 7 follows
the relevant expression for the critical determinant in the MFLG model for
binary systems:

$$J_c^{\,*} = \begin{vmatrix} \dfrac{\partial J_s}{\partial \phi_1} & \dfrac{\partial J_s}{\partial \phi_2} \\[3mm] \Lambda_1 & \Lambda_2 \end{vmatrix}$$

where:

$$\Lambda_1 = \phi_1 \left(\frac{\partial A_1}{\partial \phi_1}\right)_T + \phi_2 \left(\frac{\partial A_1}{\partial \phi_2}\right)_T = \phi_1 A_{11} + \phi_2 A_{12}$$

and:

$$\Lambda_2 = \phi_1 \left(\frac{\partial A_2}{\partial \phi_1}\right)_T + \phi_2 \left(\frac{\partial A_2}{\partial \phi_2}\right)_T = \phi_1 A_{12} + \phi_2 A_{22}$$

Using equations 2 and 3 together with $J_c^* = 0$ in the adjustment of α_m and g_m (T) to eight experimental critical points leads without any further complications to the values:

$\alpha_m = 1.101$;

$g_{mo} = -0.5591$;

$g_{m1} = 39.172 \cdot 10^{-4}$;

$g_{m2} = -6.214 \cdot 10^{-6}$.

The calculated critical line is drawn in figure 5. The above values for the parameters of the mixture were used to predict the P, T, x - space model for the system ethylene-naphtalene including (solid (S) = L), (L = V) and (S = L = V) equilibria[16].

CONCLUSION

The introduction of a relevant expression for the critical determinant in the mean-field lattice gas model for binary systems is discussed here. It leads to an alternative and thermodynamic consistent method of adjusting two-particle interaction functions to experimental critical binary liquid-vapour densities. The present approach might lead to new developments in the determination of MFLG parameters for the mixture in small-molecule mixtures and in polymer solutions and polymer mixtures (blends). These relevant critical conditions appear because of the extra constraint, which is the equation of state, put on the hole model, and are therefore not questioning or contradicting the method of Gibbs determinants[17,18].

Figure 5: MFLG calculated p (T) and T (x) projections of the critical curve (L = V) for the system ethylene/naphtalene, using the relevant critical deter-minant J_c^*.

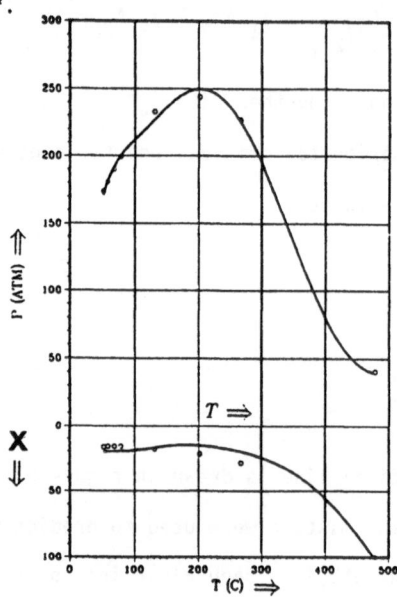

ACKNOWLEDGEMENT

The author wants to thank prof. dr. Koningsveld, prof. Šolc and dr. Kleintjens for helpful discussions, and L. van Opstal for assistance in some of the calculations. I thank DSM for a post-doc. grant in the Universitaire Instelling Antwerpen.

REFERENCES

1. Kleintjens, L.A., Ph. D. Thesis, Essex, UK, 1979.

2. Staverman, A.J., Recl. Trav. Chim., Pays-Bas, 56 (1937) 885.

3. R. Koningsveld, L.A. Kleintjens and A.-M. Leblans-Vinck, J. Phys. Chem., 91 (1987) 6423.

4. E.J. Beckmann, R.S. Porter and R. Koningsveld, J. Phys. Chem., 91 (1987) 6429.

5. L.A. Kleintjens and R. Koningsveld, Colloid α Pol. Sci., 258 (1980) 711.

6. L.A. Kleintjens and R. Koningsveld, Sep. Sci. and Techn., 17 (1) (1982), 215.

7. L.A. Kleintjens, Fluid Phase Eq., 10 (1983) 183.

8. A.-M. Leblans-Vinck, R. Koningsveld, L.A. Kleintjens and G.A.M. Diepen, Fluid Phase Eq., 20 (1985) 347.

9. L.A. Kleintjens, R. Koningsveld and M. Gordon, Macromolecules, 13 (1980), 303.

10. L.A. Kleintjens, in 'Integration of Polymer Science and Technology', Eds. L.A. Kleintjens and P.J. Lemstra. Applied Sci. Publishers, London, 1985.

11. N.J. Trappeniers, J.A. Schouten and C.A. Ten Seldam, Chem. Phys. Letters, 5 (1970) 541.

12. J.D. Van der Waals and Ph. Konstamm, 'Lehrbuch der Thermodynamik', vol. II, Leipzig 1912.

13. R. Koningsveld and G.A.M. Diepen, Fluid Phase Eq., 10 (1983) 159.

14. R. Koningsveld, L.A. Kleintjens and G.A.M. Diepen, Ber. Bunsenges. Phys. Chem., 88 (1984) 848.

15. G.S.A. Van Welie and G.A.M. Diepen, Recl. Tzar. Chim., Pays-Bas, 80 (1961) 659.

16. R. Van der Haegen, R. Koningsveld, L.A. Kleintjens and L. van Opstal, accepted for publication in Fluid Phase Eq.

17. 'The Scientific Papers of J. Williard Gibbs'; Dover Publications: New York, 1961; vol. I.

18. Koningsveld, R. and Staverman, A.J., J. Polymer Sci., Polym. Phys. Ed., 6 (1968) 325.

STATE DIAGRAMS OF SYSTEMS WITH WATER SENSITIVE POLYMERS AND WATER

W. Borchard, E. Dobnik, B. Luft, P. Reutner

Angewandte Physikalische Chemie der Universität-GH-Duisburg, FRG

ABSTRACT

In principle it is possible to draw state diagrams of binary systems by means of calorimetric-measurements. Such state diagrams of low molecular systems are well known, but there is only little knowledge of systems of which at least one component is a polymer. First results were obtained from the systems gelatin-water and polyethylene oxide-water. Measurements of the system x-carrageenan-water are not yet completed.

INTRODUCTION

With the results of calorimetric investigations in principle state diagrams of binary systems may be drawn. Such state diagrams are well known of systems, which contain low molecular components, but only little has been found in the literature concerning phase diagrams of systems of which at least one component is a polymer [1,2]. The reason is that systems with high molecular components show a phase behaviour where effects like a glass transition or several are superimposed. Therefore it is sometimes difficult to analyse the caloric data. In fig. 1 a thermogram with some of these effects is shown schematically.

Recently it has been shown how phase diagrams of polymer/solvent systems can be determined [3,4,5,6]. The complete diagram of the system gelatin-water has successfully been constructed. Therefore the systems polyethylene oxide (PEO)-water and partly x-carrageenan-water were investigated too.

MATERIALS AND METHODS

The gelatin was an industrially produced and dialyzed product, a granu-
lar mixture of α-, β-, and γ-gelatin, which was also used in earlier
papers[5]. The PEO with M = 9 125 g/mol was synthesized in the group
of Prof. Heidemann in Mainz. The substance x-carrageenan was of the
pectin factory in Copenhaven. It was the potassium salt. All the substances were mixed and swollen with water over night. The quantity
of added water was corrected by the moisture content of the polymers
which was determined by drying the substance in an oil-free high
vacuum.

The calorimetric measurements have been performed with a differential
scanning calorimeter (DSC-2C) from Perkin Elmer equipped with a modified Texas Instrument calculator and combined with a low temperature
equipment from the firm Keltorr.

RESULTS AND DISCUSSION

The complete diagram of the system gelatin-water which has successfully been constructed is shown in fig. 2. This system is of an eutectic type. At the invariant eutectic temperature ice, nearly pure water,
and mixed crystal with an over-all concentration of 67 % by wt. gelatin
coexist. The graph of the experimentally determined specific eutectic
enthalpy of transformation versus mass fraction of gelatin y_2 confirm
this value. The liquidus and solidus curves have been determined up
to temperatures very close to the melting point of pure gelatin at
510 K. The glassy solidification curve which has been measured between
a solution of 40 and 100% by wt. of gelatin is situated below but very
close to the solidus. From the change of the cooling rate different
non-equilibrium states can be verified caused by the crystallization of
pure water and an incomplete mixed crystal formation of gelatin. This
leads to a second, morphologically stipulated glass transition T_g^Φ,
which is independent of concentration.

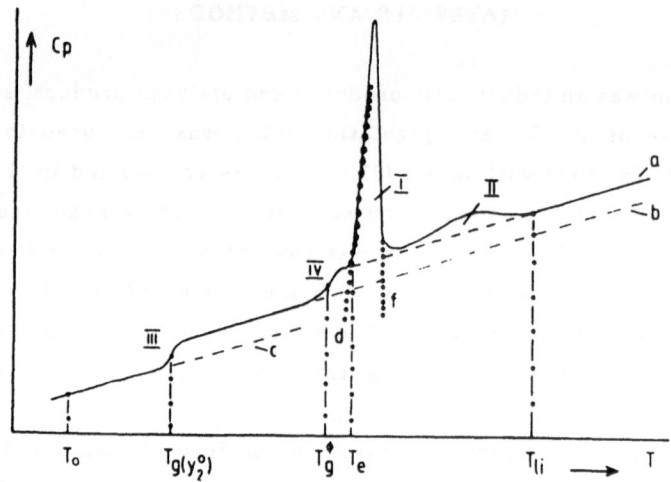

Fig. 1: A schematic thermogram showing various possible transitions of a system with a polymer component.

I, II, III, IV different phenomena; $T_{g(y_2^o)}$ first glass transition temperature; T_g^Φ second glass transition temperature; T_e eutectic temperature; T_{li} liquidus temperature.

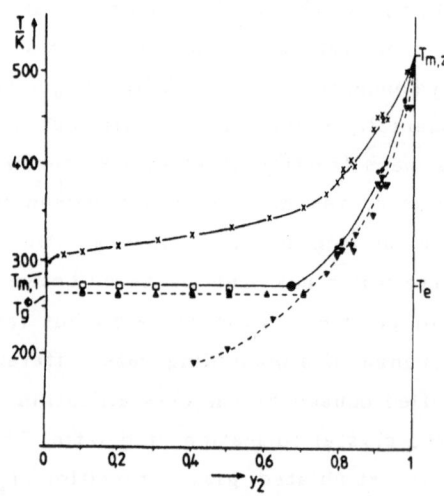

Fig. 2: State diagram of the system gelatin-water

x liquidus curve

• solidus curve

□ eutectic curve

● eutectic mixed crystal

▲▼ glassy solidification

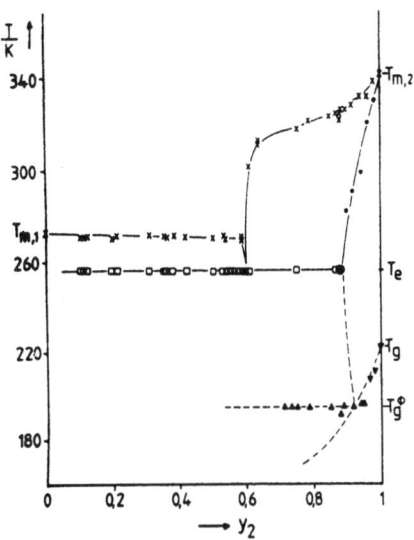

Fig. 3: State diagram of the system PEO-water

Symbols see fig. 2

Fig. 3 presents the state diagram of the system PEO-water. It is of an eutectic type exhibiting mixed crystal formation and glass transition phenomena like the system mentioned. The graph of the experimentally determined eutectic enthalpy of transformation versus mass fraction of PEO verifies the eutectic composition.

The preliminary state diagram of x-carrageenan-water is shown in fig 4. This diagram looks very similar to both other systems which were found. The eutectic point is also shifted to vanishing polymer concentrations which is practically equal to zero.

CONCLUSIONS

State diagrams of polymer/water systems of which the polymer is water sensitive essentially seem to be of the eutectic type known from low molecular binary systems but including some peculiarities caused by non-equilibrium phenomena and incomplete phase transitions. The

90

present results of calorimetric investigations by means of a DSC-2C demonstrate that it is possible to draw such state diagrams of polymer-solvent systems.

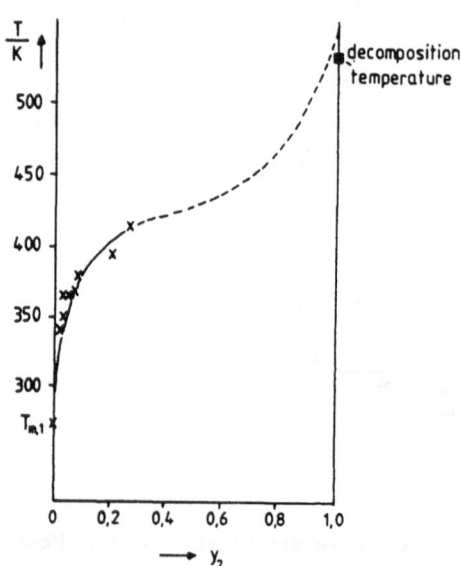

Fig. 4: State diagram of the system x-carrageenan-water

ACKNOWLEDGEMENT

The financial support by the "Deutsche Forschungsgemeinschaft" is greatfully acknowledged.

REFERENCES

[1] Smith, P. and Pennings, A.J., Polymer, 1974, 15, 413
[2] Smith, P. and Pennings, A.J., Polymer Sci. Polymer Phys. ed., 15, 523
[3] Borchard, W., Bremer, W., and Keese, A., Colloid & Polymer Sci., 1980. 258, 516
[4] Borchard, W., Luft, B. and Reutner, P., Ber. Bunsenges., 1984, 88, 1010
[5] Reutner, P., Luft, B., Borchard, W., Colloid & Polymer Sci., 1985, 263, 519
[6] Borchard, W. Luft, B. and Reutner, P., J. of Photographic Sci., 1986, 34, 132

PURIFICATION OF POLYMERS BY SUPERCRITICAL
FLUID EXTRACTION IN PROCESSING MACHINES

L.A. Kleintjens
DSM Research
P.O. Box 18
6160 MD Geleen
The Netherlands

ABSTRACT

Many organic (low-molecular) components can be dissolved in a variety
of simple fluids under near- or supercritical thermodynamic conditions.
The phase behaviour a such mixtures is nowadays well-understood and
several extraction procedures based on supercritical dissolution processes
are on stream. Laboratory equipment, as well as bench-scale units for such
high-pressure research, are commercially available and thermodynamic
models have been developed that can direct and correlate the experimental
research.

The high diffusion speed of small molecules (even under nearcritical
conditions) in comparison with ordinary solvents and the dissolution
selectivity of such systems can be used in the extraction of low-molecular
organic products out of amorphous polymeric materials. Some results for
acrylonitrile containing (co-) polymers/blends and thermoplastic polymers
like ethylene-propylene-diene terpolymers are reported.

First extruder experiments show that this technique can be successfully applied to polymers in processing machines.

INTRODUCTION

The ability of small molecular fluids under nearcritical conditions to
dissolve low-vapour-pressure solid materials was first discovered by
Hannay et al. (1). Scheffer and coworkers (2) investigated extensively the
solubility of naphthalene in near- and supercritical ethylene. Since then
many researchers have started to study the possibilities of supercritical
solvents and within the past two decades several research institutes have
investigated and developed the principles and technology of supercritical
fluid separations. Commercial application can be found in areas as diverse
as spice extraction, monomer purification, coal extraction, nicotine and
caffeine extraction, fractionation of (co-) polymers or the extraction of
oils from all kinds of natural products. Reviews of most of this work are

given eg. by Schneider et al. (3) and by McHugh et al. (4).
The basis of the success of this new kind operation is twofold, viz: the
high selectivity of supercritical fluids by fine-tuning the extraction
conditions and thereby the density, the interactions, polarity of the
fluid etc., and secondly the high diffusivity of such small molecules,
even in natural products like coffee beans or in coal. Furthermore, the
extracted products often can easily be obtained from the fluid by a simple
change in pressure and/or temperature.
    For a successful application of supercritical fluid extraction (SFE)
to a given system it proved to be necessary to:
a) Understand the phenomena going with supercritical phase behaviour.
b) Determine the actual phase equilibria by experiment. To reduce the
number of laborious experiments, an adequate thermodynamic correlation
model for nearcritical mixtures can be helpful.
c) Carry out bench-scale separation at proces conditions in the chosen
processing machine(s) and find the optimal separation procedure.
    In the following these three steps are discussed for the purification
of polymer material by SFE.

PHASE BEHAVIOR OF SUPERCRITICAL MIXTURES

Usually engineers try to stay away from the gas-liquid critical conditions
when they design processroutes or separation conditions. In SFE use is
made of the detailed phase behaviour of near- and supercritical fluid mixtures.
Principles of phase equilibria after Gibbs (5) and Van der Waals (6)
supply the ordering procedure for these phenomena. The application of
these principles has been greatly stimulated by the work of Bakhuis
Roozeboom (7).
The $p(T,x_b)$ diagram in Fig. 1 schematically shows the various Solid-Liquid
(S-L) and Liquid-Liquid (L_1-L_2) equilibria occurring in a binary system
with a heavy component $\underline{b}$ and a fluid $\underline{a}$. The nonvariant triplepoint of pure
component b,(Δ), is located at a higher temperature than the vapour/liquid
(V/L) critical point of solvent $\underline{a}$, C_a. We also assume the solubility of $\underline{b}$
in $\underline{a}$ to be small.
Addition of low-molecular component $\underline{a}$ makes the S_bLV equilibrium ($\bullet$) mono-
variant and leads to a depression of the melting temperature of $\underline{b}$. At the
same time the pressure must rise because of the high volatility of the
solvent and we have an S_bLV curve starting from the triplepoint of $\underline{b}$.
There are three branches representing the vapour (V), liquid (L) and solid
(S_b) phases respectively which, when projected on the front $p(T)$ plane,
form a single S_bLV curve. Quite typical for supercritical dissolution is
the phenomenon that at the second critical endpoint Ce_2, $(V = L)_2S_b$, the V
and L branches interfere with the $V = L$ critical curve C_aC_b which connects
the VL critical points of $\underline{a}$ and $\underline{b}$. In Ce_2 vapour and liquid become identical and we have a critical fluid in nonvariant equilibrium with a solid
phase. At temperatures below Ce_2 we only have two-phase fluid/solid
equilibria until we reach the first critical point Ce_1. Here we have the
other intersection between the S_bLV curve and the critical curve coming
from Ca. Below T_{Ce_1} the V and L phases are no longer identical. The vapour
pressure of solutions saturated with $\underline{b}$ is only slightly smaller than that
of the pure solvent because of the small solubility of $\underline{b}$ in $\underline{a}$.
Supercritical extraction of solids is preferably carried out at a temperature between T_{Ce_1} and T_{Ce_2}.

If the component to be extracted is a liquid, we have another class of
supercritical separations where only fluid phases play a role. In such

systems only gas-gas, gas-liquid and liquid-liquid phase separations will
appear in a Bakhuis Roozeboom diagram.
Depending on the sign of the excess volume ΔV^e miscibility in the liquid
phase increases or decreases with rising pressure (negative or positive
ΔV^e, respectively). Various possible types of phase behaviour have been
comprehensively reviewed by Schneider (8).

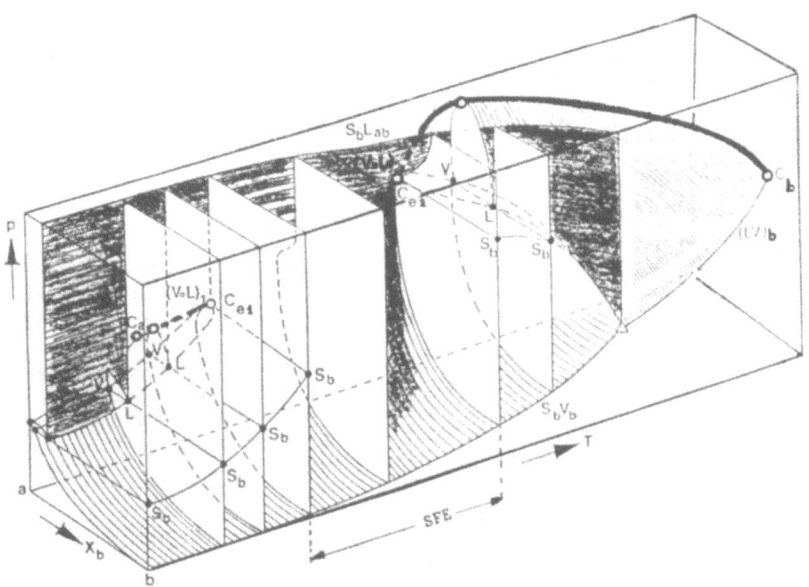

Figure 1. Bakhuis-Roozeboom (pTx) diagram for a binary system with two
critical end-points, showing the temperature range within SFE
preferably is carried out (see text).

THERMODYNAMIC MODELS FOR SFE

Since Van der Waals, many thermodynamic models have been proposed. Some of
them can be applied more or less successfully to nearcritical systems.
Such models can be classified in semi-empirical methods (9-13),
corresponding states approaches (14-17), pertubation theories (18-20) and
equation of state models (21-28), of which the latter were shown to be
well able to correlate experimental solubilities in the nearcritical
region.

For polymer systems it is common to use rigid lattice models (29-32).
However, since in SFE the phases differ in density such treatments are
unsuitable. Lattice gas description (32-36) can circumvent this problem
and combine the advantage of lattice statistics with the flexibility of
equation of state models. Fig. 2 shows an example of the phase behaviour
for the system linear polyethylene/n-hexane at nearcritical conditions,
calculated with the mean-field lattice-gas model (35).

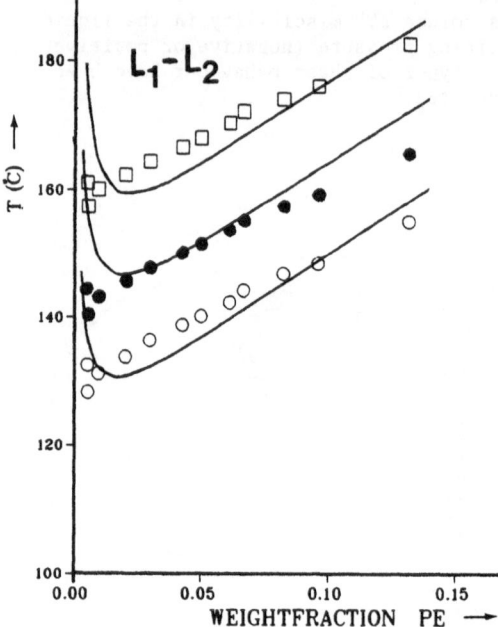

Figure 2.

Comparison between experimental
cloud points and spinodals
calculated with the mean-field
lattice-gas model for
n-hexane/linear polyethylene
at 6 bar (o), 25 bar (●) and
50 bar (□) (L. van Opstal,
private communication).

EXPERIMENTAL

A. Equipment

Phase boundaries and phase compositions at nearcritical conditions
can be measured today in commercially available stirred window-autoclave
systems for temperatures up to 200 °C and a pressure up to 1000 bar. We
developed a system in which we further can take samples at equilibrium
under constant pressure and temperature. This was possible via an involved
piston system that compensates for the decrease in volume during the
sampling operation.
To study the relation between extraction time and structure of the
material as a function of p and T, a semi-continuous extraction system
with a wide range in flow and a sampling system is required. We opted
for a (4 1) double autoclave system with a gas-compressor, a high-pressure
liquid pump and a sampling system.
To optimize supercritical extraction under processing conditions we made
use of:

i. A bench scale reactor with:
- a feed system, in which we have in parallel a gas compressor (400 bar) and
a liquid pump (350 bar) connected to a dissolution vessel (850 bar,
250 °C). The latter contains two sapphire windows, through which the
system can be inspected via a TV system.
- an extraction system; we have two possibilities, viz. a 4 1 batch
autoclave (400 bar, 200 °C, opened and loaded again in a few minutes) or a
two metre long extraction column (200 bar, 200 °C).

- a separation system, in which we decompress the extraction stream and collect the extracted material in a 4 l autoclave (200 bar, 200 °C).
- a recycling system for the extraction fluid.
     The bench-scale unit is equipped with an on-line gas chromatograph for the characterisation of process streams.

ii. Single-screw extruder

Although in general both a single- and a double-screw extruder can be used for SFE experiments we choose a Schwabenthan single-screw (D = 30 mm) extruder, equipped with a hopper, a three zone screw (L = 35 D), an injection unit for the extraction fluid at 20 D from the hopper and a capillary spin opening (∅ 1.7 mm).
The screw consisted of a feeding-zone (L = 10 D), a compression zone (L = 8 D) and a pumping zone (L = 15 D), including one mixing part (L = 5 D).
The pressure distribution in the extruder is set in such a way that the polymer and the gas will be transported together to the spin opening. Thus the extruder does not need any device for separating the polymer from the extraction fluid. The latter just expands to atmospheric pressure, leaving the purified polymer for further processing.

B. Results

i. Solubility of impurities in supercritical fluids

The possibilities of SFE were studied on glassy polymers and on thermoplasts, viz. acrylonitrile containing (co-) polymers and ethylene-propylene-diene terpolymers, respectively. We first investigated, in window-autoclave systems, the solubility of the impurities in CO_2 and/or ethylene. Fig. 3 shows some typical examples. Most organic impurities of the studied polymers could be dissolved at pressures and temperatures well within the reach of the standard processing equipment.

ii. Extraction in bench-scale reactor

The extraction of impurities of three types of acrylonitrile containing polymers were investigated in our bench-scale unit.
The polymer was loaded as spheres of about 2 mm diameter.
The extraction was carried out with CO_2 as SFE-fluid. Extraction conditions and experimental results are reported in table I.

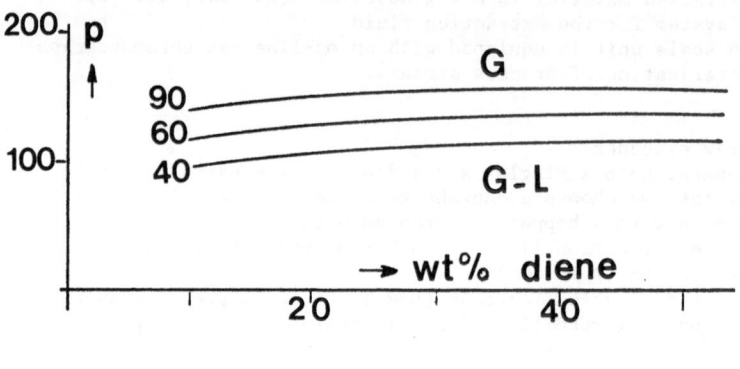

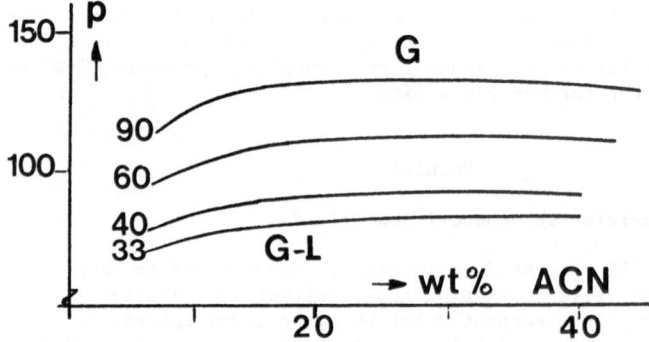

Figure 3.

Solubility in CO_2 at indicated temperatures.

TABLE I
Result SFE of ABS and SAN with CO_2

	temp. °C	pressure bar	ext. time min.	ACN in mass %	α-MSt in mass %	St in mass %
original ABS A				< 0.01	0.03	0.30
A1	40	180	30	< 0.01	< 0.01	< 0.01
A2	40	180	60	< 0.01	< 0.01	< 0.01
A3	40	180	15	< 0.01	< 0.01	< 0.01
A4	40	300	15	< 0.01	0.01	0.17
A5	35	180	15	< 0.01	< 0.01	0.03
original ABS B				0.0077	0.19	0.16
B1	40	180	90	0.0004	0.01	< 0.01
B2	40	180	300	0.0003	0.01	< 0.01
original SAN				0.53	< 0.01	2.27
S1	40	180	30	< 0.01		< 0.01
S2	40	180	60	< 0.01		< 0.01
S3	40	300	60	< 0.01		< 0.01
S4	35	300	15	< 0.01	< 0.01	0.58

In the same set-up we performed extractions of impurities out of EPDM (crumb) with CO_2 as SFE-medium. Extraction conditions and results are reported in table II.

TABLE II
Results SFE of EPDM with CO_2

Test	T (°C)	P (bar)	time (min.)	C_6 mass %	Diene mass %	H_2O mass %
original EPDM-I				5.3	0.94	28
EI-1	60	200	30	2.1	0.57	19
EI-2	60	200	30	1.9	0.58	20
EI-3	60	200	60	1.6	0.52	23
EI-4	60	200	210	0.03	0.03	16
EI-5	60	200	120	0.02	0.03	17
original EPDM-II				0.60	1.35	1.4
EII-1	60	200	30	≤ 0.01	0.01	0.01
EII-2	60	200	15	< 0.01	0.03	0.02
EII-3	60	200	5	< 0.01	0.27	0.24
EII-4	40	150	30	< 0.01	0.04	0.11
EII-5	40	150	15	< 0.01	0.15	0.2
EII-6	35	80	30	< 0.01	0.29	0.4
EII-7	35	200	15	< 0.01	0.11	0.33

Extraction with ethylene as SFE-fluid resulted in comparable quality of purification. The extraction conditions had to be slightly modified for this extraction medium.

iii. Extraction in single-screw extruder

First experiments were carried out to investigate the possibilities of SFE in this type of processing equipment. An EPDM rubber containing: 1.2 mass % solvent (C_6), 0.95 % diene and 54 % H_2O is extracted in a single-screw extruder (15.2 grams/min.) at 72 °C and 165 bar for 2 min. resulting in an impurity content of the processed material of 0.2 % C_6, 0.5 % diene and 2 % H_2O.

Another EPDM sample contained: 0.7 mass % C_6, 0.95 % diene and 22 % H_2O. The sample was processed at a speed of 16.8 grams/min. in a single-screw extruder and extracted with CO_2 at 75 °C and 175 bar for 2.4 min. The resulting rubber contained 2 % H_2O, 0.2 % C_6 and 0.5 % diene.

DISCUSSIONS AND CONCLUSIONS

As was shown before crystalline and glassy polymers (37) can be purified from low-volatile organic monomer rests and additives by SFE. In this work we show that also in amorphous polymers as well as in polymer melts SFE can be applied suprisingly succesfull for the removal of monomer rests. The estimated diffusion coefficient for supercritical gases in this type of polymers is orders of magnitude higher than that of liquid solvents. Nevertheless, the usual extraction time in processing is not long enough

for complete purification (38, 39). Extended extraction-time leads to surprisingly clean polymer materials.
Such materials may have improved performance compared with standard commercial products e.g. the heat distortion temperature of ABS proved to be raised by over 6 °C by a simple SFE extraction with CO_2.
First extruder experiments on EPDM rubber demonstrate that standard equipment can easily be modified for SFE (40). For successful extractions an extraction time of 2 minutes is not long enough.
Better results will be obtained when the mixing part of the extruder is extended and longer extraction times can be achieved.
Extruders can be equipped with such a screw arrangement and experiments with longer extraction times are planned.

REFERENCES

1. J.B. Hannay and J. Hogarth, Proc. R. Soc. London <u>29</u>, 324 (1879).

2. e.g. G.A.M. Diepen and F.E.C. Scheffer, J. Am. Chem. Soc. <u>70</u>, 4081, 4085 (1948).

3. G.M. Schneider, E. Stahl, G. Wilke, 'Extraction with supercritical gases', Weinheim, W-Germany, Verlag Chemie (1980).

4. M. McHugh, Ber. Bunsenges, Phys. Chem. <u>88</u> (1984).

5. J.W. Gibbs, Collected Works, Vol. I Dover Publ., Reprint, New York 1961.

6. J.D. van der Waals, Ph. Kohnstamm, 'Lehrbuch der Thermodynamik', Barth, Leipzig 1912, Vol. II.

7. H.W. Bakhuis Roozeboom, 'Die heterogenen Gleichgewichte vom Standpunkte der Phasenlehre', Vieweg, Braunschweig 1913.

8. G.M. Schneider, Chem. Thermod. Vol. II, Specialist Period. Repts., Chem. Soc. London 1978, p. 105.

9. P.L. Chueh and J.M. Prausnitz. AIChE J. <u>13</u>, 1107 (1967).

10. J.J. Czubryt, M.M. Meyers and J.C. Giddings, J. Phys. Chem. <u>74</u>, 4260 (1970).

11. E.U. Franck, Z. Phys. Chem. <u>6</u>, 23 (1956).

12. G.L. Rössling and E.U. Franck, Ber. Bunsenges. Phys. Chem. <u>87</u>, 882 (1983).

13. N. Gangoli, Ind. Eng. Chem. Prod. Res. Div. <u>16</u>, 209 (1977).

14. J.S. Rowlinson, 'Liquids and Liquid Mixtures', 2nd ed. Butterworths, London 1969.

15. I.R. McDonald, 'Statistical Mechanics', Vol. I, p. 134, Spec. Period. Reports, The Chem. Society, London 1973.

16. A.S. Teja and J.S. Rowlinson, Chem. Eng. Sci. <u>28</u>, 529 (1973).

17. e.g. J.W. Leach, P.s. Cheppelear and T.W. Leland, Proc. Am. Petrol. Inst. 46, 223 (1966) and AIChE J. 14, 568 (1968).

18. G.A. Mansoori, W.F. Carnahan, K.E. Starling and T.E. Leland, J. Chem. Phys. 54, 1523 (1971).

19. J.D. Weeks, D. Chandler and H.C. Anderson, J. Chem. Phys. 54, 5237 (1971).

20. K.E. Gubbins and C.H. Twu, Chem. Eng. Sci. 33, 363, 879 (1978).

21. M.L. McGlashan, K. Stead and C. Warr, Int. Conf. on Chem. Thermod. Vienna 1973, J. Chem. Soc. Faraday II 73, 1889 (1977).

22. R.L. Scott, Int. Conf. on Chem. Therm., Vienna 1973.

23. D. Peng and D.B. Robinson, Ind. Eng. Chem. Fundam. 15, 59 (1976).

24. G. Soave, Chem. Eng. Sci. 27, 1197 (1972).

25. O. Redlich and J.N.S. Kwong, Chem. Rev. 44, 23 (1949).

26. R.T. Kurnik, S.J. Holla and R.C. Reid, J. Chem. Eng. Data 26, 47 (1981).

27. S. Peter and H. Wenzel, Ber. Bunsenges. Phys. Chem. 76, 331 (1972).

28. M.E. Mackay and M.E. Paulaitis, Ind. Eng. Chem. Fundam. 18, 149 (1979).

29. A.J. Staverman and J.H. van Santen, Rec. Trav. Chim. 60, 76 and 640 (1941).

30. P.J. Flory, J. Chem. Phys. 10, 51 (1942); 12, 425 (1944).

31. M.L. Huggins, Ann. N.Y. Acad. Sci. 43, 1 (1942).

32. G. Kanig, Kolloid Z. Polym. 190, 1 (1963); 233, 829 (1969).

33. H.G. Killian, Kolloid Z. & Z. Polym. 252, 353 (1974).

34. I.C. Sanchez and R.H. Lacombe, J. Phys. Chem. 80, 2352 and 2568 (1976).

35. L.A. Kleintjens, Ph.D. Thesis, Essex Univ. UK, 1979.

36. L.A. Kleintjens, R. van der Haegen and R. Koningsveld in 'High Pressure Chemistry and Biochemistry', ASI-Series C197, Reidel Publ. Co. 1987, pag. 157.

37. G. Braun and R. Steiner, Preprints: 'High Pressure Chem. Engineering', GVC Symposium Erlangen, W-Germany 1984, pag. 297.

38. Eur. Patent Appl. 183.314.

39. US Patent 4.725.667.

40. Eur. Patent Appl. 233.661.

DETERMINATION OF THE THERMODYNAMIC PROPERTIES OF THE SYSTEM POLYETHYLENE OXIDE/WATER

A. Michalczyk and W. Borchard

Angewandte Physikalische Chemie der Universität-GH-Duisburg, FRG

ABSTRACT

The thermodynamic properties of the system polyethylene oxide (PEO)/ water were investigated with osmotic measurements in a temperature interval ranging from 283 - 323 K. To carry out these experiments in a large concentration range a differential technique has been used by measuring the osmotic pressure difference between two solutions of different concentrations. Measurements of osmotic vapour pressure with a low molecular polymer and those of membrane osmosis with a high molecular polymer indicate that the polymer associates in solution as a function of temperature. The Flory-Huggins-parameter is independent of concentration and dependent on temperature in the investigated temperature and concentration range.

INTRODUCTION

The understanding of the thermodynamic behaviour of PEO in aqueous solutions is important for predicting the phase behaviour of these compounds and many of polyoxyethylene derivates. PEO has a wide application in technics, surface and colloid chemistry, medicine and pharmacy. The purpose of this work is, to determine the thermodynamic functions in a large range of concentration and temperature with the aid of vapour pressure (VPO) and membrane osmotic (MO) measurements. Furthermore we have used a differential measuring technique in VPO, which is a modified steady state method we have developed [1, 2]. A similar method was introduced in the membrane osmosis by Rehage and Meys [3].

THEORETICAL CONSIDERATION

In the Flory-Huggins theory for polymer solutions the difference of the chemical potentials of the solvent in the solution and pure state $\Delta\mu_1$ reads, taking the base molar fraction of the solvent x_2^* as the generalized concentration variable,

$$\Delta\mu_1/RT = \ln x_1^* + (1 - 1/r_n)x_2^* + \sum_{i=1}^{N} \chi_i x_2^{*(i+1)} , \qquad (1)$$

where r_n is the number average degree of polymerisation and T the temperature, χ_i are the Flory-Huggins-parameters independent of concentration and dependent on temperature. The osmotic pressure Π is related to the difference of the chemical potentials and that of the free enthalpy of mixing ΔG by the expression

$$\left(\frac{\partial \Delta G}{\partial n_1}\right)_{T,P,n_{i\neq 1}} = \Delta\mu_1 = - \Pi V_1 \approx - \Pi V_{01} , \qquad (2)$$

where it is assumed that the partial molar volume V_1 of the solvent is independent of pressure and concentration. This means that V_1 may be represented by the molar volume of the solvent V_{01}. Differentiation of eq. (1) with respect to x_2^* at constant temperature and pressure and considering eq. (2) leads to the expression

$$1/(1 - x_2^*) - (1 - 1/r_n) - \frac{V_{01}}{RT}\left(\frac{\partial \Pi}{\partial x_2^*}\right)_{T,P} = \sum_{i=1}^{N} (i+1)\chi_i x_2^{*i} . \qquad (3)$$

If small values for the difference Δx_2^* of the concentrations of two polymer solutions are chosen, one can write

$$\lim_{\Delta x_2^* \to 0}\left(\frac{\Delta \Pi}{\Delta x_2^*}\right)_{T,P} = \left(\frac{\partial \Pi}{\partial x_2^*}\right)_{T,P} . \qquad (4)$$

With eq. (3) and (4) it is possible to determine the χ_i-parameters over a large range of concentration by measuring the osmotic vapour pressure difference of two solutions with slightly different concentrations.

Expansion of the logarithmic term in eq. (1) into a power series of x_2^* up to i = 1 and using eq. (2) leads to the well known expression

$$\frac{V_{01}}{RT} \frac{\Pi}{x_2^*} = \frac{1}{r_n} + \left(\frac{1}{2} - \bar{\chi}\right)x_2^*, \tag{5}$$

where $\bar{\chi}$ is a mean value of the interaction parameter, which is independent of concentration and molar mass. In a plot of the reduced osmotic pressure $\frac{V_{01}}{RT} \frac{\Pi}{x_2^*}$ against x_2^* the ordinate section should be a constant, independent of temperature. r_n is given by $r_n = M_n/M_0$. The quantities M_n and M_0 are the mean molar mass of the polymer and the molar mass of the monomer.

MATERIALS AND METHODS

PEO 2000 for synthesis (Merck-Schuchardt) was chosen for osmotic vapour pressure measurements. Fructose p.a. (Fa. Merck) was used as calibrating substance. For the membrane osmosis PEO 35000 (Hoechst AG) was taken. This product was fractionated by slowly cooling a solution (10 wt.% PEO) of a mixture water/acetone (15/85 wt./wt.). This procedure was carried out to make sure, that the sample doesn't contain portions of low molecular polymers < 20000 g/mol. The membranes (Sartorius, Göttingen, regenerated cellulose Cat.-Nr. 11539) have a lower limit for determination of molar masses of 20000 g/mol.

The vapour pressure osmometer (Dr. Knauer, Berlin) was provided with a universal-two-thermistor-probe for all kinds of liquids. The probe is one part of a Wheatstone-bridge, which is connected with an external amplifier, an analog meter and a recorder. The membrane osmometer is from the same producer.

RESULTS AND DISCUSSION

In fig. 1 we have plotted a mean $\bar{\chi}$-parameter versus the reciprocal temperature [4]. The temperature dependence of the $\bar{\chi}$-parameter in the range

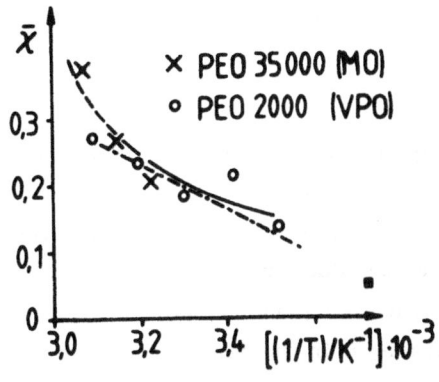

This work for PEO 2000/water:

$$\begin{cases} \alpha = 1,427 \\ \beta = -374,05 \text{ K} \\ r_n = 33 \end{cases}$$

$\beta = -709$ K from calorimetric measurements [4]

■ $\bar{\chi}$-value from [5]

Fig. 1: Mean $\bar{\chi}$-parameter for PEO 2000 and PEO 35000 in the concentration range up to 10 wt.% versus reciprocal temperature

Fig. 2: Plot of the reduced osmotic pressure $\frac{V_{01}}{RT}\left(\frac{\Pi}{x_2^*}\right)$ versus x_2^* for PEO 2000

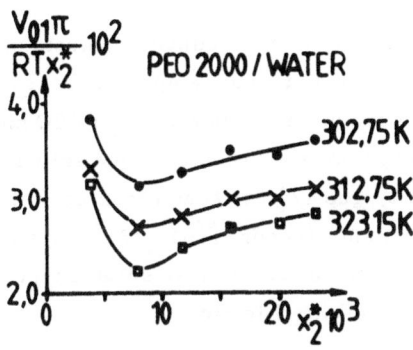

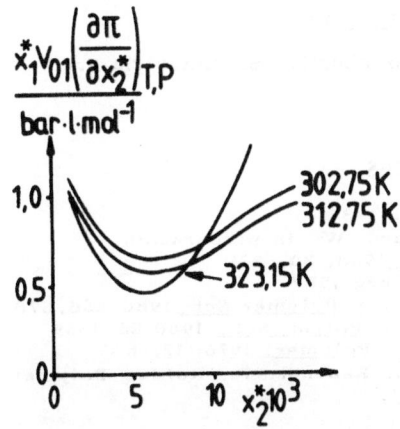

Fig. 3: Plot of the quantity $x_1 V_{01}\left(\partial\Pi/\partial x_2^*\right)_{T,P}$ versus x_2^* of PEO 2000 for different temperatures

of 283 - 323 K for the system PEO 2000/water is represented by the well known expression [3]:

$$\bar{\chi} = \alpha + \beta/T .\qquad(6)$$

where α is the entropy and β the enthalpy term of the mean $\bar{\chi}$-parameter. The value for β (PEO 2000) is in agreement with the results of direct calorimetric measurements in the same temperature interval [5]. For temperatures T>323 K eq. (6) does not hold and the additional term $\gamma \ln T$ has to be added in order to describe the temperature dependence of the enthalpy and entropy of mixing. The mean $\bar{\chi}$-parameter in the concentration range up to 10 wt.% PEO is consistent with the phase diagram of Hager and Macrury [6]. They have fitted their experimental data of the liquidus curve in the system PEO/water also with a mean interaction parameter of $\bar{\chi} = 0.05$, shown in fig. 1. The evaluation of the state diagram PEO/water presented on this symposium will be published later [8].

The plot of the reduced osmotic pressure versus the base molar fraction for the different temperatures (fig. 2) does not show a constant ordinate section. This indicates, that PEO molecules tend to associate with increasing temperature in aqueous solution. As the system has a closed miscibility gap [7] for temperatures well above 373 K, we expect the thermodynamic factor $x_1^*\left(\partial\Delta\mu_1/\partial x_1^*\right)_{T,P} = x_1^* V_{01}\left(\partial\Pi/\partial x_2^*\right)_{T,P}$ to tend to zero values at the lower critical solution point. This is indicated by our findings represented in fig. 3.

ACKNOWLEDGEMENT

The financial support by the "Fonds der chemischen Industrie e.V." is greatfully acknowledged.

REFERENCES

[1] Adames. W., Diplomarbeit, Duisburg 1985
[2] Adames, W., Michalczyk, A., Borchard, W., in preparation
[3] Rehage. G., Meys. H., J. Polym. Sci.,1958, **30**, 271
[4] Michalczyk. A., Diplomarbeit, Duisburg 1987
[5] Schönert, H., Monshausen, F., Colloid & Polymer Sci., 1980, **258**, 578
[6] Hager. S.L., Macrury, T.B., J. Applied Polym. Sci., 1980 **25**, 1559
[7] Saeki. S., Kuwahara, N., Kaneko, M., Polymer, 1976, **17**, 685
[8] Borchard, W., Dobnik. E., Luft, B., Reutner, P., Rolduc Polymer Meeting 3. 1988, this volume

THERMODYNAMICS OF POLYMER MIXTURES: FACTORS
AFFECTING MISCIBILITY IN POLYMER BLENDS

CONSTANTINOS G.PANAYIOTOU
Chemical Process Engineering Research Institute
and Department of   Chemical Engineering
University   of Thessaloniki
Thessaloniki,540 06, GREECE

ABSTRACT

Some thermodynamic aspects of polymer-polymer compatibility are
discussed in this work.the analysis is made with the Lattice-
Fluid theory of polymer solutions as modified recently by the
author.the theory has been extended to multicomponent and mu-
ltigroup systems and is,thus,applicable besides others to mixtures of polydisperse polymers and mixtures of random copoly-
mers.the effect of pure component properties,of pressure and
of polydispersity on the critical behavior of polymer mixtures
are examined.theoretical estimations are compared with experimental data whenever available.A satisfactory agreement is
observed between theory and experiment.

INTRODUCTION

Polymer-polymer miscibility is one of the fields in Polymer
Science and Technology which has attracted particular attention in the last fifteen years(1-3).Numerous pairs ot polymers
(many more than one could imagine in the early seventies)have
been found showing very interesting price-performance-composi-
tion profiles.A low degree of miscibility is,sometimes,suf-
ficient for the blend to have important properties.Intimately
associated with polymer miscibility is their lower critical
solution temperature(LCST) behavior and,thus,methods for
its prediction are of both theoretical and technological im-
portance.Although LCST behavior is the rule for homopolymers,
mixtures of random copolymers may,equally well,show upper
critical solution temperature (UCST)behavior (4-6).Besides
this,judiciously chosen random copolymers,when added to a
blend of two (co)polymers, may have beneficial effects on
the properties of the blend by enhancing their miscibility.In
view of all this, it is understandable that successful

theoretical frameworks for correlating and predicting the thermodynamic behavior of (co)polymer blends must,at least, be able to predict both the LCST and the UCST behavior. In addition,their formalism must be fairly general,so that they can be applied to various cases of mixtures including mixtures of homopolymers or random copolymers,binaries or multicomponent, consisting of monodisperse or polydisperse polymers.

A thermodynamic model meeting all the above requirements is presented in the next section. It is based on the Lattice-Fluid theory of Sanchez and Lacombe(7) as modified recently by the author (8-12).So far the model has been applied to solvent-homopolymer and homopolymer-homopolymer(both monodisperse)mixtures(10),to the gas solubility in polymeric liquids (8),to the glass transition temperatures in polymer mixtures (9),to mixtures of two random copolymers(6), to mixtures of associated substances(11) and very recently(12) it has been reformulated in terms of contributions of functional groups. In this work we will discuss some further aspects of the thermodynamics of polymer mixtures such as the effect of pure component properties on polymer-polymer miscibility, the influence of pressure on the LCST, and the influence of polydispersity on the phase behavior of polymer mixtures.

THEORY

In this section we will present only the essentials of the Lattice-Fluid(LF) theory.Further details of the formalism may be found in the literature(6-11). According to the LF model, each molecule consists of a number of constitutional repeat units (CRU) which may or may not be the same. In the case of dissimilar CRU's they are assumed distributed randomly along the molecule (random copolymer).Each such unit of type i is characterized by three scaling constants, namely an interaction energy $\varepsilon_i^* = s_i \varepsilon_i / 2$, a number of segments r_i, and a hard core volume per segment υ_i^*. The molecular parameter s_i is the number of intermolecular contacts per segment. An equivalent set of parameters are the characteristic temperature $T_i^* = \varepsilon_i^* / R$, the characteristic pressure $p_i^* = \varepsilon_i^* / \upsilon_i^*$ and the close-packed density $\rho_i^* = M_i / (r_i \upsilon_i^*)$,where R is the gas constant and M_i the molar mass per repeat unit.Entirely analogous scaling constants $(\varepsilon^*, r, \upsilon^*, T^*, P^*, \rho^*)$ may be defined,in the one-fluid approach, for the mixtures in terms of the volume fraction φ_i and the surface fractions ϑ_i of each CRU of type i. Thus υ^* in the mixture may be obtained from the classical quadratic mixing rule

$$\upsilon^* = \sum_i \varphi_i^2 \upsilon_i^* + 2 \sum_i \sum_{j>i} \varphi_i \varphi_j \upsilon_{ij}^* \qquad (1)$$

One may, quite generally, write for the intersegmental hard core volume υ_{ij}^*.

$$\upsilon_{ij}^* = \xi_{ij} \left(\frac{\upsilon_i^{*1/3} + \upsilon_j^{*1/3}}{2} \right)^3 \qquad (2)$$

ξ_{ij} being a binary parameter for the pair i-j. By assuming $\upsilon_{ij}^{*}=0.5(\upsilon_{i}^{*} + \upsilon_{j}^{*})$ one may recover the simple mixing rule, $\upsilon^{*}=\Sigma \; \varphi_{i}\upsilon_{i}^{*}$, proposed by Sanchez and Lacombe(7).

The interaction energy ε^{*} in the mixture is given by

$$\varepsilon^{*}=\Sigma_{i} \; \varphi_{i}\varepsilon_{i}^{*} - RT \; \Sigma_{i} \; \Sigma_{j>i} \; \varphi_{i}\vartheta_{j}X_{ij} \qquad (3)$$

where

$$X_{ij} = \frac{\varepsilon_{i}^{*} + \dfrac{s_{i}}{s_{j}} \varepsilon_{j}^{*} - 2\zeta_{ij}\sqrt{\dfrac{s_{i}}{s_{j}}\varepsilon_{i}^{*}\varepsilon_{j}^{*}}}{RT} \qquad (4)$$

ζ_{ij} is a binary interaction parameter for the pair of units i-j. The ratio s_{i}/s_{j} may be obtained from molecular size and shape characteristics of the pair i-j(10).

Phase equilibrium may be dealt with through the equality of chemical potentials of each component in all phases while phase stability may be studied through the appropriate spinodal conditions and the conditions for the critical points(13). The present LF model uses an "entropic" correction term in the expression for the chemical potential entirely analogous to the corresponding correction term of the Equation-of-State theory of Flory and coworkers(14) and which is equal to $-r_{i}\vartheta_{j}^{2}q_{ij}$ for a binary mixture i-j. The "entropic" parameter q_{ij} is a unitless adjustable binary parameter. This correction term does not appear in the expressions for the excess volume and the excess enthalpy of the mixture.

APPLICATIONS AND DISCUSSION

The first application we will discuss in this section is the influence of the pure component properties (basically the volumetric properties) on the phase behavior in a mixture of two homopolymers. A mixture for which all necessary information is available(7,10) is the mixture of Polystyrene(PS) with the Poly(vinyl methyl ether)(PVME). Scaling constants for PS are obtained from Reference 7 while scaling constants for PVME and binary parameters are obtained from Reference 10. In Figure 1 is shown the influence of the thermal expansion coefficient α of PS on the spinodal curve of the mixture of PS and PVME considered both monodisperse with molecular weight 51000. As shown in the Figure a small increase in the thermal expansion coefficient of PS of the order of 1% enhances the compatibility of PS with PVME by raising the spinodal curve and the critical temperature by about 30 degrees. An increase of α of PS of the order of 4% causes a shift of the spinodal curve by about 115 degrees centigrade. This dramatic influence of α on the phase behavior of polymer mixtures may easily be explained by realizing that a change in α causes primarily a change in T^{*} of PS with a concomitant change in X_{12}. Thus an increase of 1% in α of PS causes a decrease of (the negative) X_{12} of the order of 10% while an

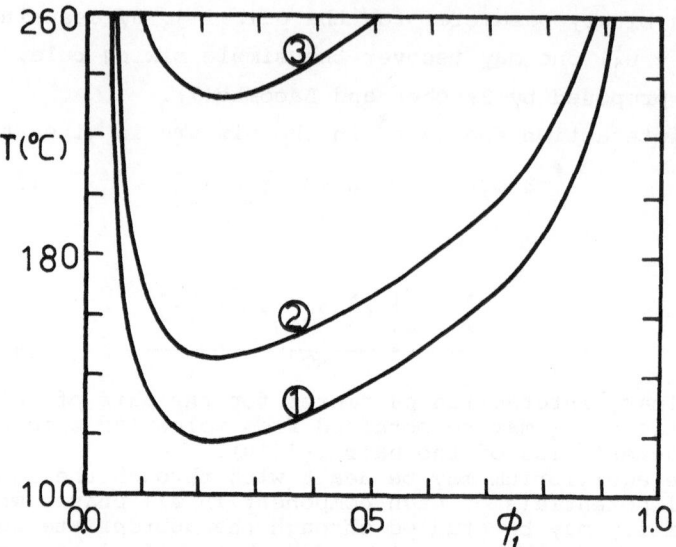

Figure 1. Spinodal curves for the system PS(1)-PVME(2).Curve 1: $\alpha_1=4.03\times10^{-4}K^{-1}$; 2:$\alpha_1=4.07\times10^{-4}K^{-1}$; 3:$\alpha_1=4.19\times10^{-4}K^{-1}$

increase of 4% in α causes a decrease of 38% in X_{12}.

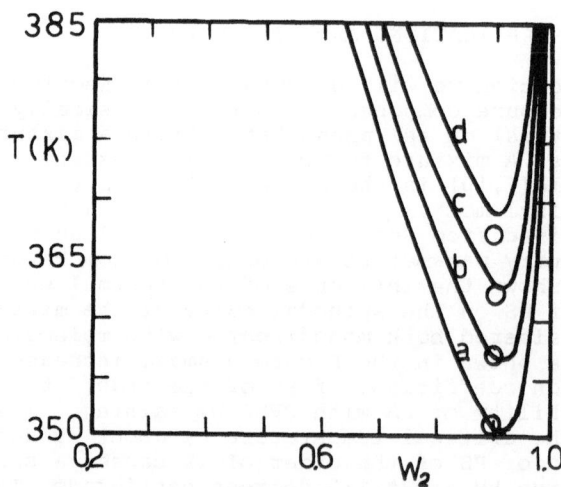

Figure 2. Spinodals for PES(1)-PEO(2).o:Exper cloud points(15). Curves a,b,c & d at 1,108,221 & 348 atm respectively

A change in the thermal pressure coefficient,γ,of one of the components causes,primarily,a change in the characteristic pressure P^* of this component.Such a change in P^* has no significant effect on the critical behavior of the polymer mixture at ordinary pressures,but it becomes significant at high pressures.

As a second application we will discuss the influence of pressure on the critical behavior of a polymer mixture.In Figure 2 are compared the theoretical estimations with the experimental data of Walsh and Rostami(15)for the system Poly(ether sulfone)(PES)(component 1)and Poly(ethylene oxide)(PEO). Scaling constants for the two polymers have been obtained from the known volumetric data(15) and are:$T_1^*=926K,P_1^*=432MPa,\rho_1^*=1393Kg/m^3,T_2^*=644K,P_2^*=402MPa,\rho_2^*=1155Kg/m^3$. The binary parameters ζ_{12} and ξ_{12} have been determined from the data on volumes of mixing and heats of mixing of the low molecular weight analogs and are $\zeta_{12}=1.1114$, $\xi_{12}=1.010$. s_1/s_2 was set equal to 0.83 as proposed by Walsh and Rostami(15). The entropic parameter q_{12} was estimated equal to 0.19 from the requirement that the calculated LCST of the system at 1atm be equal to the experimental LCST (15). On the basis of these binary parameters we calculated the spinodal curves shown in Figure 2.A rather satisfactory agreement between theory and experiment is observed for both critical temperatures and critical compositions. In addition, the shape of the spinodals closely resembles the shape of the experimental cloud-point curve at 1 atm(15).The enhancement of miscibility with the pressure which is observed in Figure 2 is easily understood to be caused not by a change in the interactional term X_{12} but by the increase in the density of the system and the change in the compressibility term in the equation for the spinodal (6,10).

As a third application we will examine the influence of polydispersity on the phase behavior of the system PS(1)-PVME(2). In Figure 3 are compared the theoretical estimations(spinodals) with the experimental(16) cloud points of a mixture of PS+PVME 50% by weight in PS. PVME is monodisperse with molecular weight equal to 51,500, while PS is a mixture of two monodisperse samples a and b with molecular weights a:10,000 and b:110,000. Calculations in Figure 3 have been made on the basis of the pure component and binary parameters of the literature(7,10).The abscissa in Figure 3 is the weight fraction of the a component in the polydisperse PS.As shown, the agreement between theory and experiment is satisfactory.

The model has,also,been applied to mixture of three random copolymers. Conditions for the enhancement of compatibility in polymer blends by the use of a third copolymer are examined as well as the influence of temperature in the phase behavior of ternary mixtures whose constituent binaries may show either UCST or LCST. These findings are, as yet, only theoretical and for space limitation they are not reproduced here. They will be, however, the subject of a forthcoming publication.

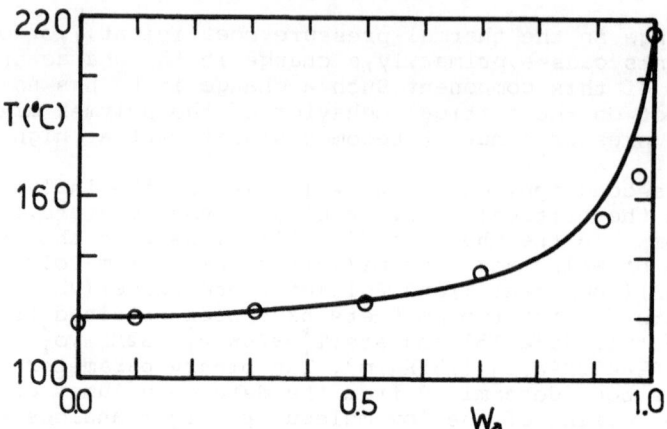

Figure 3. Influence of MW and polydispersity on the phase behavior of PS(1)-PVME(2) as explained in the text

REFERENCES

1. Paul,D.R.and Newman,S.,Eds.,Polymer Blends,Academic Press, New York,1978

2. Olabisi,O.,Robeson,L.M.and Shaw,M.T.,Polymer-Polymer Misci- bility,Academic Press,New York,1979

3. Solc,K.,Ed.,Polymer Compatibility and Incompatibility: Principles and Practice,Harwood,New York,1982

4. Ueda,H. and Karasz,F.E.,Macromolecules,1985,18,2719

5. Kamer,H.W.,Acta Polym.,1986,37,1

6. Panayiotou,C.,Makromol.Chem.,1987,188,2733

7. Sanchez,I.C. and Lacombe,R.H.,Macromolecules,1978,11,1145

8. Panayiotou,C.,Makromol.Chem.,1986,187,2867

9. Panayiotou,C.,Polymer J.,1986,18,895

10. Panayiotou,C.,Macromolecules,1987,20,861

11. Panayiotou,C.,J.Phys.Chem.,In Press

12. Panayiotou,C.,Polymer J.,Submitted for Publication

13. Kurata,M.,Thermodynamics of Polymer Solutions,Harwood, New York,1982

14. Flory,P.J.,Orwoll,R.A. and Vrij,A.,J.Am.Chem.Soc.,1964,86, 3515

15. Walsh,D.J. and Rostami,S.,Macromolecules,1985,18,216

16. Nishi,T. and Kwei,T.K.,Polymer,1975,16,285

EXTRA HIGH TEMPERATURE ACRYLICS AND THEIR ALLOYING POTENTIAL

L.C. SEDEREL
Rohm and Haas European Laboratories
Sophia Antipolis
06565 Valbonne, France

ABSTRACT

The properties of a new class of high temperature resistant imide-copolymers are discussed. Some of the most notable characteristics of these clear thermoplastic copolymers are, a very good weatherability, an exceptional stiffness, high heat capabilities and very good electrical, optical and barrier properties. The fact that the imide-copolymers are miscible and compatible with a large number of other polymers, make them prime candidates for alloying and coextrusion.

INTRODUCTION

PARALOID EXL 4000 series is a new class of high temperature resistant, clear thermoplastics developed by Rohm and Haas Company. The EXL 4000 systems are copolymers which, in their simplest form, can be thought of as N-methyl-dimethylglutarimide/methyl methacrylate copolymers.

Other variants include materials which can be described as terpolymers in which the third unit contains polar groups. EXL 4000 series bear some of the good properties of PMMA such as, very good weatherability (see Figures 1 and 2), and excellent electrical and optical properties (90% TWLT and 2% haze). In addition to the above mentioned properties, which are very similar to those of acrylics, they also have notable characteristics which differ strongly from those of acrylics, such as high heat capabilities, exceptional stiffness, a much better resistance to common solvents, a lower coefficient of linear thermal expansion, and considerably better oxygen barrier properties (two times better than bi-oriented PET). Figure 3 shows the barrier properties of EXL 4000 vs PC and of SAN. Some of the main properties of the EXL 4000 series are summarized in Table 1.

TABLE 1
Key properties of EXL 4000 series

PROPERTY	ASTM D	UNITS	EXL 4240	EXL 4150	EXL 4260	EXL 4170
Refractive index	542	–	1.53	1.53	1.54	1.54
Specific gravity	792	–	1.21	1.21	1.22	1.22
Flexural modulus	790	GPa	3.7	4.1	3.8	4.3
Izod impact	256	J/m	21	21	21	21
Rockwell hardness	785	–	M98	M98	M100	M100
Vicat-B	1525	°C	134	142	150	163
HDT-A (annealed)	648	°C	132	140	148	160
Coeff. of linear thermal expansion	696	°C-1	5.2E-5	R.2E-5	4.9E-5	4.7E-5
Water abs. (23°C)						
- 24 hrs immersion	570	%	0.3	0.3	0.3	0.4
- equilibr. 50% RH		%	1.2	1.3	1.2	1.6

The ability of the EXL 4000 series to wet pigments and fillers, and the fact that the EXL 4000 series prove to be miscible/compatible with a large number of polymers, make them prime candidates for alloying and coextrusion. Fowler, Paul, Cohen and Freed (1) investigated the miscibility of EXL 4000 systems with SAN and PVC. Most EXL 4000 polymers appeared to be miscible with PVC over the entire composition range. For optimal use in PVC, a specific imide-copolymer has been developed which is now marketed under the PARALOID HT 510 designation. In the case of the imide/SAN blends, miscibility appeared to be dependant upon the acrylonitrile content of the SAN copolymer and on the type of EXL 4000 polymer (see Figure 4). Miscible blends of SAN and imide copolymers were only obtained with imide copolymers which did not contain polar groups. In this paper the preparation and properties of other EXL 4000/polymer blends are described.

EXPERIMENTAL

Materials
The following EXL 4000 polymers were used for the study as described in this paper:- PARALOID EXL 4240, 4150, 4260 and 4170. The second digit indicates the presence/absence of polar groups in the imide copolymer.
    - 1: copolymer with polar groups.
    - 2: copolymer without polar groups.
The third and fourth digit indicate the Tg plus 100°C. Example: PARALOID EXL 4260, a copolymer without polar groups with a Tg of 160°C. Blends were prepared with polyamide 6, polyamide-12 and with clear ABS. In some cases a newly developped reactive, all acrylic Rohm and Haas core/shell impact modifier (PARALOID EXL 3386) was also included in the evaluation (2,3).

Methods

All EXL 4000/polymer blends were prepared on a Werner and Pfleiderer ZSK-30 twin screw extruder with a L/D of 24 and at a screw speed of 200 rpm. The melt was degassed. Polymers, EXL 4000 systems and modifier were all added in the throat of the extruder. After compounding, the blends were dried and injection moulded into test specimen on an Arburg Allrounder injection moulding machine. Properties of the polyamide blends were measured either 'dry as moulded' or after conditioning until equilibrium at 23°C and 50% RH.

RESULTS AND DISCUSSIONS

For several reasons the alloying of EXL 4000 with clear ABS is of interest. First of all, the RI of clear ABS (Terluran 2802 TR from BASF, RI=1.535) is very close to that of the EXL 4000 polymers, and secondly, ABS contains a large amount of SAN, of which the blending with the EXL 4000 series already has been investigated in the past (1) (see Figure 4). In Figure 5, the optical properties of Terluran 2802/EXL 4000 blends are shown.

The best optical properties were obtained with EXL 4260. Total white light transmission and haze remained on the original level of the clear ABS. Good optical properties were also obtained with EXL 4240 and 4170, whereas a significant amount of haze was generated in the EXL 4150/clear ABS blends. The addition of 20% EXL 4260 to clear ABS resulted in a modulus increase of 400 MPa, an HDT-A increase of 6°C, whereas impact was only very slightly reduced (Izod impact dropped from 79 to 71 J/m). Similar results were obtained with EXL/4000 polyamide blends. In all cases an increase of the modulus (especially in the conditioned state), an improved dimensional stability, and a higher HDT-A were found. In order to improve the slighly reduced impact strength, a newly developped reactive all acrylic core/shell impact modifier was added to some of the PA-6/EXL 4000 blends.

Figure 6 shows the influence of the addition of the EXL 4000 polymers and/or EXL 3386 on the tensile modulus of PA-6. Although both EXL 4150 and 4240 have the same influence on the modulus improvement of the PA-6/EXL 3386 blends, a marked difference is found between the effect of these two imide-copolymers on the Izod impact strength. As can be seen from the data presented in Table 2, the addition of EXL 4150 to the PA-6/EXL 3386 blends only has a small negative effect on the impact strength, whereas the addition of EXL 4240 appears to be detrimental for the impact strength of the impact modified polyamide blends. The PA-6/EXL 4150/EXL 3386 show a very interesting property profile, especially in the conditioned state; a very high impact strength in combination with a modulus and HDT-A which is higher than that of unmodified PA-6.

TABLE 2
Impact strength of PA-6/EXL 4000/EXL 3386 blends

EXL 4000 Type	%	% EXL 3386	IZOD IMPACT STRENGTH J/m) Dry as moulded	Conditioned
–	0	0	43	200
–	0	10	220	1131
EXL 4150	17.5	0	32	173
EXL 4150	17.5	10	125	960
EXL 4240	17.5	0	33	80
EXL 4240	17.5	10	52	117

ACKNOWLEDGEMENT

The author wishes to acknowledge Dr. Didier Leblanc and Mr, Augusto Negri of Rohm and Haas European Laboratories, and Mr. Herman Lippers of Twente University of Technology (The Netherlands), for their valuable contributions to this paper.

REFERENCES

1. Fowler, M.E., Paul, D.R., Cohen, L. and Freed, W.T., Blends of Imidized Acrylic Polymers with SAN Copolymers and with PVC, to be published.

2. Sederel, L.C., Mooney, J. and Weese, R.H., Super Tough Blends based on LV Nylon and an Acrylic Core/Shell Modifier, Proceedings of the European Symposium on Polymer Blends, May 25-27, 1987, Strasbourg (France).

3. Sederel, L.C., Impact Modification of High Performance Engineering Plastics, Proceedings of the High Performance Additives Conference, The British Plastics Federation, May 10-11, 1988, London (UK).

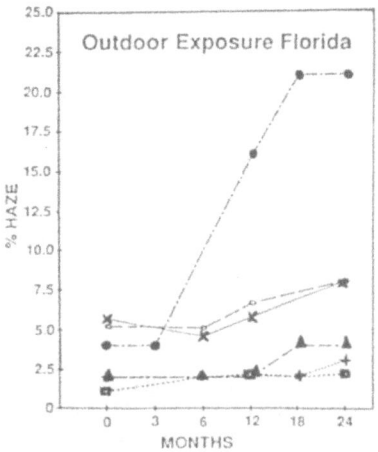

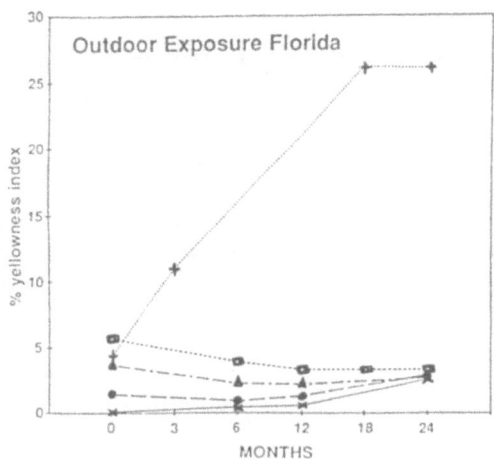

Figure 1. Haze vs outdoor exposure time in Florida (x,o: high impact PMMA, □: standard PMMA, ▲: EXL 4150, +: EXL 4240, ●: UV stab PC)

Figure 2. Yellowness index vs outdoor exposure time in Florida (x,●: high impact PMMA, □: EXL 4150, ▲: EXL 4240, +: UV stab PC)

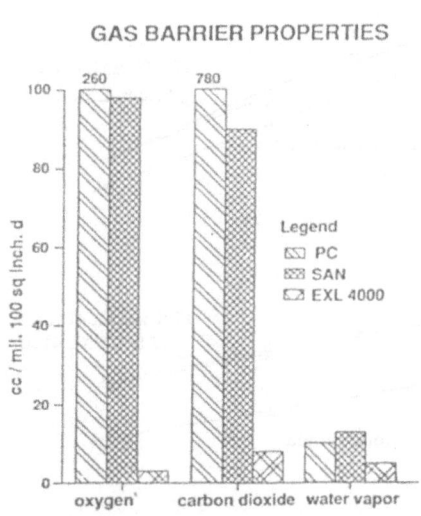

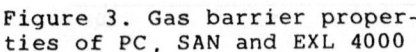

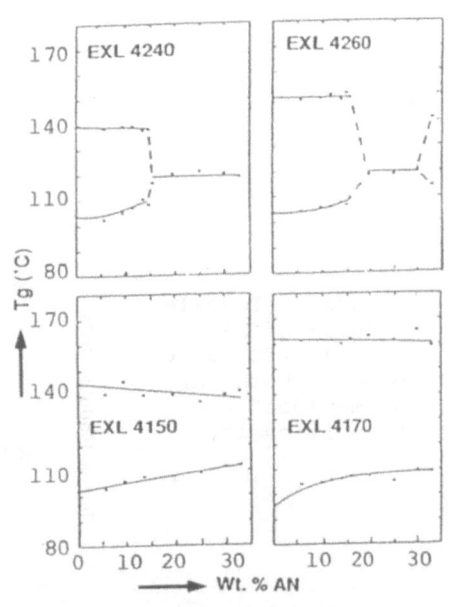

Figure 3. Gas barrier properties of PC, SAN and EXL 4000

Figure 4. Glass transition temperature for 50/50 blends vs AN content of SAN copolymer

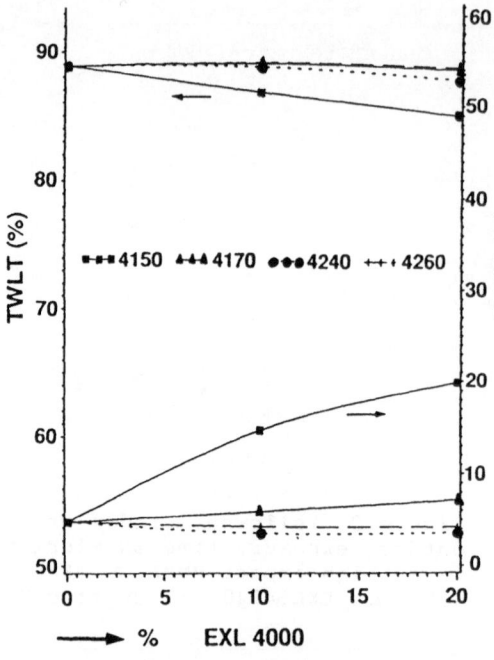

Figure 5. Optical properties of EXL 4000/ clear ABS blends.

Figure 6. Tensile Modulus of PA-6/ EXL 4000/EXL 3386 blends, both "dry-as moulded" and "conditioned".

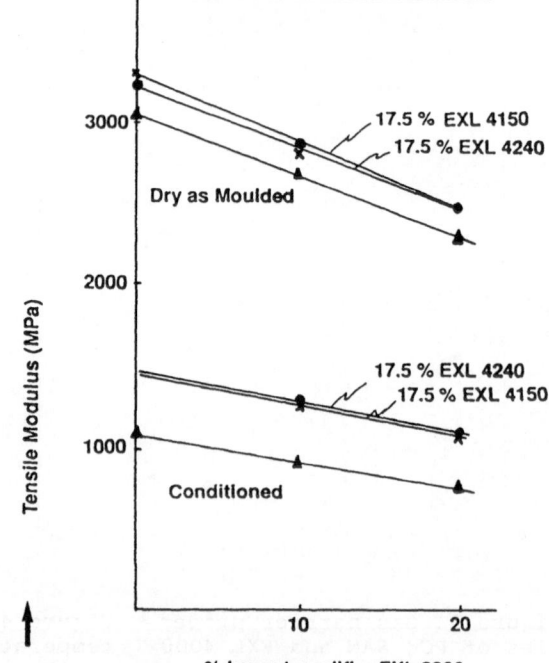

SUBINCLUSION MORPHOLOGY IN COMPATIBILISED POLYMER BLENDS

L. L. Ban and M. J. Doyle
Exxon Chemical Company, Polymers Business Group, Linden, NJ, USA.

M. M. Disko
Exxon Research & Engineering Co, Corporate Research Laboratories,
Annandale, NJ, USA.

G. Braun and G. R. Smith*
Exxon Chemical International Inc., Brussels, Belgium.

INTRODUCTION

The effect of polymer blend morphology on the stiffness and toughness of blends of ethylene-co-propylene rubber (EPR) in polyamides (PA) has been the subject of a number of recent papers and patents (1-7). In these toughened blends, the rubber exists as a separate phase of discrete particles. It has been shown that both the rubber particles size and total volume fraction have an effect on toughness and, especially, the ductile-to-brittle transition temperature (DBTT).

The effects of appropriate chemical functionality of the elastomer phase on its ability to form a refined morphology when dispersed in the nylon have also been related to the formation of graft copolymer during mixing. This graft copolymer can act as a dispersant to form a thermodynamically stable dispersed morphology (4). Through steric stabilisation, these graft copolymer molecules are also effective during the mixing process to reduce the efficiency of droplet coalescence, i.e. reagglomeration of the dispersed rubber particles as the mixing process progresses, or in subsequent processing (8).

Previous work (3, 6) has suggested that the spacing between the rubber domains is the critical parameter which correlates the DBTT with rubber content and particle-size. Work to date, however, appears to have calculated this interparticle spacing from the measured particle-size distribution and the known volume fraction of rubber added to the blend, rather than from a direct stereological analysis of electron micrographs. The former method assumes that there is no occlusion of the nylon continuous phase within the rubber particles during mixing.

Transmission electron microscopy (TEM) and quantitative Electron Energy Loss Spectroscopy (EELS) have been employed in the work presented here. It is shown that a subinclusion morphology of nylon particles within the rubber domains can occur if EPR is used which contains reactive sites capable of grafting to the nylon. In several blends of maleic anhydride grafted EPR and PA 66, blended in a twin screw compounding extruder, the rubber domains were found to contain subinclusions amounting to 40% of the domain volume. The corresponding blend using an unmaleated EPR showed little evidence of a subinclusion morphology.

Dynamical mechanical thermal analysis (DMTA) results also indicate that the effective rubber volume fraction is greater in the blends containing maleic anhydride grafted rubber.

Other results using optical microscopy are also presented. These illustrate how the incorporation of dispersed rubber alters the crystalline microstructure of the nylon matrix. Differential scanning calorimetry (DSC) shows that the rubber, whether functionalised or not, can accelerate the rate of nylon crystallisation.

RESULTS AND DISCUSSION

The elastomers used for this study were an amorphous EPR (44%wt ethylene, 50 ML Mooney viscosity) and a reactive elastomer formed by maleic anhydride grafting this same polymer (0.7%wt grafted maleic anhydride). The PA 66 used is Zytel 101 (DuPont Company). Blends of nylon and rubber in a 80:20 weight ratio were prepared by mixing in a 30 mm non-intermeshing, twin screw extruder with a compounding screw configuration. Barrel temperature in the mixing and conveying sections was 280 deg C.

For microscopy studies, thin films (less than 100 nm thick) were prepared by microtomy of individual pellets at -150 deg C. Different techniques were used to image the polymer phases in the transmission electron microscope (TEM). In the first, the microtomed films were exposed for 20 minutes to the vapour above a 0.5% aqueous solution of ruthenium tetraoxide (9). Figures 1 and 2 show TEM micrographs of the blends with unmodified and maleic anhydride grafted EPR respectively in the nylon blends. In both figures, the darker phase, which contains the ruthenium stain, is the nylon. In Figure 1, the blend containing the unmodified EP rubber is seen to have a phase size several microns in diameter and the rubber particles contain very few subinclusions. However, the blend containing maleated EPR, shown in Figure 2 has particles of less than 1 micron in size and all contain multiple subinclusions which individually range in size from 0.5 microns to less than 50 nm in diameter. From random lines drawn on the micrograph, the lineal fraction of intercepts of the rubber domains leads to a value of approximately 40 % vol. of subinclusions within each domain. A TEM similar to Figure 2 has been recently published (10), although the presence (or not) of graft copolymer in the blend was not mentioned.

In a second TEM technique, a thin layer of gold was evaporated under high vacuum onto a microtomed thin film of the blend containing the maleated elastomer. TEM in Figure 3, shows small islands of gold (of order 5 nm) on the surface of the specimen which are more densely packed on the crystalline polyamide than on the amorphous rubber. The presence of subinclusions within the rubber domain is clearly revealed by the gold decoration.

Transmission electron energy loss spectra (EELS) were obtained along the line A-B in Figure 3. The spectra were measured with 300 kV incident electrons and the single inelastic scattering distribution was obtained (11, 12). The spectra from the polyamide matrix and subinclusion were found to be very similar, indicating that the domain consists of nylon subinclusions within EPR rather than of two different mixed miscible phases of different composition (13).

Quantitive EELS analysis was conducted at fourteen probe positions along the line A-B in Figure 3. These results are shown in Figure 4. The nitrogen concentration in the polyamide matrix is 12 +/- 2 atomic % (expected value 14.3% for nylon 66), zero in the EPR and near 10% in the subinclusions. Lower values in the subinclusion could arise because the film thickness and subinclusion diameter are of similar size. Thus some EPR may be overlapping the subinclusion. Relative mass thicknesses (figure 4b) were computed from the elastic and inelastic intensities (12). Observation that the nitrogen concentration and thickness is similar in the subinclusion and in the matrix shows clearly that the subinclusion is chemically similar to the nylon matrix and passes nearly all the way through the sample volume.

Previous studies of high impact polystyrene (HIPS) found that increases in rubber volume fraction in HIPS materials due to subinclusions of polystyrene had the effect of decreasing shear modulus and increasing the loss tangent in the glass transition region of the rubber (14). Thus DMTA results (Polymer Laboratories equipment in flexure at 3 Hz) of the unmodified and chemically modified EPR/PA 66 blends are presented in Figure 5. For the nylon based blends studied here, a similar increase in loss tangent in the region of the rubber transition is found for the blend containing subinclusions, although the modulus values remain unaffected. It is worth noting that in the study of HIPS, subinclusions could account for up to 80% of the rubber phase volume. However, the TEM results above showed nylon subinclusions in these blends to be no more than 40% of the rubber particles phase, hence less significant DMTA variations are understandable.

Polarised light optical microscopy has been used to determine the crystalline structure of the PA 66 matrix in the blends. The micrographs shown in Figure 6 depict a finer PA 66 spherullitic structure in the blend containing the maleated EPR rubber. As was found previously (15), and as seen in figure 6b, the blend with unmodified EPR has a coarser crystal structure.

Kinetics of crystallisation of the blends have also been studied by following crystallisation isotherms using DSC. Figure 7 presents an Avrami plot (16) for PA 66 and for the nylon matrix blends containing either EPR or functionalised EPR. From these data, half-times for crystallisation at 240 deg C were calculated. For PA 66, the half time is 1.6 minutes. Unmodified EPR in the nylon lowers this to 1.4 minutes, while with functionalised EPR in the blend, the value is 1.3 minutes. As reproducibility is only +/- 0.1 minutes, the difference between the two blends may be insignificant. However, it appears that the presence of EPR may increase the crystallisation rate of the nylon matrix.

CONCLUDING REMARKS

Extruder compounded blends of polyamide 66 and functionalised EP rubber have a more complex morphology than has been previously assumed. TEM, EELS and DMTA data, presented here, lead to the conclusion that the rubber domains in these blends contain typically 40% volume of occluded nylon as multiple subinclusions. By contrast, blends of nylon with unmodified EPR have less than 1% of subinclusions. These occlusions probably arise due to steric stabilisation effects of EPR-g-nylon compatibiliser molecules which may encapsulate a nylon droplet within the EPR particle and effectively prevent coalescence of that subinclusion with the nylon matrix phase. It is suspected that this phenomenon also occurs in other, similarly compatibilised, polymer blends. Its occurrence is likely to be influenced by a number of factors, including the relative compatibility of the polymers being blended, the rate of compatibiliser formation versus the rate of dispersion during mixing, and the effectiveness of the compatibiliser as a steric stabiliser of the subinclusions.

Rubber incorporation into the nylon also affects crystallinity. Although the overall extent of PA 66 crystallinity is not greatly altered (15), the rate of crystallisation and final nylon spherullite size is affected by the rubber particles. Presumably due to smaller rubber particles, functionalised EPR also leads to a more refined nylon crystal structure than does unmodified EPR in the blends studied here.

In 1959 (17) it was found for nylon that a finer spherullite size led to higher yield strength. Later work (17), however, has suggested a more complex relationship of properties and crystallity for polypropylene (which may similarly also apply to polyamides).

Current micro-mechanical understanding of functional elastomer-nylon blends presumes a continuum behaviour of the nylon matrix and rubber particle inclusions, i.e. that the properties of the matrix are unaffected by the inclusion of rubber toughening particles and that the rubber particles do not contain subinclusions (2-7). Results presented here, however, suggest additional complexity in nylon-rubber blends. The effects of nylon crystallinity and rubber particles subinclusion morphology on yield properties, toughness and the ductile-brittle transition temperature clearly require further mechanistic consideration.

REFERENCES

1. US patent 4,174,358.
2. Wu, S., J. Poly Sci., Phys. Edit., 1983, 21, 699.
3. Wu, S., Polymer, 1985, 26, 1855.
4. Wu, S., Poly. Eng. & Sci., 1987, 27, 335.
5. Borggreve, R.J.M., Gaymans, R.J. and Luttmer, A.R., Macromol. Chem., Macromol. Symp., 1987 (in Press).
6. Borggreve, R.J.M., Gaymans, R.J., Schuijer, J. and Ingen Housz, J.F., Polymer 1987, 28, 1489.
7. Cimmino, S., D'Orazio, L., Greco, R., Maglio, G., Malinconico, M., Mancarella, C., Martuscelli, E., Palumbo, R. and Ragosta, G., Poly. Eng. & Sci., 1984, 24, 48.
8. Elmendorp, J.J. and Van der Vegt, A.K. Poly. Eng. & Sci., 1986, 26, 1332.
9. Trent, J.S., Scheinbeim, J.I. and Couchman, P.R., Macromolecules, 1983, 16, 589.
10. Sawyer, L.C., and Grubb, D.T. "Polymer Microscopy" Chapman and Hall, New York, 1987, p. 211.
11. Leapman, R.D., Fiori, C.E., and Swyt, C.R., J. Microscopy, 1984, 133, 239.
12. Egerton, R.F., "Electron Energy Loss Spectroscopy in the Electron Microscope", Plenum Press, New York, 1986.
13. Ban, L.L., Doyle, M.J., Disko, M.M. and Smith, G.R., Polymer Comm., 1988 (in press).
14. Bucknall, C.B. and Hall, M.M., J. Materials Sci., 1971, 6, 95.
15. Hahn, M.T., Hertzberg, R.W. and Manson, J.A., J. Materials Sci., 1983, 18, 3551.
16. Avrami, M., J. Chem. Phys., 1939, 7, 1103; 1940, 8, 212; 1941, 9, 177.
17. Starkweather, H.W. and Brooks, R.E., J. Appl. Poly. Sci., 1959, 1, 236.
18. Way, J.L., Atkinson, J.R. and Nutting, J., J. Materials Sci., 1974, 9, 293.

121

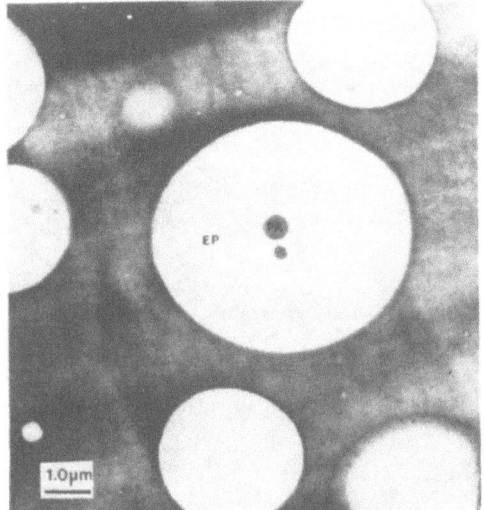

FIGURE 1 TEM OF PA66/EPR (20 WT %) BLEND.

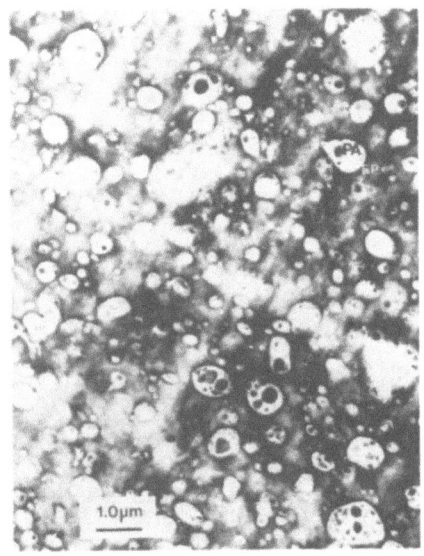

FIGURE 2
TEM OF PA66/MALEATED EPR (20 WT %) BLEND.

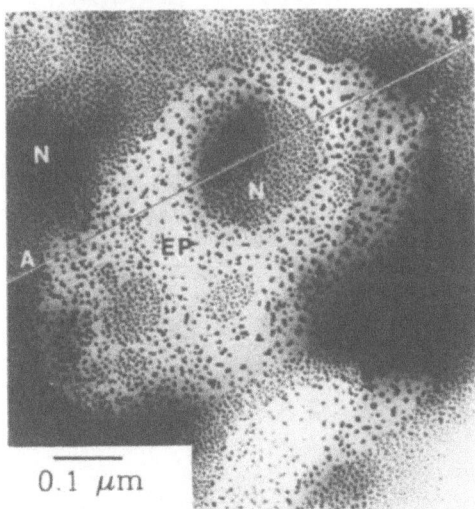

FIGURE 3 TEM OF A GOLD DECORATED THIN FILM FROM THE PA66/ MALEATED EPR BLEND.

FIGURE 4
(A) NITROGEN CONCENTRATION AND (B) MASS-THICKNESS PROFILES FOR THE A-B LINE SCAN SHOWN IN FIGURE 3.

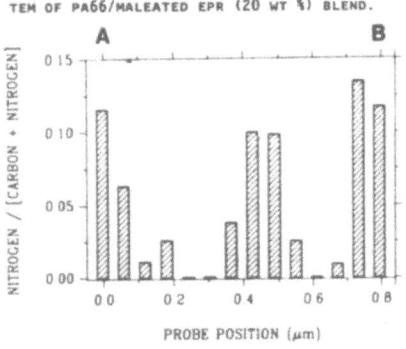

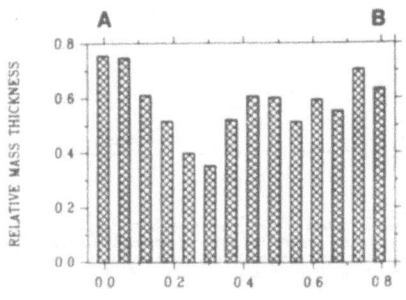

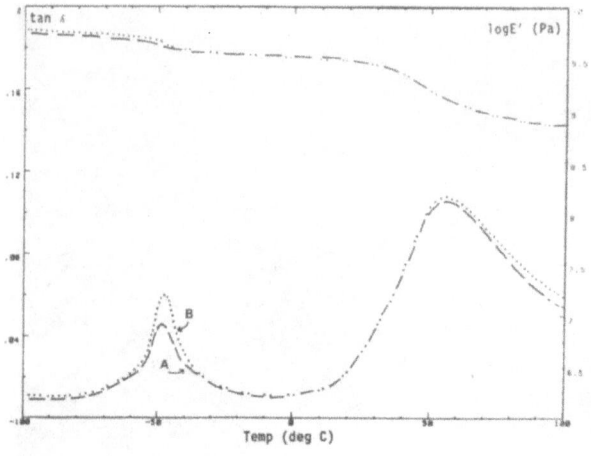

Figure 5
Shear modulus and loss tangent versus temperature for PA66 blends
containing unmodified (A) or functionalised EPR (B).

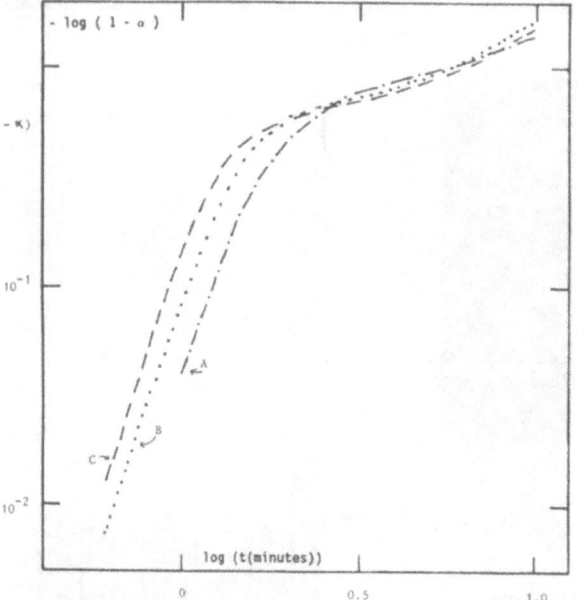

FIGURE 6
POLARISED LIGHT OPTICAL
MICROGRAPHS OF PA66 BLENDS
CONTAINING UNMODIFIED (A)
OR FUNCTIONALISED EPR (B).

Figure 7
Avrami plots for unmodified PA66 (A) and the PA66 blends
containing unmodified (B) or functionalised EPR (C). α is the wt.
fraction of uncrystallised material after time, t, at isothermal
conditions (240 deg C).

THE MODIFICATION OF EPDM-RUBBER WITH MALEIC ANHYDRIDE BY REACTION BLENDING

A.J. Oostenbrink, R.J.M. Borggreve, R.J. Gaymans
University of Twente
Department of Chemical Technology
PO Box 217, 7500 AE Enschede, The Netherlands

ABSTRACT

Modification of EPDM rubber was carried out in a corotating twin screw extruder. The effect of process conditions, initial maleic anhydride (MA) and initiator concentration on the total amount of grafted MA was studied. Quantitative infrared analysis and potentiometric titration were used to determine the amount of grafted MA. Within the operating limits of the extruder, process conditions had only little effect on the total amount of grafted MA. Increasing throughput rates at high set temperatures resulted in a small decrease in grafted MA. The initiator concentration had a strong effect on the amount of grafted MA. Higher initiator concentrations led to a higher grafting efficiency. With increasing initial MA concentration, a sharp decrease in grafting efficiency was established.

INTRODUCTION

Polyamide-6 has a fair impact resistance but can be made super tough by modifying with rubber. For this purpose the rubber (10-20%) has to be present as a very fine dispersion, with a particle size less than 0,5 μm [1]. A fine dispersion of ethylene-propylene-dicyclopentadiene ([EPDM]) rubber in nylon can be obtained if the EPDM is modified with maleic anhydride (MA)[1]. MA acts as an interfacial agent, i.e. during blending with PA-6, grafted MA reacts with the NH_2 end groups or amide groups of PA-6. In this paper we report on the modification of EPDM-rubber with MA in a twin screw extruder.

In literature several authors reported on the grafting of MA on PP, PE and EP rubber in solution [2]. For EP copolymers the following reaction mechanism is supposed [2]. Primary radicals formed by peroxide decomposition abstract hydrogen atoms from the polymer chains. MA molecules may add to these macro radicals thus forming succinic anhydride radicals covalently bonded to the rubber. These rubber-MA-radicals have a low tendency to homopolymerize. Cimmino [2] observed no homopolymerization. Propagation reaction takes place when these radicals abstract protons from the rubber. Gaylord [3] suggested that grafting of EP rubber with maleic anhydride under conditions at which peroxides have short half life times (these conditions are reached in the extruder) results in the formation of reactive MA-species. These excited MA species increase the radical generation on the polymer beyond that attributable to the radicals from peroxide. Formation of poly-MA grafted rubber is also possible by repetitive coupling of an MA excimer followed by ionic coupling.

EXPERIMENTAL

The functionalization of EPDM with MA is carried out in a Berstorff 25 mm corotating twin screw extruder. EPDM (Keltan 740; DSM), Perkadox 14 (bis(tert. butyl peroxy isopropyl) benzene; AKZO; half life times: 5 min at 175 °C; 10 sec at 210 °C [4]) and Nourymix(50/50 wt% masterbatch of MA on polyethylene; AKZO) were mixed and starved fed to the first section of the extruder. A nitrogen atmosphere was maintained over the reaction mixture throughout the extruder. In the last sections free maleic anhydride and other volatile components were removed by applying vacuum on the last vent. Standard extrusion conditions were: throughput rate: 1 kg/h; residence time: 5 minutes; rotor speed: 150 rpm; Mass temperatures in the die: 220-260 °C. The MA-content in the rubber was determined with potentiometric titration and infrared analysis.

RESULTS AND DISCUSSION

Relation between λ and MA content

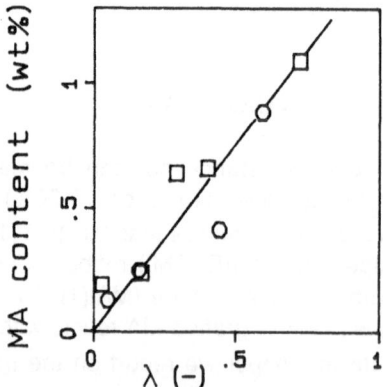

Figure 1. MA content values obtained from potentiometric titration versus λ obtained from infrared analysis.

Quantitative infrared analysis of functionalized rubber films is a far more simpler, faster and cheaper method than potentiometric titration. The quotient (λ) of the absorbance peak of the carbonyl stretching vibration of the anhydride group at 1785 cm^{-1} and the C-H stretching vibration of the CH$_2$-group at 715 cm^{-1} should be directly proportional to the MA content. In figure 1 the results of relating λ to the wt% MA grafted measured with potentiometric titration, are given. λ and MA content showed a linear relationship given by : MA content (wt%) is 1,6 times the value of λ. Deviations from this line are introduced by local inhomogeneities in the samples.

Influence of processing variables
The effect of throughput on the MA-coupling was studied at relatively low (140-195 ºC) and high(135-220 ºC) barrel temperature settings (figure 2a). From figure 2a it can be concluded that with the low temperature profile, throughput has no effect on the reaction efficiency (34%). When a high temperature profile was applied, however, the efficiency decreased from 34% at low throughput rate to 23% at the highest throughput rate.

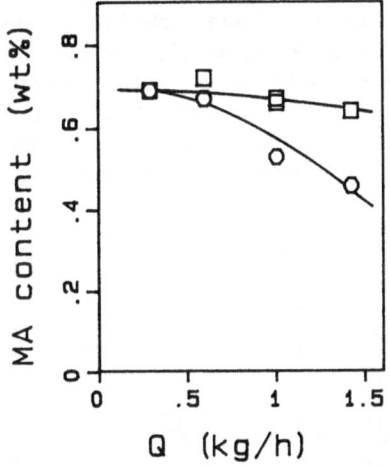 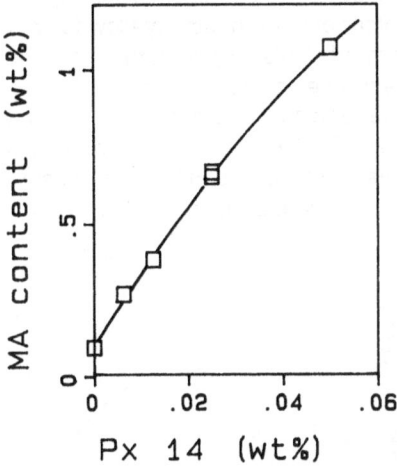

Figure 2 a) Effect of throughput rate at low (□) and high (o) barrel set temperatures (mass temperatures of the rubber 223 and 260ºC).
b) Influence of peroxide concentration on grafted MA content. (initial MA concentration : 2 wt%)

Effect of initiator concentration
At constant processing conditions and constant initial MA concentration (2 wt%) the influence of initiator concentration was studied (fig 2b). At fixed MA concentration (2 wt%), increasing the peroxide concentration results in a higher amount of polymer radicals generated and/or a higher concentration of excited

MA. According to figure 2b this is, as expected, leading to a higher grafted MA content. Rheology measurements showed that the melt viscosity of the functionalized rubber increased with peroxide concentration. A concentration of 0.025 wt% Perkadox 14 increased the melt viscosity in comparison with the virgin EPDM rubber over a wide shearrate range by 50% .

Effect of initial MA concentration

At constant initiator concentration (0.025 wt% Px14) and constant processing conditions the effect of the initial MA concentration on the total amount of grafted MA was studied(fig. 3a).

After an initial increase of the grafted MA content the amount of grafted MA shows a decrease with increasing initial MA concentration. Increasing the MA concentration at fixed peroxide concentration should increase the probability of a reaction between MA and an elastomer radical. So this decrease in absolute amount of grafted MA with higher initial concentration can not be explained by the radical propagation reaction mechanisms.

According to the reaction mechanism by Gaylord, however, the MA content in the elastomer could decrease with increasing initial MA concentration owing to the quenching of excited MA by an increasing amount of ground state MA. The colour of the products (yellow to light brown) indicates that some polyMA has been formed. This suggest that the mechanism proposed by Gaylord holds. The decrease in the total amount of grafted MA can also be a result of another phenomenon. From earlier experiments with a single screw extruder it is known that with high MA concentrations (4 wt%) phase separation of MA and rubber exists. Part of the initiator is possibly lost in this separated MA phase leading to a less efficient grafting reaction.

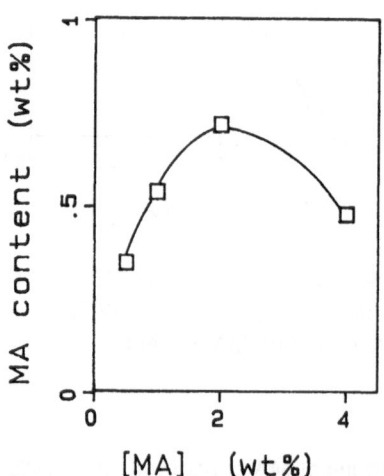 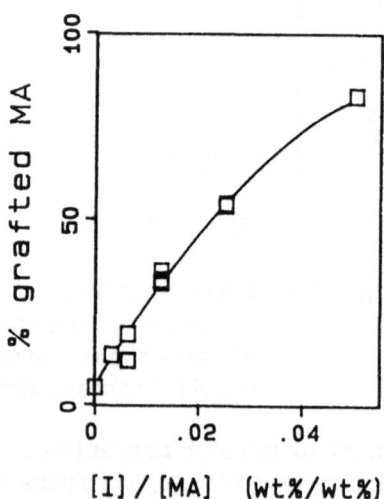

Figure 3   a) Total amount of grafted MA versus initial MA concentration.
b) Grafting efficiency versus peroxide/MA quotient.

The viscosity of the functionalized rubber was independent of the initial MA content, as could be concluded from the negligible differences in die pressure during extrusion of the different compositions. Rheology measurements showed also no difference.

Figure 3b shows that in the range of initiator and MA concentrations studied, the ratio initiator/MA determines the grafting efficiency. This result is in accordance with Borggreve [5] who produced modified rubbers with a single screw extruder. He observed that if the initiator/MA ratio is kept constant, the grafting efficiency is independent of the MA concentration.

CONCLUSIONS

Functionalization of EPDM by grafting with maleic anhydride in a twin screw extruder is possible in a controlled way. Within normal operating limits of the extruder good grafting efficiency could be achieved with Perkadox 14 as initiator. It is shown that at constant process conditions the grafting efficiency is determined only by the initiator/MA ratio, and that the melt viscosity is influenced by the initiator concentration.

REFERENCES

1. R.J.M. Borggreve, R.J. Gaymans, J. Schuijer and J.F. Ingen Housz, Brittle-tough transition in nylon rubber blends: effect of rubber concentration and particle size, Polymer, 1987, **28**, 1489-96.

2. S. Cimmino, L. D'orazio, R. Greco, G. Maglio, M. Malinconico, C. Mancarella, E. Martuscelli, R. Palumbo and G. Ragoista, Morphology-Properties Relationships In Binary Polyamide 6/Rubber blends: Influence of the Addition of a Functionalized Rubber, Polymer. Eng. and Sci, 1984, **24**, no 1, 48-56.

3. N.G. Gaylord, M. Mehta and R. Mehta, Degradation and Crosslinking of EPR on reaction with MA and/or Peroxides, J. Appl. Polym. Sci, 1987, **33**, 2549-58.

4. W. Hofmann, Crosslinking agents in Ethylene-Propylene Rubbers, Progress in Rubber and Plastics Technology, **21**, no 2 (1985), 18-55.

5. R.J.M. Borggreve, R.J. Gaymans, D. Heikens, submitted to Polymer.

Irradiation of Isotactic Polypropylene and
Polypropylene/Ethylene-Propylene-(diene-monomer) Blends

J.G.M. van Gisbergen, J.I. Meijerink and N. Overbergh
Department of Polymer Technology
Eindhoven University of Technology
P.O. Box 513
5600 MB Eindhoven
The Netherlands

Abstract

The influence of electron beam irradiation on rheological properties and morphology of polypropylene and polypropylene/ethylene-propylene rubber blends was studied. Electron beam irradiation of isotactic PP causes pronounced chain scission (degradation) at dosis $\leq$ 100 kGy. Melt viscosity can be controlled easily up to this dose for pure PP. For the blend, however, an anomalous rheological behaviour is observed.

Attempts have been made to induce a special morphology and to retain this morphology during subsequent shaping processes via fixation by radiation induced crosslinks.

Radiation crosslinking of the dispersed EPDM phase and a simultaneous grafting of PP onto EPDM is proposed to be the explanation for both the anomalous rheological and an enhanced morphology fixation. This idea is supported by gelfraction measurements.

Introduction

High energy irradiation (electron beam or gamma) of polymeric materials results in a multitude of chemical reactions[1,2]. The two main reactions are crosslinking, as observed in polyethylene and chain scission, for example polyisobutylene. In polypropylene with its structure in between polyethylene and polyisobutylene, both reactions are observed. The relative importance of chain scission (degradation) over crosslinking depends on the physical state, the irradiation parameters, presence of oxygen etc. Irradiation at room temperature in the presence of oxygen mainly results in chain-scission up to doses of about 500 kGy[3].

Degradation of materials is less desirable, but controlled scission can be of industrial importance like e.g. the use of degraded irradiated polytetrafluoro-ethylene as lubricating agent. For polypropylene an attractive feature could be to polymerize one base-grade of polymer and to produce numerous subgrades via controlled degradation.

Rubber blending of PP is used to improve its impact properties. Miscibility, however, is exception rather than rule in polymer systems and consequently processing of polymer blends is quite complicated with respect to control of morphology. An induced morphology may be preserved using radiation techniques to crosslink the dispersed phase in polymer blends [4,5]. In isotactic-polypropylene/ethylene- propylene (diene monomer) blends the rubber particles will crosslink whereas the PP matrix will degrade upon irradiation [1,2,3]. Therefore, fixation of the original morphology in blends of PP and EPDM by irradiation may result in improved flow behaviour preserving the good low temperature properties as flexibility and impact strength.

The influence of dose on the chain scission reaction of the main chain of i-PP in both the pure homopolymer and the blend with EPDM has been studied. Also some preliminary experiments with respect to morphology fixation will be presented.

Experimental

Materials used, kindly obtained from DSM (the Netherlands), were PP 13E 10 (polypropylene) and two types of EPDM, Keltan 514 and Keltan 578. Before exposure to irradiation PP was first compression moulded into sheets with a thickness of 1 mm. Blends were made on a two roll mill (Schwabentahn, T = 185°C) and the blends were also compression moulded into 1 mm thick sheets. Some of the blends were made on a corotating twin-screw extruder (Berstorff ZE25) and were pelletized. Both sheets and pellets were exposed to electron beam irradiation which was performed

with a 3 MeV "Van de Graaff" electron beam accelerator (IRI, Delft).
Doses were used between 0 and 100 kGy. Both irradiated and non-irradiated
pellets were injection moulded on a Stübbe SKM 75/80 injection moulding
machine. A schematic representation of the whole process is shown in
Figure 1.

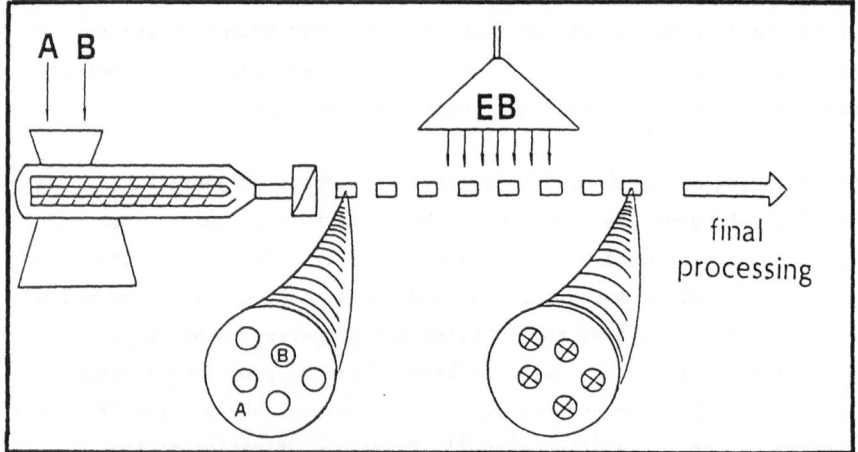

Figure 1.   Schematic view of crosslinking the dispersed phase (B)
             with (electron beam) irradiation between the blending and
             processing step.

Morphologies of the blends were characterized with a Cambridge 2000
stereoscan Scanning Electron Microscope. Samples were cut at a
temperature of –150°C and afterwards the rubber particles were extracted
during 5 seconds in boiling xylene vapour. The surface was then covered
with a thin gold layer.

Rheological experiments were performed with a Rheometrics RDS–II and a
Philips selfmade dynamic spectrometer. Circular samples for this
measurement were cut from the compression moulded sheets.

For DSC measurements the samples were first molten at a temperature of
200°C, quenched to 110°C for isothermal crystallization and afterwards
heated at a rate of 10°C/min. A Perkin Elmer DSC 7 was used.

Results

Dynamic mechanical measurements (Figure 2) show a decrease of molecular
weight upon irradiation of PP resulting in a reduced viscosity. The melt
viscosity vs frequency also shows that irradiated PP behaves more
Newtonian indicative of a narrowing of the molecular weight distribution
[6,7]. In blends of PP and EPDM, however, the latter effect is not
observed. Differences between unirradiated PP and unirradiated blends are
small, indicating that the viscosity of the blend is mainly determined by
the PP matrix. For irradiated blends a strong influence of the
crosslinked EPDM particles is observed, indicating a strong interaction
between the PP matrix and these particles.

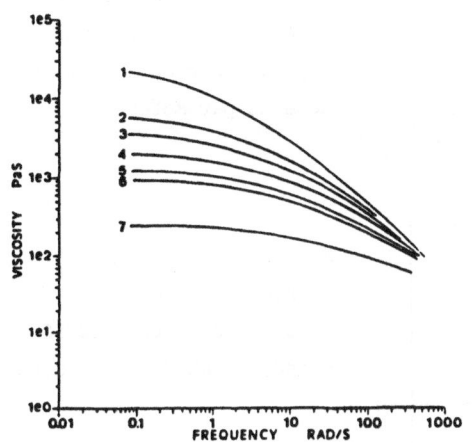

Figure 2a Dynamic viscosity
vs frequency for irradiated
Polypropylene, Temperature
180 °C:

1; 0 kGy, 2; 10 kGy, 3; 20
kGy, 4; 30 kGy, 5; 40 kGy,
6; 50 kGy, 7; 100 kGy.

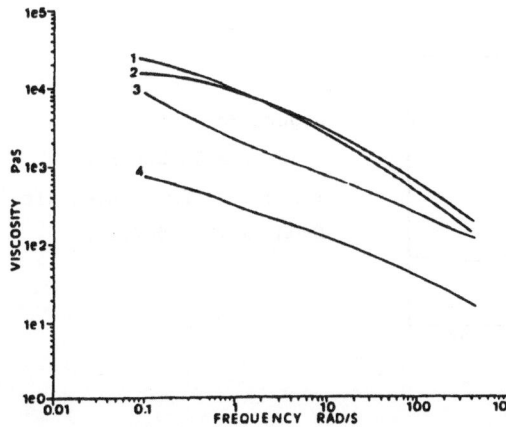

Figure 2b Dynamic viscosity
vs frequency for irradiated
Polypropylene/EPDM blends
(70/30).

1; K514, 0 kGy, 2; K578, 0
kGy, 3; K514, 43 kGy, 4;
K578 43 kGy.

Differences between irradiated blends and irradiated PP are also
visualized by DSC. Upon irradiation of pure PP two melting peaks become
visible (Figure 3a). This is explained in terms of a faster
reorganization from imperfect to perfect crystals [8], as a result of
reduced melt viscosity. For the blend this behaviour is observed only at
higher dosis (Figure 3b). When irradiated PP (with a dose of 50 kGy) is
blended with unirradiated EPDM again clearly two peaks are visible
(Figure 3c). The behaviour shown in Figure 3 for K578 acting as the
dispersed phase was also observed for K514.

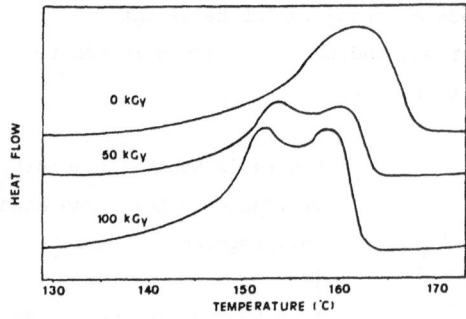

Figure 3a
DSC curves of irradiated
polypropylene: subsequent
heating at a rate of 10
°C/min after isothermal
crystallization at 110 °C,
dose as parameter.

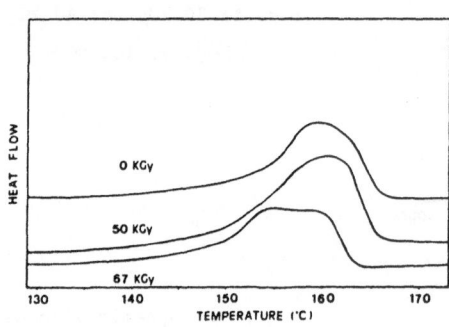

Figure 3b
as figure 3a but for
irradiated PP/EPDM 50/50
blends. EPDM K578 was used.

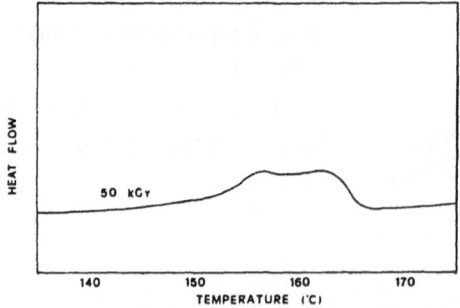

Figure 3c
As figure 3a but for
irradiated PP blended with
EPDM (K578) (50/50).

Irradiation of PP/EPDM blends causes pronounced differences in rheological behaviour compared with irradiated PP, which cannot be explained only by crosslinking of the EPDM phase. A possible explanation is the formation of grafted copolymers from PP onto the EPDM rubber particles. In blends a higher crosslinked fraction (based on 100% EPDM) is found than in irradiated EPDM (Table 1). The authors, however, are well aware that this is not an absolute prove of the presence of a graft copolymer and certainly is not a quantitative method to measure the amount of grafting. In order to get more evidence IR and NMR experiments will be performed in the future.

Table 1.   Crosslinked fractions in EPDM and PP/EPDM blends based on 100% EPDM. Soxhlet extraction in xylene.

PP/K578	dose		PP/K514		dose
	20	50 kGy	20	50 kGy	
0/100	34	70	38	69	
50/50	40	90	48	82	
70/30	41	70	44	74	

The morphologies of some of the blends are shown in Figures 4a to 4c. Very small EPDM particles are obtained when using a corotating twin-screw extruder (Figure 4a). After injection moulding a cocontinuous structure is observed (Figure 4b). Irradiation of the blend after extrusion but before injection moulding preserves the original morphology to a large extent. The absence of coalescence is even more surprising keeping in mind that the viscosity of the PP matrix is substantially lowered upon irradiation. The enhanced morphology fixation may be explained by the formation of a graft copolymer in addition to the crosslinking of the dispersed phase.

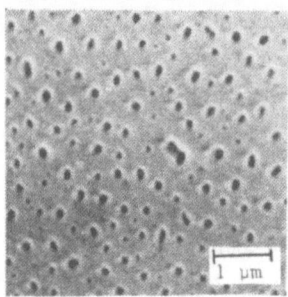 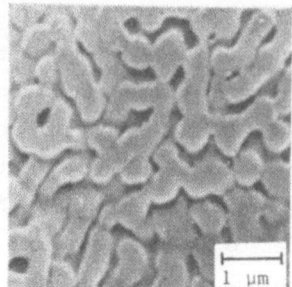 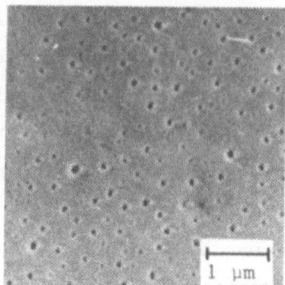

Figure 4a
SEM micrograph of a
PP/EPDM (80/20) blend
extrudate parallel to
the direction of
extrusion

Figure 4b
As Figure 4a but after
injection moulding

Figure 4c
As Figure 4b, but
irradiated with a dose of
43 kGy before injection
moulding

Conclusions

Electron beam irradiation of PP reduces both the molecular weight and
molecular weight distribution resulting in controlled flow properties. In
PP/EPDM blends only a reduction in melt viscosity is observed and the
viscosity is still strongly dependent on shear-rate. Grafting of PP seems
to be the major explanation. The formation of this graft-copolymer in
addition to crosslinking of the dispersed EPDM particles may enhance the
morphology fixation. Preliminary experiments on morphology fixation seem
to be successful but optimization is needed.

Acknowledgment

The authors wish to thank Dr. v. Dijk (Fasson) and Dr. v.d. Brule
(Philips) for enabling the performance of the rheological experiments and
their continuous interest. They also appreciate the assistance of Prof.
Hummel and Dr. Hom with the irradiation experiments.

Literature

1. Charlesby A., Atomic Radiation and Polymers, Pergamon Press, Oxford, 1960.
2. Makhblis F.A., Radiation Physics and Chemistry of Polymers, Keter Publishing House, Jerusalem, 1975, translated from Russian.
3. Black R.M. and Lyons B.J., Proc. Roy. Soc., London, A 253, 1959, 322-330.
4. Meijer H.E.H., Lemstra P.J. and Elemans P.H.M., Makromol. Chem. Macromol. Symp., 16, 1988, 113-135.
5. Elemans, P.H.M., Van Gisbergen, J.G.M., Meijer, H.E.H., in the proceedings of the 3rd Rolduc Polymer Meeting, Elseviers Appl. Science Publishers, 1988.
6. C.R. Zeichner, P.D. Patel, "A Comprehensive Evaluation of Polypropylene Melt Rheology", Report Hercules Inc., Wilmington, 1985.
7. M. Pahl, Praktische Rheologie der Kunststoff Schmelzen und Lösungen, VDI-Verlag, Düsseldorf, 1983.
8. Fujiwara Y , Col. Pol. Sci., 253(4), 1975, 273-282.

TOUGH INTERPENETRATING POLYMER NETWORKS

ROBERT P. BURFORD
Department of Polymer Science
University of New South Wales
P.O. Box 1, Kensington, NSW, 2033, Australia
and YIU-WING MAI
Department of Mechanical Engineering
University of Sydney
Sydney, NSW, 2006, Australia

ABSTRACT

Interpenetrating polymer networks have been prepared by polymerization of styrene in polybutadiene (BR) and styrene butadiene rubbers (SBR), of varying crosslink density. The resulting transparent or translucent sheets were characterized by Charpy impact and 3 point bending tests. Fracture surface morphologies were revealed by SEM and microstructure by TEM. They vary enormously in fracture toughness, with G_c ranging from < 1 to > 10 kJm^{-2}. Factors leading to high fracture toughness will be discussed.

INTRODUCTION

Research in interpenetrating polymer networks (IPN's) over the past ten years has been reviewed by Sperling [1,2]. Those IPN's consisting of poly- styrene and SBR have been described in some detail [3-5] and a SANS study of BR/PS IPN's has also appeared [6]. The emulsion SBR's contained either 5 or 23.5 wt. % styrene, and crosslink densities were low.

Semi-IPN's were shown by TEM to contain PS domains, with size increasing with polystyrene content, whereas true IPN's had a more continuous microstructure.

The subsequent paper [4] details dynamic mechanical, tensile and impact properties. Impact strength increased with SBR content above 10%, whereas other factors, were less important. A SEM study of uncrosslinked SBR, 2% DVB crosslinked PS and 20% SBR content revealed at high magnification nodules which were correlated with the domains found in the TEM. Crazing is proposed as the primary failure mode, although it is influenced by the degree of crosslinking of polymer I. We supplement this previous substantial study by using three elastomer classes, a BR and random and block SBR's, as polymer I, and higher crosslink densities are used. We also employ more rigorous fracture mechanics principles.

MATERIALS AND METHODS

Materials. Dicumyl peroxide ("Hercules Dicup 40C") was dispersed into the elastomer on a two-roll mill and cured at 145°C for 60 min, according to ASTM D3182-81. Crosslink densities were obtained by equilibrium swelling in n-heptane. The cured rubber was equilibrium swollen in inhibitor-free styrene monomer containing 1% benzoyl peroxide, with or without 5% divinyl benzene (DVB). After oven ageing hard transparent or translucent plastic sheets about 4 mm thick were separated from the polyester film-lined frames. Prolonged soxhlet extraction with chloroform revealed the percentage of soluble (linear) polystyrene, whilst total polystyrene content was determined gravimetrically. Details of elastomer and IPN compositions are summarised in Table 1.

TABLE 1

Compositional properties of the elastomer/polystyrene semi and full IPN's

Rubber Type	Dicumyl Peroxide	M_c	Divinyl Benzene	Wt. % Polystyrene	% Crosslinked
Austrapol BR 1220	0.2	7000	0	84	58
	0.2	7000	5	85	96
	1.0	3200	0	73	83
	1.0	3200	5	74	98
Austrapol 9702	0.2	9200	0	89	41
Emulsion SBR	0.2	9200	5	89	97
Random, 20% styrene	1.0	3700	0	77	67
	1.0	3700	5	77	98
Solprene 1205	0.2	5400	0	86	46
Anionic SBR	0.2	5400	5	87	93
A-B linear block	1.0	1900	0	70	80
25% styrene	1.0	1900	5	71	96

In all cases the levels of polystyrene in the highly crosslinked elastomers range from 70 to 77% whereas for the lower crosslinking the range is 84-89% PS. The high amount of crosslinked material for the semi-IPN's reflects significant grafting of polystyrene onto the crosslinked elastomer. With DVB, negligible linear polymer exists.

Mechanical property and fracture toughness measurement

Flexural experiments were carried out in three-point loading geometry with inner span being 20 mm, and crosshead speed 1 mm/min. The impact fracture tests were conducted in a Zwick impact tester using Charpy specimens 4x10x50 mm. They contained razor-sharp initial cracks of 1,2,3,4 and 5 mm in the depth direction. The impact fracture toughness (G_c) was obtained using the equation of Plati and Williams [7].

$$U_c = G_c BD\phi + U_k \tag{1}$$

where U_c is the energy expended after impact fracture, U_k is the kinetic energy loss and ϕ is defined by:

$$\phi = C/dC/d(a/D) \tag{2}$$

with C equal to the specimen compliance. Plati and Williams computed ϕ for a range of 1/D values at varying a/D ratios for both Charpy and Izod specimens. Here 1/D = 4 and 0.1 < a/D < 0.6.

An example of impact data is that for the highly crosslinked 1205 semi IPN shown in Figure 1, and this represents maximum scatter. Satisfactory linear fits were found for all other IPN's.

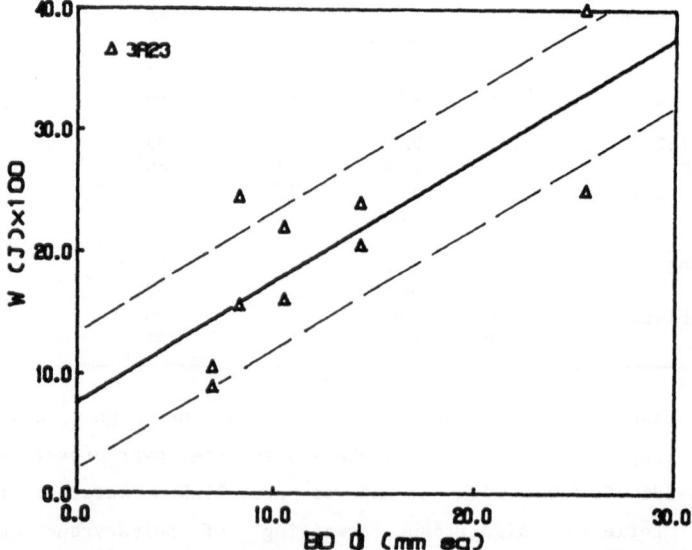

Figure 1.   Impact data for 1205/PS semi IPN.

Preliminary tensile evaluation was conducted by straining $\approx$ 3x10x40 mm specimens in an Instron 1115 testing machine using an intergrip distance of 25 mm and a crosshead speed of 20 mm min^{-1}.

Electron microscopy

Carbon coated fracture surfaces were examined using a JEOL 840 SEM. Network structures were recorded by cutting 50 nm thick sections using a diamond knife, floating the sections on 5% aqueous OsO_4 for 0.5 h, and examination in a Philips 300 TEM.

RESULTS

The IPN's varied widely in mechanical and fracture behaviour, as summarised in Table 2. Included in this Table is an indication of 3 point mode of bending, type I being a single crack, II being multiple cracking, III being multiple crazing followed ultimately by fracture and IV being ductile bending with no evidence for crazing or cracking. Opacity data is also provided, 0 being water-clear and 4 being hazy/translucent.

TABLE 2
Mechanical, physical and toughness properties of IPN's

Polymer	Dicup (%)	DVB (%)	E_b GPa	G_c (kJ m^{-2})	Mode	Opacity	σ_y (MPa)	E_b (%)
1220	0.2	0	1.4	1.7	II	2	21	30
	0.2	5	1.5	2.6	III	1	29	30
	1.0	0	0.70	2.5	IV	2	15	>500
	1.0	5	1.03	2.1	IV	1	19	250
9702	0.2	0	1.1	0.5	I	4	18	5
	0.2	5	2.0	1.5	III	1	43	15
	1.0	0	1.02	1.5	II/III	3	16	120
	1.0	5	1.50	1.5	III	1	33	70
1205	0.2	0	1.15	1.4	I,III	1	21	15
	0.2	5	1.37	0.7	III	0	22	25
	1.0	0	0.37	11.4	IV	1	14	70
	1.0	5	0.60	4.1	IV	1	21	140

The lightly crosslinked rubbers, containing high levels of polystyrene, are more brittle than those with higher crosslink density and less polystyrene. The latter also have a lower bending modulus, E_b, than the low crosslinked rubber counterparts. The mode of fracture for the high polystyrene/low crosslinked elastomers is generally either brittle cracking, crack/crazing or multiple crazing, whereas all the BR 1220 and SBR 1205 IPN's with high elastomer crosslinking showed no evidence of localized deformation or rupture in bending. Divinyl benzene crosslinker increases stiffness and clarity, without severely reducing toughness or ductility.

The fracture morphologies of selected impact samples are shown at low and high magnification in Figure 2. It can be seen that three basic types of failure occur.

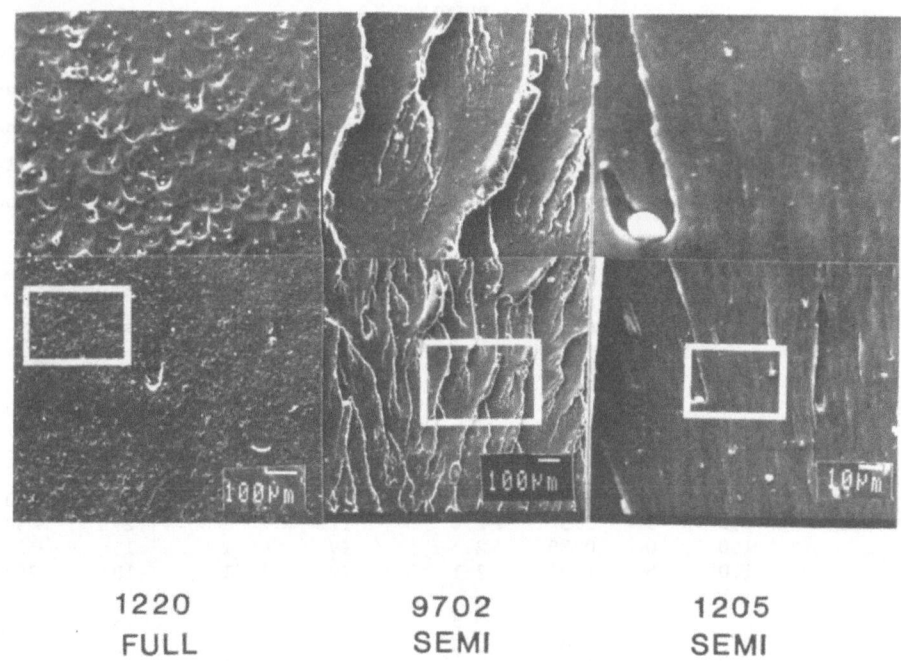

1220 FULL   9702 SEMI   1205 SEMI

Figure 2. Fracture morphologies of three IPN's.

The high fracture toughness of the highly crosslinked 1205 is notable. Shown in Figure 3 are transmission electron micrographs of high and low crosslinked semi IPN's from this elastomer. With low elastomer crosslinking (i.e. high PS content), domains of uniform size are found, whereas for high crosslinking a more dispersed structure, with a finer, layered morphology, consistent with higher toughness, is observed. This is in good general agreement with the random SBR [3-5] and BR [6] microstructural trends found elsewhere.

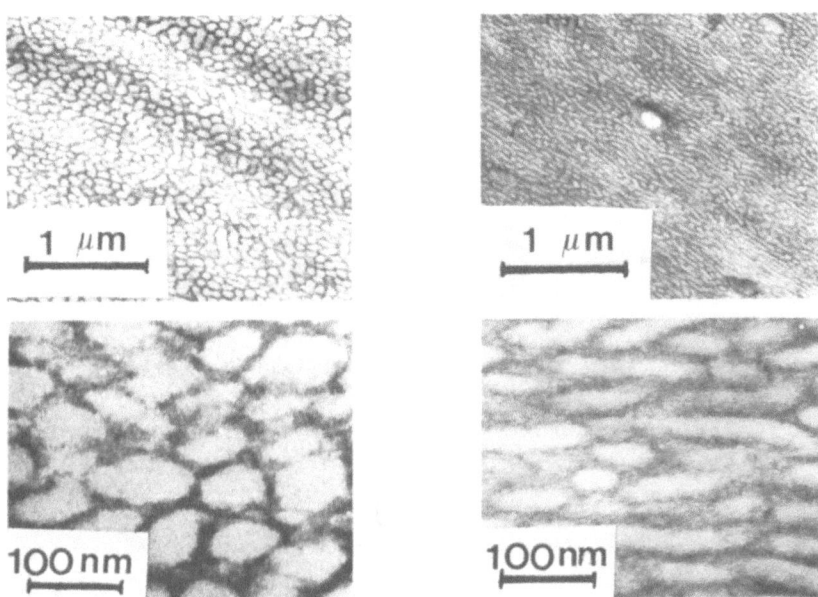

Figure 3. Transmission electron micrographs of 1205 semi-IPN's. A = 0.2% Dicup, B = 1.0% Dicup.

REFERENCES

1. Sperling, L.H., **Interpenetrating Polymer Networks and Related Materials.** Plenum, New York, 1981.

2. Sperling, L.H., Pure and applied research on interpenetrating polymer networks and related materials. In **Polymer Blends and Mixtures,** eds., D.J. Walsh, J.S. Higgins and A. Maconnachie, NATO ASI Series. Series E, Applied Sciences, No. 89, Martinus Nijhoff, Dordrecht, 1985, pp.267-287.

3. Donatelli, A.A., Sperling, L.H. and Thomas, D.A., Interpenetrating polymer networks based on SBR/PS. 1. Control of morphology by level of crosslinking. **Macromolecules,** 1976, **9,** 671-675.

4. Donatelli, A.A., Sperling, L.H. and Thomas, D.A., Interpenetrating polymer networks based on SBR/PS.2. Influence of synthetic detail and morphology on mechanical behaviour. **Macromolecules,** 1976, **9,** 676-680.

5.   Donatelli, A.A., Sperling, L.H. and Thomas, D.A., A semiempirical
     derivation of phase domain size in interpenetrating polymer networks.
     J. Appl. Polymer Sci., 1977, **21**, 1189-1197.

6.   Fernandez, A.M., Wignall, G.A. and Sperling, L.H. Interpenetrating
     polymer networks based on polybutadiene and polystyrene. Morphology
     and phase dimensions by small-angle spectron scattering and electron
     microscopy. In **Multicomponent Polymer Materials**, ed., D.R. Paul and
     L.H. Sperling, ACS Advances in Chemistry Series, 211, ACS, 1986.

7.   Plati, E. and Williams, J.G., The determination of the fracture
     parameters for polymers in impact. **Polym. Eng. Sci.**, 1975, **15**,
     470-477.

DISPERSIVE MIXING ON A TWO-ROLL MILL

C. VAN DER REIJDEN-STOLK, J. VAN DAM and H. BOERSTOEL
Department of Polymer Technology,
Delft University of Technology,
Julianalaan 136, 2628 BL Delft,
The Netherlands

ABSTRACT

Despite the fact that the flow field between the rolls of a two-roll mill is rather complex, it is quite possible to determine where the greatest shear and elongational rates are present. These rates determine the maximum forces exerted upon the droplets of the dispersed phase when a two-phase liquid system is mixed on the mill and therefore the maximum drop size. For mixtures of two Newtonian fluids good agreement between experimental and theoretically predicted drop diameters is obtained.

INTRODUCTION

Polymer blends are produced by mixing incompatible polymers in the liquid state. In blenders such as two-roll mills and internal mixers shear as well as elongational flows occur and in most cases the flow field is rather complex.

In this contribution a description is given of the flow field in a two-roll mill and its influence on the resulting particle size of the dispersed phase when a two-phase system is blended. For the components of the two-phase system Newtonian fluids were chosen because for these break-up criteria are available.

THEORY

A complete description of the flow field between the rolls of a two-roll mill can only be obtained by numerical methods. Comparison with numerical and experimental results shows, however, that for a Newtonian fluid the relevant characteristics of the flow field are very well described by an approximation due to Gaskell [1]. For a calender having equal sized rolls (radius R) rotating at the same speed (surface velocity U) the velocities in the gap in the x- and y-directions are given by:

$$u_x = U[1 + \frac{3}{2} (\frac{h_\infty}{h} - 1)(1 - \frac{y^2}{h^2})]$$

$$u_y = -\frac{3}{2} U \frac{x}{R} [\frac{y^3}{h^3} (\frac{h_\infty}{h} - \frac{2}{3}) - \frac{h_\infty}{h} \frac{y}{h}]$$

(1)

in which h is the half gap width at position x and $2h_\infty$ the gap width at position x_{max} where the liquid detaches itself from one of the rolls (see also Figure 1); here the velocity profile is assumed to be flat.

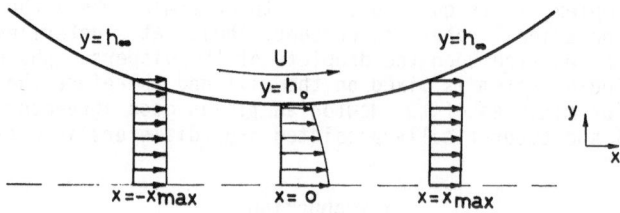

Figure 1. Coordinate system and velocity profiles between the rolls in a symmetrical calender.

From Equations (1) the shear and elongational rates at every position in the nip area of the rolls can be calculated. These rates determine the forces exerted on the dispersed particles. If these hydrodynamic forces exceed the forces due to interfacial tension (which tend to keep the drops together) break-up will occur. The ratio of the two forces is expressed by the Weber number:

$$We = \frac{\eta_c \, G \, r}{\sigma}$$

(2)

where η_c is the viscosity of the continuous phase, G the rate, r the radius of the droplet and σ the interfacial tension. For a wide range of viscosity ratios p = η_d/η_c, where η_d is the viscosity of the dispersed phase, Bentley and Leal [2] measured critical Weber numbers for drop burst, We_{cr}. By using a computer-controlled, four-roller apparatus they were also able to vary the flow parameter α which is a measure of the relative strength of the strain rate and vorticity in the flow:

$$\nabla \underline{u} = \frac{1}{2} G \begin{pmatrix} 1+\alpha & 1-\alpha & 0 \\ -1+\alpha & -1-\alpha & 0 \\ 0 & 0 & 0 \end{pmatrix} \tag{3}$$

For simple shear $\alpha = 0$, $\alpha = 1$ corresponds to pure straining motion. In Figure 2 Bentley and Leal's results are summarized:

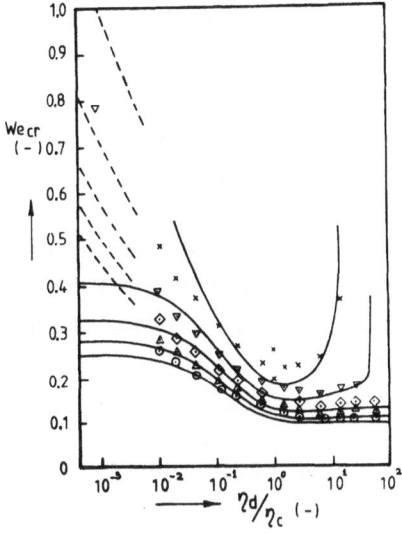

Figure 2. Critical Weber number for drop burst as a function of viscosity ratio for different values of the flow parameter: $\alpha = 0.2$ (upper curve), 0.4, 0.6, 0.8 and 1, respectively.

The velocity field according to Equations (1) can be expressed in the form of Equation (2), giving G and α at every position between the rolls. Then from Figure 2 We_{cr} can be obtained and thus $r_d = We_{cr}\sigma/(\eta_c G)$, the critical drop radius at that position. As as example, in Figure 3 the drop size calculated in this way is given at several cross-sections in the nip area for two widely different values of the viscosity ratio, p = 0.017 and p = 4.3:

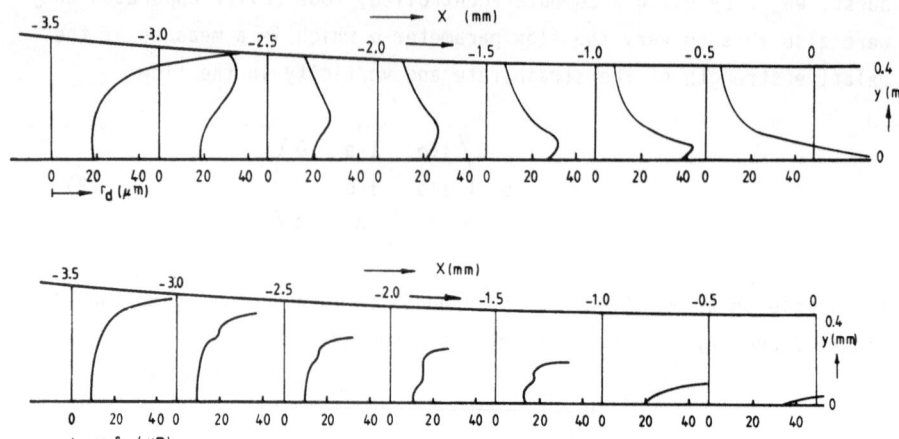

Figure 3. Calculated critical drop radii at several cross-sections
between the rolls. Upper diagram: p = 0.017, lower diagram:
p = 4.3.

So, if we suppose that the system is ideally mixed which means that all
drops have passed through every point within the gap, including the point
where the critical drop radius is at minimum, we can predict the maximum
drop size in the system.

EXPERIMENTAL RESULTS

Mixing experiments were carried out on a Collin laboratory mill (diameter
of the rolls 10 cm), at room temperature. As a continuous phase corn
syrup was used. Silicon oil was mixed in as the dispersed phase at very
low concentrations (about 0.5%) in order to eliminate the influence of
coalescence. Silicon oils are available in a wide range of viscosities,
the viscosity of corn syrup can easily be varied by changing the water
content. Thus it was possible to obtain viscosity ratios between 0.017
and 4.5. In order to obtain different shear and elongational rates the
roll speed and the roll separation were varied, from 5.6 to 16 r.p.m. and
from 0.2 to 1.6 mm, respectively. In all cases a mixing time of 15
minutes was sufficient to reach an equilibrium drop size distribution.

After each mixing operation the material was removed from the roll surface and the size of the largest droplets present in the blend was determined under a microscope. In Figure 4 the experimental radii are compared with the theoretically predicted maximum drop radii for three different viscosity ratios.

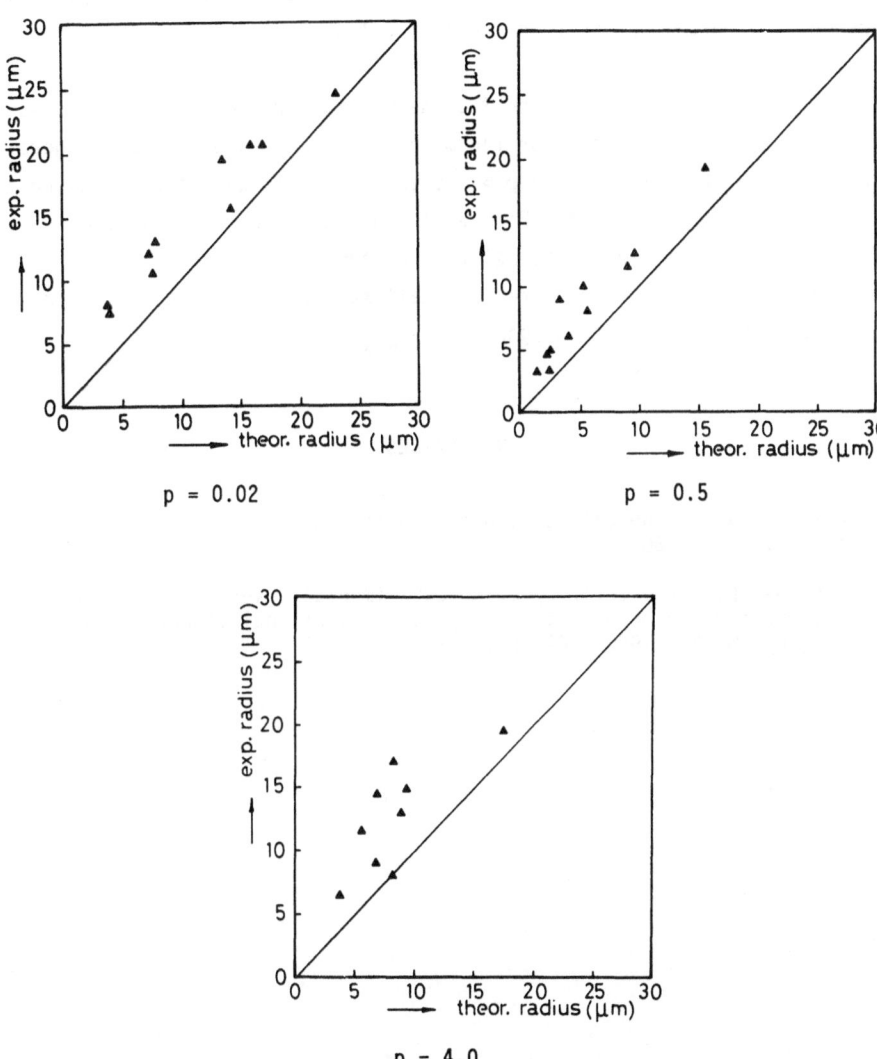

Figure 4. Comparison between experimental and calculated drop radii for three different viscosity ratios.

DISCUSSION AND CONCLUSION

The agreement between experimental and theoretical values of the drop radii is quite satisfactory. The observed drop diameters are always greater than those calculated. One reason for this deviation is our assumption that the system is ideally mixed. Furthermore it is implicitly assumed that drops larger than the critical size break up instantaneously.

For viscosity ratios below 3.5 the minimum critical drop radius is found at the roll surface near the midpoint (x = 0, see Figure 3). Here we have pure shear flow. When p > 3.5 shear flow is unable to break up the dispersed particles (see Figure 4). In the latter case the drop size is determined by the elongational flow component. The maximum in elongational rate and therefore the smallest drops occur in the centre plane at the entrance of the gap (x = $-x_{max}$).

REFERENCES

1. Gaskell, R.E., The calendering of plastic materials. J. Appl. Mech., 1950, 17, 334-336.

2. Bentley, B.J. and Leal, L.G., An experimental investigation of drop deformation and breakup in steady, two-dimensional linear flows. J. Fluid Mech., 1986, 167, 241-283.

CONTINUOUS EMULSION ABS PRODUCTION IN A 2-STAGE REACTOR CASCADE: OPERATING CONDITIONS AND POLYMER PROPERTIES

R. Arenós, P. Centrich, L. Puigjaner and F. Recasens
UNIVERSITAT POLITECNICA DE CATALUNYA
Diagonal, 647 08028 Barcelona, Spain

ABSTRACT

Emulsion ABS resins produced in continuous stirred-tank reactors in series, are compared in performance with a standard polymer prepared in a conventional batch reactor. Polymer specifications are best met by using a 2-stage CSTR with interstage feed of the faster reacting monomer.

INTRODUCTION

Reaction engineering of polymerisation processes is a promising field where new advances are expected in the next few years. Early work by Denbigh (1) shows that when chain propagation time is well below the residence time, as is the case of polyadditions, continuous stirred-tank reactors (CSTR) insure a product with a narrower spread in statistical properties. For the case of continuous emulsion polymerisation, sound theoretical studies are available (2-5). Also, the dynamics of such systems is well reviewed in the literature (6). The mathematical understanding of emulsion systems is a challenging problem in polymer reaction engineering (7). However, while polymer quality can be predicted from kinetic models in a few cases (7,8), very scarce data are available in the open literature about the influence of reactor type and mode of operation on polymer properties. In this paper we study the effects of reactor operating conditions on the performance of the finished polymer. Our reacting system is the graft emulsion polymerization of a modified Styrene (MS) and Acrylonitrile (AN) onto rubber in a CSTR system.

MATERIALS AND METHODS

Batch polymerisation runs, intended to establish a starting recipe for continuous operation, were carried out with 15% rubber and 85% copolymer. Monomer composition was azeotropic.The reaction was carried out at 70°C and 20% total solids, using standard types of redox initiator, chain-transfer agent and soap. Monomer conversion (X_p) was above 92% as measured by GC. Industrial grade monomers were used. After devolatilization, coagulation and drying, glass transition temperatures (Tg) were routinely measured (Perkin Elmer DSC-4 apparatus). Preliminary continuous runs were carried out in a 1-litre, stainless-steel, temperature-controlled reactor at weight hourly space velocities (WHSV) of 4.3 to 0.25. In order to prepare significant amounts of polymer, a production-scale

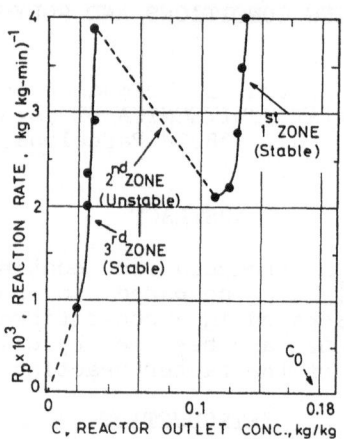

Fig. 1. Reaction rates vs. monomer concentration at 70ºC.

CSTR system, with either one or two reactors in series, were used. Reactor holdup was about 8 kg for the two-stage process. The ABS latices produced were first steam-stripped, then coagulated, dried and processed by extrusion and pelleting. Test specimens were prepared by injection molding and evaluated mechanically and thermally . In addition to the Tg, the measurements included the Izod impact strength (ISO R 179, with notched test specimens) and the Vicat softening temperature (ISO 306, B/120). Occasionally, electron microscopy observations of thin sections of chips were made using the $Os \ O_4$ staining technique on a Philips 301 transmission apparatus.

RESULTS AND DISCUSSION

Polymerisation in a Single CSTR: Conversion Stability

Typical specifications of batch polymer are: X_p > 0.92, Izod = 18 kg-cm/cm, Tg = 136 ºC, VST = 118 ºC. The formulation of the batch polymer was used to carry out a preliminary CSTR run at a very low space velocity (WHSV = 0.25) in order to reach high conversion. We used two separate feed streams (a monomer premix and an aqueous latex premix). The ratio of streams was such that the A:B:S mass fraction was 25:15:60. The properties of this preliminary polymer dropped markedly with

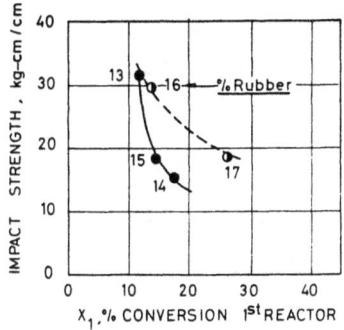

Fig. 2. Effect of monomer conversion X_1 on impact strength.

respect to those of the batch polymer (X_p = 0.87, Izod = 14 kg-cm/cm, Tg = 127 ºC, VST = 112 ºC). In order to understand this behaviour, we decided to examine the reaction kinetics in terms of rates vs. conversions to see which operating zones were more favourable. Results are given in Fig. 1 where polymerisation rates are plotted vs. steady-state monomer concentrations. In the plot, three well-defined regions are seen. In the first zone, for conversions X_p < 0.28, the reactor is stable but strongly underdamped oscillations are observed at small conversions. The second zone is inaccessible and clearly unstable except at the extreme conversions of the dotted line which act as steady-state attracters. In the third zone, for conversions above 80 %, the reactor is subject to overdamped oscillations such that the steady-state regime is approached exponentially.

Operation of Two Continuous Reactors in Series

From these kinetic features it is important to carefully choose those operating zones that are both stable and at the same time provide a polymer of acceptable quality. In the low conversion range, the polymer contains so much rubber that it cannot be evaluated without further conversion. On the other hand, if a single reactor working in the third zone is used, the wetting efficiency of the rubber particles by the monomer is limited, hence a high degree of grafting cannot occur. It is thus clear that a first CSTR that provides a sufficient monomer-rubber contacting is necessary. Thus we chose two CSTR reactors in series, the first tank operating at low X_p and low temperature (70°C) and the second tank to complete the polymerisation. Hence a higher temperature (80°C) and a low WHSV are necessary in the second unit. Note that the space velocity of the first tank is critical. This is so because if X_1 is very small, the wetting efficiency will be good but the reactor will be subject to oscillations. On the other hand, if X_1 is too high, stability will improve but the degree of rubber grafting will be low. Table 1 summarizes the effects of the conversion in the first reactor on polymer performance.

TABLE 1

Polymerisation Conditions of a 2-stage Reactor System.

Polymer	WHSV$_1$	X_1	WHSV$_2$	X_2	Izod	VST,^{o}C	Tg,^{o}C
A	1.58	0.26	0.40	0.82	17.8	111	123
B	2.15	0.14	0.37	0.83	29.1	113	124
C	2.14	0.15	0.36	0.83	29.1	112	124

(1) = 1st Reactor, (2) 2nd Reactor, Izod in kg-cm/cm

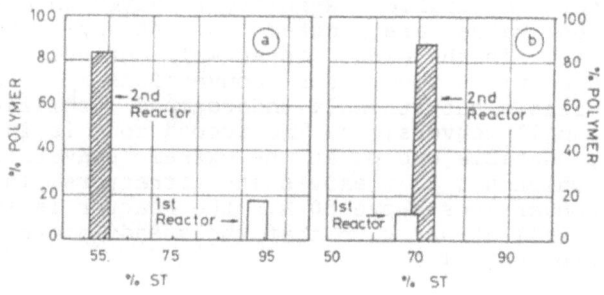

Fig. 3. Copolymer compositions in 2-Stage CSTR without (a) and with (b) insterstage feed of Styrene

As seen in Table 1, the lower the conversion in the 1st reactor, the higher the impact strength and, presumably, the grafting efficiency. This conclusion is reached when the electron micrographs of polymers A and C are compared (not shown). In Polymer C the typical white-patched structure of grafted rubber is seen whereas in polymer A the rubber particles appear uniformly stained by the Osmium. Figure 2 summarizes our results about the impact properties. For decreasing conversions in the first reactor , the impact strength values increase sharply. Results of Table 1 indicate that the Tg and VST of polymers A, B and C are substantially lower than those of the batch polymer (with 136 ºC and 118 ºC for Tg and VST). Fig. 3 shows the copolymer composition distribution for the fraction of polymer produced in each reactor. Note that the final product is a mixture of two polymers with very different compositions (90% and 56% MS, resp.). However since the second fraction (with 56% MS) is

more abundant (85% of total copolymer), it will almost
entirely determine the thermal properties of the ABS matrix
polymer. This is probably why the Tg and VST are so low. In
order to obtain a more uniform reaction of Styrene in the two
reactors, some of the total MS was withheld from the feed to
the 1st tank and fed later in the second reactor as a separate
stream, thus as an interstage feed. This operating policy has
been recommended previously (10) as a means of controlling the
spread of composition in copolymer reactor in series when
monomer reactivities differ widely. By employing this
technique it is possible on one hand to maintain a high impact
strength (as it depends only on the conversion in the first
reactor) and on the other hand, achieve a higher Tg and VST
due to a more uniform copolymer composition. Fig.3b
illustrates the difference in copolymer composition when the
intermediate feed strategy is adopted. In this case polymer
performance is greatly improved to Tg= 135 oC and VST= 117 oC.

ACKNOWLEDGMENT

The financial support for this project provided by ARCO
Chemical Company and continued by ATOCHEM ESPAÑA , Barcelona,
is gratefully acknowledged.

REFERENCES

1. Denbigh,K.G.,J. Appl. Chem., 1951,1, 227.

2. Gershberg, D.B. and Longfield,J.E., 54th AIChE Meeting, New
   York, 1961.

3. Nomura,M., Kojima,H., Harada,M., Eguchi,W. and Nagata, S.
   J. Appl. Pol. Sci., 1971,15, 675.

4. Brooks,B.N., Kropholler,H.W. and Purt,S.N.,Polymer, 1978,
   19, 193.

5. Kiparissides,C., MacGregor,J.F. and Hamielec,A.E. Can. J.
   Chem. Engng., 1980, 58, 56.

6. Penlidis, A., MacGregor, J.F. and Hamielec, A.E.,AIChE J,
   1985,31, 881.

7. Ray, W.H., ACS Symposium Series, 1983,226, 101.

8. Villermaux, J., Blavier,L. and Pons,M. in Polymer Reaction
   Engineering, Hanser Publishers, Munich, 1983.

9. Lin, C.-C. and Chiu, W.-Y., J.Appl.Pol.Sci., 1982, 27, 1977

10.Recasens, F., Alonso, M. and Puigjaner, L., Chem. Eng.
   Process., 1986, 20, 85

Part 3

FIBRES/COMPOSITES

VISCOELASTICITY OF ARAMID FIBRES

M.G. NORTHOLT, J.H. KAMPSCHREUR, S. VAN DER ZWAAG
Akzo Research Laboratories, P.O. Box 9300,
6800 SB Arnhem, The Netherlands.

ABSTRACT

Both the low creep and stress relaxation rates of aramid fibres are outstanding properties which are presumably related to the rigid nature of the chains, the high crystallinity of the structure and the hydrogen bonding between the chains. For a better understanding of the viscoelasticity of the fibre, the structural changes during creep and stress relaxation have been investigated. A useful tool is provided by the dynamic compliance, because in well-oriented fibres it is linearly related to the second moment of the chain orientation distribution. Experiments have shown that during creep and stress relaxation a progressive contraction of this distribution occurs, which is explained by shear deformation of the crystallites. This phenomenon can be explained by a single-phase series model which incorporates viscoelasticity.

INTRODUCTION

In many applications of industrial fibres the creep and the stress relaxation play an important role. Users often want to know creep and stress relaxation rates to enable them to predict long-term behaviour of these fibres. However, since the processes underlying these relaxation phenomena are still not well understood, the prediction of long-term behaviour is rather precarious.

The creep of polymer fibres is composed of the primary creep, which is recoverable with time, and of the non-recoverable or secondary creep. The secondary creep is almost negligible if a fibre has been mechanically pre-conditioned, i.e. when it has first been loaded with a stress considerably higher than the stress applied in the subsequent creep experiment.

In this report we present a model for the primary creep of well-oriented aramid fibres. It has been shown for well-oriented fibres of PET, cellulose and poly-(p-phenylene terephthalamide), abbreviated here as PpPTA, that the dynamic compliance, S, is a linear function of the second moment of the orientation distribution of the chains $\langle \sin^2\varphi \rangle$ /1/ /2/ /3/. By measuring S during creep and relaxation of a fibre, the changes in the orientation distribution can be followed. As shown here, such an experiment offers a valuable tool for the investigation of the viscoelasticity in polymer fibres.

EXPERIMENTAL

The creep and stress relaxation of two PpPTA fibres (Twaron $^{\circledR}$) were investigated. The initial dynamic compliances, S_o, of the unconditioned fibres were 11.9×10^{-3} and 8.24×10^{-3} m²/GN; these values correspond to moduli of 84 and 121 GNm^{-2} respectively. The creep and stress relaxation of both unconditioned and mechanicallly conditioned fibres were measured. In the latter case the fibre was preloaded with a stress of 1.3 GNm^{-2} for 24 minutes, followed by a recovery period at almost zero stress of 16 hours, after which the actual creep experiment was started. The dynamic compliance, S, was measured during creep and stress relaxation by means of a pulse-echo technique /1/.

Figures 1a and 1b show the creep strain, ε, and S plotted against $^{10}\log t$ for unconditioned and conditioned fibres. Both graphs show linear functions with log t. This was observed up to loads of 2.8 GNm^{-2}. Creep and recovery experiments show that changes in ε correspond directly with the changes in S, as is depicted in Figure 2. This is also observed in the stress relaxation experiments, as shown in Figure 3. The creep rate of unconditioned fibres increases with σ, approximately as a linear function of log σ, from about 2.5×10^{-4} for 0.4 GNm^{-2} to about 7.5×10^{-4} per decade for a creep stress of

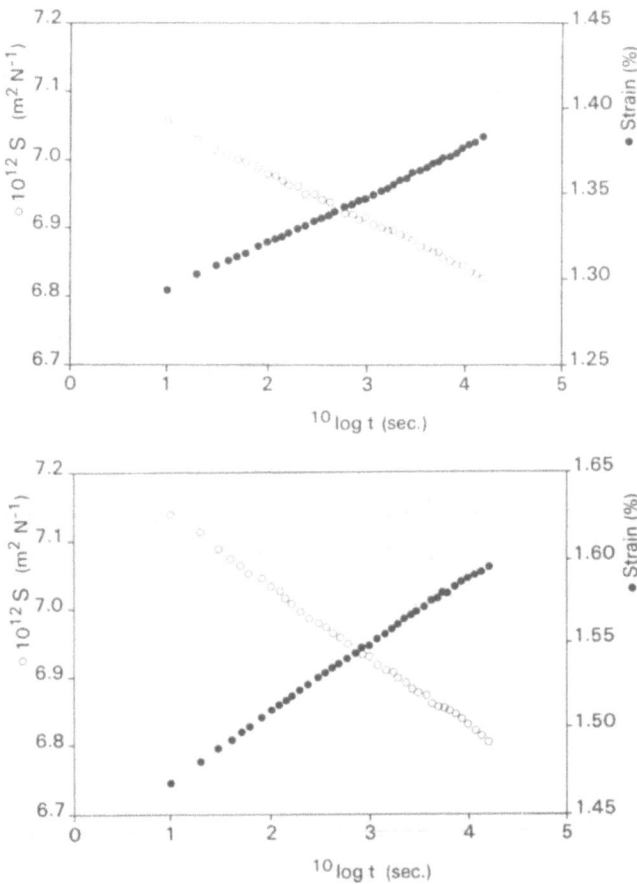

Figure 1. Extension ε (•) and dynamic compliance S (o) during
creep for σ_o=1.1 GNm^{-2}, unconditioned fibre (top)
and mechanically conditioned fibre (bottom).

2.8 GNm^{-2}. For the fibre with the lower initial compliance dε/dlog t is
smaller by a factor of 0.6. Figure 4 shows that the creep rate of
conditioned fibres is almost independent of the creep stress for
stresses between 0.5 GNm^{-2} and the preloading stress. The lower stress
limit is related to the strain of 0.5% in the tensile curve at which a
weak yield effect is observed. This apparent yield strain is
independent of the fibre compliance, which has also been observed for
cellulose based fibres /2/.

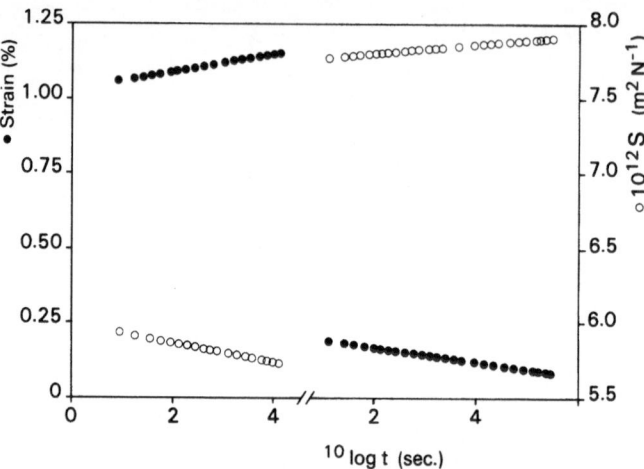

Figure 2. Creep and recovery of an unconditioned fibre for
for $\sigma_o=1.1$ GNm^{-2}: creep strain (•) and dynamic
compliance (o).

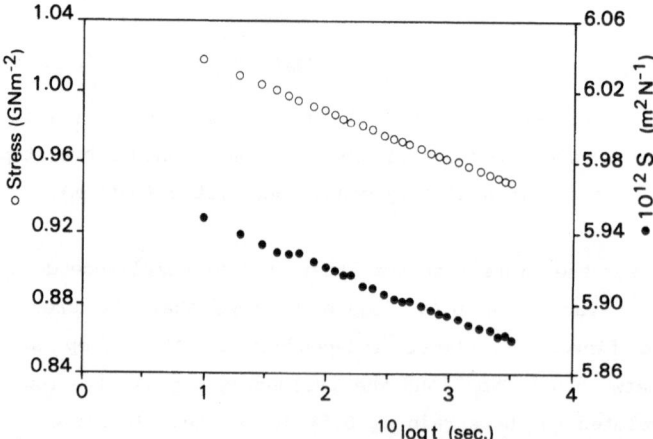

Figure 3. Unconditioned fibre: stress relaxation (o) and dynamic
compliance (•) for $\varepsilon_o=1\%$.

THEORY

A modified series model was developed for the interpretation of the elastic extension of well-oriented, crystalline polymeric fibres /3/. According to this model the elastic strain-stress curve for limited strains is given by

$$\varepsilon \simeq \frac{\sigma}{e_3} + \frac{\langle \sin^2\varphi_o \rangle}{2} \left[1 - \exp(-\sigma/g_o) \right] \tag{1}$$

and the compliance by

$$S = \frac{1}{e_3} + \frac{\langle \sin^2\varphi \rangle}{2g_o} \tag{2}$$

where e_3 is the tensile modulus of the chain, g_o the modulus for shear parallel to the chain axis of the crystallite and $\langle \sin^2\varphi_o \rangle$ the initial orientation parameter.

From measurements of S and $\langle \sin^2\varphi \rangle$ during tensile deformation of PpPTA fibres we found for $(2g_o)^{-1}$ a value of $0.26\ m^2/GN$, or $g_o = 1.92\ GNm^{-2}$ /3/. This result is in good agreement with the shear modulus determined from torsional experiments on single filaments $G = 1.83 \pm 0.17\ GNm^{-2}$ /4/.

Expressions (1) and (2) represent the response of a serial arrangement of two springs with spring constants e_3 and $2g_o/\langle \sin^2\varphi \rangle$, as shown in Figure 5. The macroscopic strain is the sum of two contributions, namely one due to the stretching of the chain, the other due to the shear deformation of the crystallites which is accompanied by rotation of the chains towards the stress direction /5/.

The model is now further developed to incorporate the viscoelastic extension of the fibre. To this end, we make the simple asssumption that the time-dependent part of the creep strain of the fibre arises solely from the rotation of the chains towards the fibre axis direction as result of the shear deformation of crystallites. This yields for the fibre extension as a function of the time t during creep caused by a stress σ_o

$$\varepsilon(t) \simeq \frac{\sigma_o}{e_3} + \frac{1}{2}(\langle \sin^2\varphi_0 \rangle - \langle \sin^2\varphi(t) \rangle) \tag{3}$$

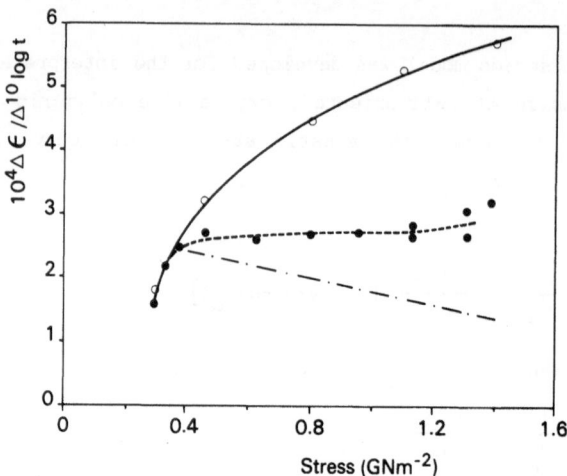

Figure 4. Creep rate versus stress for an unconditioned fibre (o)
and a mechanically conditioned fibre (•), relation (10)
is indicated by a straight line.

or

$$\varepsilon(t) \simeq \frac{\sigma_o}{e_3} + g_o[S_o - S(t)] \qquad (4)$$

Similar relations can be derived for the the stress relaxation /1/ and
we obtain irrespective of the functional dependence on the time the
following expressions

for creep: for stress relaxation:

$$\frac{\varepsilon(t) - \varepsilon(t_o)}{S(t_o) - S(t)} = -g_o \quad (5) \qquad\qquad \frac{\sigma(t_o) - \sigma(t)}{S(t_o) - S(t)} = g_o e_3 \quad (6)$$

Relations (5) and (6) imply a progressive contraction of the
orientation distribution in the fibre during creep and stress
relaxation.

On the basis of the observations presented in Figures 1 and 2 we
introduce a relaxation function for the orientation parameter.
Employing

$$u(t) = \langle \sin^2\varphi(t) \rangle$$

163

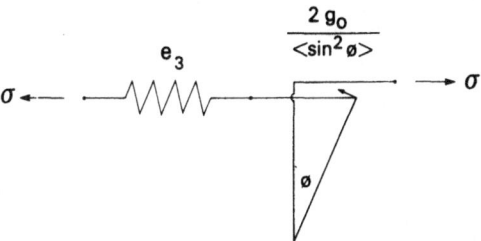

Figure 5. Schematic presentation of the model for elastic
extension of a fibre: a serial arrangement of a
"chain spring" and a "shear spring".

the time dependence of u(t) is given by

$$u(t) = u_o \exp\left[-\frac{\sigma_o}{g_o} - \frac{\lambda}{g_o}\ln(\frac{t}{t_o})\right] \tag{7}$$

where u_o is the value for $\langle \sin^2\varphi \rangle$ before applying the creep stress and
λ a relaxation constant. For $t = t_o$ expression (7) yields the
expression for the change of $\langle \sin^2\varphi \rangle$ due to a stress σ in case of a
purely elastic deformation /3/. It should be noted here that expression
(7) does not describe transient phenomena.

For $\lambda/g_o \ll 1$ we derive from (1) and (7)

$$\varepsilon(t) \simeq \frac{\sigma_o}{e_3} + \frac{u_o}{2}\left[1 - \exp(-\sigma_o/g_o)\right] + \frac{\lambda u_o}{2g_o} \exp(-\sigma_o/g_o) \ln(t/t_o) \tag{8}$$

and

$$S(t) \simeq \frac{1}{e_3} + \frac{u_o}{2g_o} \exp(-\sigma_o/g_o)\left[1 - \frac{\lambda}{g_o}\ln(t/t_o)\right] \tag{9}$$

which implies that

$$\exp(\sigma_o/g_o) \frac{d\varepsilon}{d(\ln t)} = \frac{\lambda u_o}{2g_o} \tag{10}$$

$$\exp(\sigma_o/g_o) \frac{dS}{d(\ln t)} = -\frac{\lambda u_o}{2g_o^2} \tag{11}$$

Expressions (8) and (9) give the time dependence of the extension and of the dynamic compliance during creep. Because u_o can be calculated from S_o using expression (2), the constant λ is determined experimentally using relations (10) and (11).

As time tends to infinity expressions (3) and (7) yield a finite extension of

$$\varepsilon(t \to \infty) = \frac{\sigma}{e_3} + \frac{1}{2} u_o \tag{12}$$

which is attained when all chains are oriented parallel to the stress directed along the fibre axis. For the analysis of the recovery and the stress relaxation it is more appropiate to write (7) in differential form

$$d\sigma = - g_o d\left[\ln u(t)\right] - \lambda d(\ln t) \tag{13}$$

This relation describes a parallel arrangement of an elastic spring g_o and a kind of dashpot which is responsible for the stress relaxation according to $d\sigma/d(\ln t) = -\lambda$. This arrangement resembles the classical Kelvin element, as is demonstrated by comparing the creep response of this element

$$\varepsilon(t) = \frac{\sigma_o}{E} \left[1 - \exp(-\frac{t}{\tau})\right] \tag{14}$$

with expression (7) now written in a different form

$$\ln\left[\frac{u_o}{u(t)}\right] = \frac{\sigma_o}{g_o} \left[1 + \frac{\lambda}{\sigma_o}\ln(t/t_o)\right] \tag{15}$$

The contribution of this viscoelastic element to the fibre extension is given by

$$\tfrac{1}{2}\left[u_o - u(t)\right]$$

, see also relation (3). It is proposed that the primary creep, i.e. the creep of a mechanically conditioned PpPTA fibre, can be described by linking the spring e_3 in series with the viscoelastic element defined by relation (13).

RESULTS AND DISCUSSION

Typical results for $dS/d^{10}\log t$ and $\exp(\sigma_o/g_o)dS/d^{10}\log t$ versus the
creep stress are depicted in Figure 6. For $\sigma > 0.5$ GNm^{-2} the compliance
data comply with relation (11), both for conditioned and unconditioned
fibres. This clearly demonstrates that both primary and secondary creep
in PpPTA fibres are due to shear relaxation. We find for both PpPTA
fibres studied here that $\lambda/g_o = 0.01$. However, the physical
significance of λ is not yet clear. According to relation (10) the
primary creep rate is determined by u_o. Indeed, it was found that the

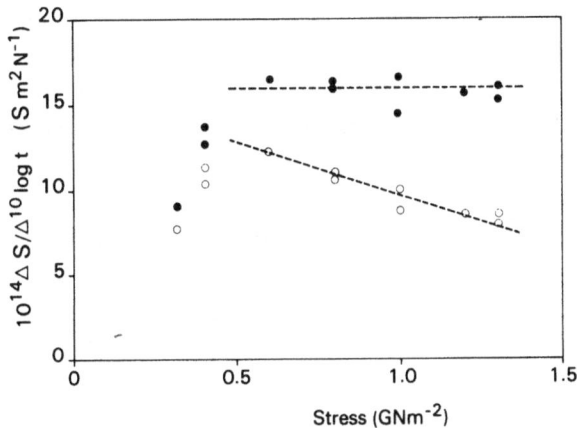

Figure 6. Unconditioned fibre: $-dS/d^{10}\log t$ (o) and
$-\exp(\sigma/g_o)dS/d^{10}\log t$ ($\cdot$) versus creep stress.

ratio of the primary creep rates of the two fibres was equal to the
ratio of the initial orientation parameters measured after mechanically
conditioning of the fibres. This was also found for the creep rates of
the unconditioned fibres. However, as shown in Figure 4 the functional
dependence on σ given by relation (10) is not observed. There is a
small extra contribution to the creep strain in the case of the
conditioned fibres, and a larger extra contribution for the
unconditioned fibres. This is also shown in Figure 7, where $-\Delta\varepsilon/\Delta S$ is
plotted versus the creep stress. According to relation (5) this

quantity should be equal to g_o = 1.9 GNm^{-2}. This discrepancy can be explained by taking into account the effect of the finite width of the chains. In conjuction with the rotation there has to be some relative displacement or slip of adjacent chains (or bundles of chains) parallel to the chain axis. This is easily visualized by a row of books that is slowly falling over. Using this model it can be shown that this slip contribution to the fibre strain increases with decreasing orientation angles of the chains, as occurs when the creep stress is increased.

Presumably the difference in creep rates between the unconditioned and the conditioned fibres can be partly attributed to a difference in size of the slipping element. In the case of first loading or secondary creep of a fibre presumably adjacent fibrils will shear, whereby the lateral hydrogen bonds between the fibrils are broken and reform in more "favourable" positions resulting in a permanent slip. On the other hand the slip is much smaller during primary creep. It is therefore probably limited to single chains in the crystallites and thus recoverable.

The logarithmic dependence of the primary creep is characteristic for a single phase structure having a constant distribution of activation energies. It implies that the logarithmic retardation spectrum $L(\ln \tau)$ is independent of $\ln \tau$ and the loss tangent is independent of the frequency.

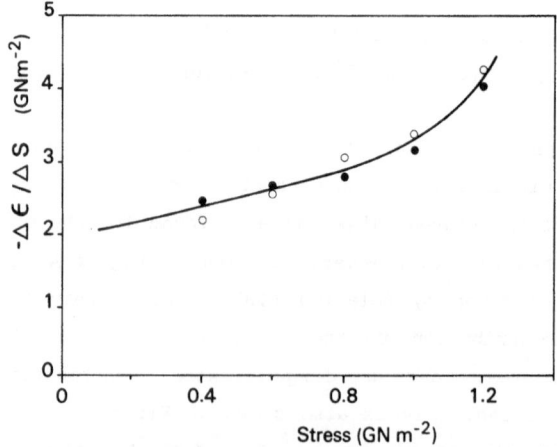

Figure 7. The quantity $-\Delta\epsilon/\Delta S$ versus creep stress measured for both fibres after mechanically conditioning.

CONCLUSIONS

Creep and stress relaxation of aramid fibres are a linear function of log(time). The creep rate of unconditioned fibres increases linearly with log σ, whereas the creep rate of mechanically conditioned fibres is for stresses above 0.5 GNm^{-2} independent of the creep stress. Creep and stress relaxation of aramid fibres are caused by shear relaxation. They can be described by a serial arrangement of an elastic spring, representing the chain modulus, and of a viscoelastic element that embodies the chain rotation due to shear deformation of the crystallites.

REFERENCES

/1/. M.G. Northolt, A. Roos and J.H. Kampschreur, submitted for publication in J. Pol. Sci., Phys. Ed..

/2/. M.G. Northolt and H. de Vries, Angew. Makrom. Chem. 133, 183 (1985).

/3/. M.G. Northolt and R. v.d. Hout, Polymer 26, 310 (1985).

/4/. W.F. Knoff, J.Mater.Sci. Lett. 6, 1392 (1987).

/5/. S. van der Zwaag, M.G. Northolt, R.J. Young, I.M. Robinson, C. Galiotis and D.N. Batchelder, Polymer Comm. 28, 276 (1987).

MAXIMUM PROPERTIES ACHIEVED BY UNIAXIAL DRAWING OF HIGH MOLECULAR WEIGHT POLYOLEFINS

Tetsuo Kanamoto[1]) and Roger S. Porter[2]),
1). Department of Applied Chemistry, Science University of Tokyo,
Kagurazaka, Shinjuku-ku, Tokyo 162, Japan
2). Department of Polymer Science and Engineering, University of
Massachusetts, Amherst, MA 01003, USA

ABSTRACT

Several flexible chain polymers of ultrahigh molecular weights (UHMW) have been uniaxially drawn from different morphologies by three techniques, including solid-state extrusion, tensile drawing, and their sequential combination (two-stage drawing). Among these, solution-grown single crystal (SGC) mats of UHMW-polyethylene, polypropylene, and poly(4-methyl-1-pentene) have been superdrawn, by two-stage drawing, to the limits in terms of mechanical and physical properties approaching theoretical limits. The morphlogies of such samples have been studied by density, DSC melting behavior, X-ray small- and wide-angle diffraction, thermal expansivity, etc. The results showed a significant difference in the microstructure between UHMW-PE and UHMW-PP superdrawn to the limits.

INTRODUCTION

The mechanical and some physical properties of linear, flexible chain polymers in an isotropic solid are far from those expected for single crystals of these polymers. Extensive studies (1-3) have been made to develope high-modulus and high-strength morphologies of these polymers by uniaxial drawing, which induces chain extension, orientation, and crystal-lization. The most successful and efficient draw has been achieved for ultrahigh molecular weight polyethylene (UHMW-PE), by flow-induced crystal-lization from solutions (4) and by ultradrawing of gels (5) prepared by quenching semidilute solutions. These techniques have been applied for a variety of polymers by many workers (1-3,6,7).

We have also found (8) that single crystal mats and powders of high density polyethylene, grown from dilute solutions, could be extrusion drawn up to a draw ratio (RD) over that reported by tensile drawing of HDPE morphologies, including single crystal mats (9). Thus, we have examined, by utilizing several techniques, the drawability of single crystal aggregates of several polar and nonpolar polymers, precipitated from dilute solutions. Among these polymers studied, only UHMW-PE (10), polypropylene (PP) (11), and poly(4-methyl-1-pentene) (P4M1P) (12) have been superdrawn to the limits in terms of mechanical and physical properties approaching limits. In this paper, we report the drawing behavior and the tensile and physical properties of superdrawn films of these polymers. Some morphological properties of such samples are also discussed.

EXPERIMENTAL

The six linear polyethylene samples used here had M_v in the range of 2-80 x 10^5. The three polypropylenes used had M_v in the range of 3-36 x 10^5 and isotactic pentad fractions ranging from 94.5-96 %. The highly isotactic UHMW-P4M1P had a M_v of 1.5 x 10^6. Solution-grown crystals of these polymers were isothermally precipitated from dilute solutions (0.05-0.2 wt%) in xylene kept near the clouding point for each of the solutions. Sedimented mats of SGC were obtained by slowly filtering the crystal suspensions at room temperature. The dried mats were drawn by three techniques, i.e., solid-state coextrusion (13), tensile drawing, and their sequential combination (two-stage drawing). Details of the preparation of SGC mats, drawing, and characterization were described elsewhere (10-14).

RESULTS AND DISCUSSION

The similarities and differences between single crystal aggregates and gels should first be summerized. The gel is usually prepared by quenching semidilute solutions (>1 wt%) giving a macroscopically coherent structure due to the formation of crystalline junctions and chain entanglements, depending on the concentration of solution from which gel is formed. This coherency of gel is a major difference in the structure from single crystals precipitated from dilute solutions, where each crystal is isolated in the crystal suspension. Although both the SGC mats and the dried gels consist of well-developed lamellar crystals, further differences in the structure are likely. The former crystals are slightly thicker (140 vs. 110 Å) and thermally more stable, and may have a more regular chain folding and less entanglements than the latter, reflecting their higher temperature and lower solution concentration for crystallization from solution.

Drawing of UHMW-PE SGC Mats
The drawability and the resultant structure and properties of drawn films were found to be sensitive to drawing technique. The mats formed from 0.05 wt% solutions were used for straight tensile drawing since those from more concentrated solutions, 0.2 wt%, exhibited poor reproducibility, likely due to the less coherent lamellar packing. At temperatures $\leq 60^{\circ}C$, the draw was found to be unstable with no effective draw possible for tensile drawing of the mats. However, at$\geq$ 90°C, mats showed excellent drawability. The draw stress initially increased rapidly, followed by a sharp drop after yielding with a sharp neck formation, depending on the draw temperature and rate. The strain hardening phenomenon, usually observed for drawing melt-crystallized polymers, was absent in tensile drawing of SGC mats. Further, the draw stress was remarkably low, compared to that for melt-crystallized UHMW-PE. These features are related to the characteristic morphology of SGC precipitated from dilute solution; the less chain entanglements and the regular chain folding in these thin lamellar crystals. They also explain the dramatic improvement in drawability of UHMW-PE.

The drawability of mats was also sensitive to drawing rate and temperature as shown in Figure 1. The complex change of the maximum DR

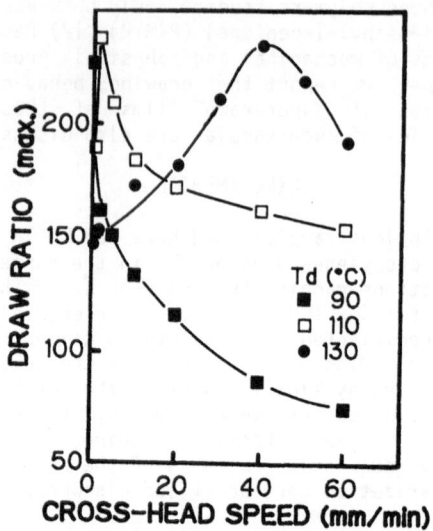

Figure 1. Maximum achievable DR. vs CHS for tensile drawing of UHMW–PE SGC mats at 90–130°C.

with CHS at 130°C indicates that two opposing factors may control the drawability at the high temperature: the increase of draw stress and the decrease of annealing, with increasing CHS. Figure 2 shows the maximum extrusion draw ratio (EDR) achieved for UHMW–PE SGC mats by solid–state coextrusion in the temperature range of 20–130°C. As the mats prepared from 0.2 wt% solution also exhibited an excellent drawability, like those

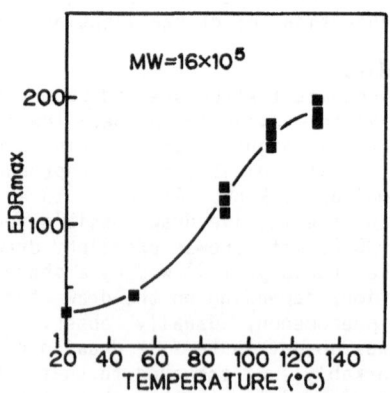

Figure 2. Maximum achievable EDR vs. temperature for solid–state coextrusion of UHMW–PE SGC mats.

from 0.05 wt% solution, the former mats were used for solid-state
coextrusion. This EDR increased with extrusion temperature from 27 at 20°C
to 200 at 130°C. The pressure of 50–240 MPa applied for extrusion
suppressed chain mobility and hence the deformability at a given temperature. This should be considered on comparing drawability by solid-state
coextrusion and by tensile drawing. A remarkable difference in drawing
behavior between two techniques was found at low temperatures. The tensile
drawing of mats at 60°C was unstable. In contrast, solid-state coextrusion
allowed effective draw, even at room temperature, up to an EDR 27. The
comparison of the drawing behavior of mats for tensile drawing and
solid-state coextrusion indicates that the coherent mats exhibit an
excellent drawability for both techniques. However, the efficiency of draw
is higher for coextrusion than for tensile drawing if the coherency of mats
is poor or on draw at lower temperatures.

Even more efficient and high draw was possible for SGC mats of several
polymers by solid-state coextrusion at a low EDR of ~6 followed by tensile
drawing at controlled rates and temperatures. Figure 3 shows the maximum

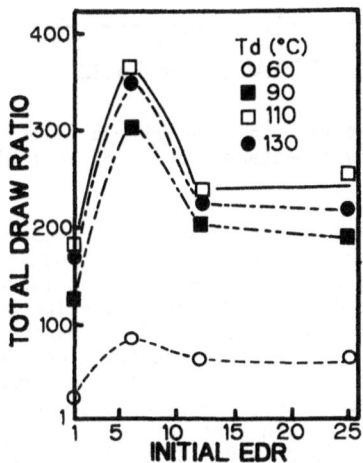

Figure 3. Maximum achievable DR_t vs. initial EDR for the second-stage
tensile drawing of SGC mats observed at constant CHS of 10 mm/min. and at
60–130°C. The sample lengths were adjusted to give the same extension rate
at any DR_t.

achieved DR_t (EDR x DR) as a function of initial EDR for two-stage drawing
at temperatures from 60 to 130°C. As the CHS was 10 mm/min, the sample
lengths were adjusted so as to give the same extension rate for the
extrudates having different EDR, as each sample passed through any DR_t.
Drawability increased markedly between 60 and 90°C, due to the crystal
softening associated with the α-dispersion of SGC mats. At temperatures
between 90–130°C, each sample exhibited a high drawability, slightly
depending on the combination of drawing temperature and rate. In general,
however, extrudates with an EDR of 6 was most deformable over a wide range
of drawing temperatures and rates. Indeed, an initial EDR 6 extrudate
could be effectively drawn, even at room temperature, up to a DR_t of 42
(10). Figure 4 shows tensile modulus as a function of DR_t for tensile

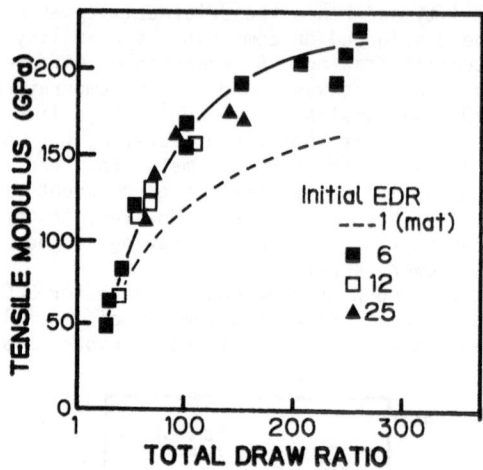

Figure 4. Tensile modulus vs. DR_t for UHMW–PE films prepared by tensile drawing of SGC mats and by two–stage drawing of initial EDR 6, 12, and 25 extrudates.

drawn films and for two–stage processed samples prepared from initial EDR of 6, 12, and 25. For two–stage drawing, the moduli fall on a single curve, independent of the initial EDR and draw temperatures ($\leq$ 130°C). The modulus increased rapidly with DR_t below a value of 150 and approached a constant of 210–220 GPa at a $DR_t \geq 250$. This maximum is close to the lowest theoretical value reported for a polyethylene crystal (220–324 GPa (7,15,16)). The modulus for straight tensile drawing also increases rapidly with DR. However, the efficiency of draw, as evaluated by tensile modulus, is remarkably lower than that for two–stage drawing of SGC mats. The lower modulus for tensile drawing is likely ascribable to the poorer coherency of the initial lamellae in the mats and also of the resultant fibrils formed by the destruction of the lamellae. The poor coherency may reduce the efficiency of molecular deformation.

The effects of molecular weight (Mv) on the drawability of poly–ethylene SGC mats and the resultant tensile properties have been studied for the Mv over the 2 to 80 x 10^5 range. The mats prepared from 0.2 wt% solutions were drawn by two–stage drawing, i.e., solid–state coextrusion at 110°C and an EDR of 6 followed by tensile drawing at the optimum temperatures (60–125 °C), which increased with sample Mv. The draw stress at these temperatures increased remarkably with increasing Mv. Never–theless, the maximum achievable DR_t increased rapidly with Mv from 80 for a Mv of 2 x 10^5 to a constant value of 300–400 for Mv $\geq$ 21 x 10^5. Figure 5 shows tensile modulus and strength at break as a function of DR_t for a range of sample Mv's. Both the tensile modulus and the strength increase rapidly with DR_t and approach constant values characteristic for the Mv at high DR_t. At the lower DR_t range, the tensile properties at a given DR_t are significantly higher for higher Mv samples. The maximum plateau moduli of high Mv samples are not affected by Mv and 220 $\pm$ 10 GPa for Mv $\geq$ 16 x 10^5. In contrast, the tensile strength of drawn films are markedly affect–ed by Mv even at high $DR_t \geq 250$ where the tensile modulus is not affected

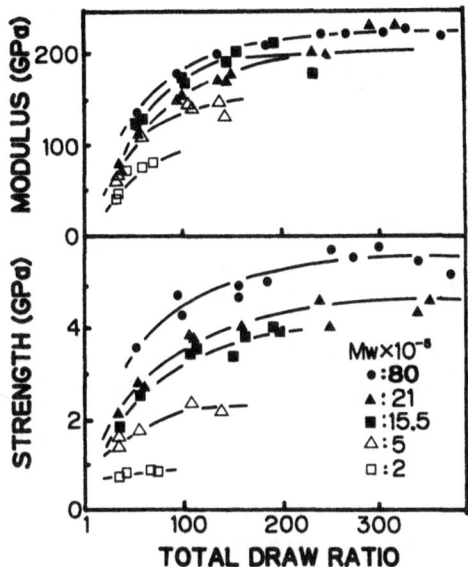

Figure 5. Tensile modulus and strength vs. DR$_t$ for PE molecular weights.

by Mv. The maximum tensile strength increased from 0.9 GPa for a Mv of 2 x 10^5 to 5.8 GPa for UHMW-PE with Mv of 80 x 10^5, the highest Mv used in this work.

Drawing of UHMW-PP SGC Mats
SGC were precipitated from 0.1 wt% solutions in xylene kept at 55°C over 20 hr. The SAXS and WAXS patterns of the mats showed lamellae of ~85 Å thickness that were well-oriented with the fold surface parallel to the mat surface and with the molecular chains perpendicular to the surface. These features are similar to those of UHMW-PE SGC mats. However, the SAXS of UHMW-PP SGC mats showed only the first order of long period scatterings, in contrast to the UHMW-PE SGC mats which exhibited up to the third and even higher order of the scatterings. This indicates a large variation in the fold length of the former crystals.

The effect of draw technique and conditions, and sample Mv on the drawability of UHMW-PP SGC mats has been studied. It was found that these mats also was most deformable by two-stage drawing, as was observed for UHMW-PE SGC mats. Furthermore, the initial EDR also had a major effect on the drawability for the second-stage tensile drawing. Thus, the SGC mats of PP with different Mv's were initially coextruded at 130°C and a low EDR of 6, followed by tensile drawing at optimum temperatures of 150-165°C, depending on the sample Mv. The maximum achievable DR$_t$ increased slightly with Mv, from 60 for a Mv of 3 x 10^5 to 80 for a Mv of 36 x 10^5. Such an increase in drawability with Mv is not so remarkable as found in PE.

Figures 6 (a) and (b) show tensile modulus and strength, respectively, as a function of DR$_t$ for three Mv's of PP. The modulus and strength increase rapidly with DR$_t$ and approach constant values characteristic of each sample Mv at high DR$_t$. Althogh the modulus at a given DR$_t$ is not affected by Mv, the maximum achievable modulus is singnficantly higher for higher Mv samples. The highest modulus of 34-37 GPa in Figure 6 (a)

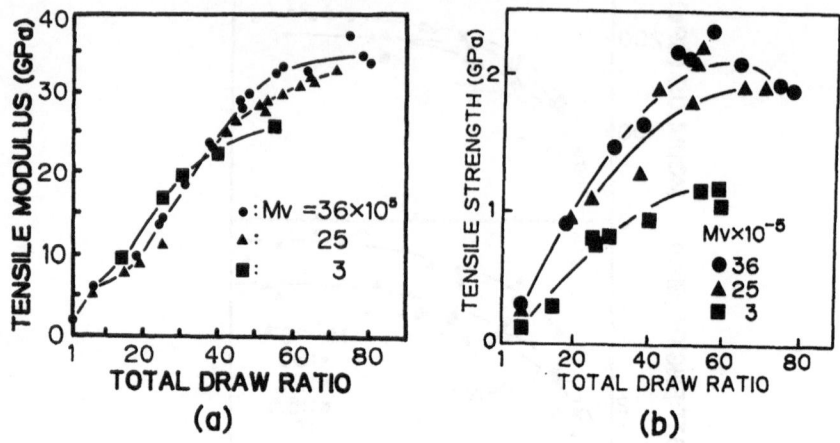

Figure 6. Tensile modulus (a) and strength (b) vs. DR_t for a range of PP molecular weights. The SGC mats were drawn by two-stage drawing.

closely approaches the reported crystal modulus of PP (35–42 GPa (15)). In contrast to the tensile modulus, the strength is substantially higher for higher Mv samples, as seen in Figure 6 (b). Thus, the effects of Mv on the

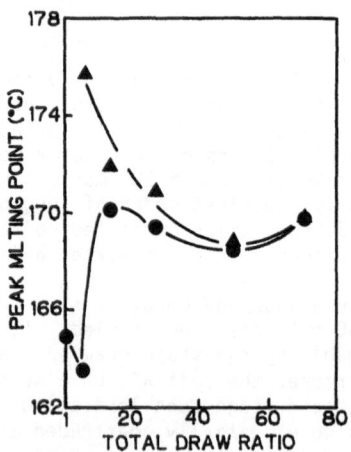

Figure 7. DSC melting peak temperature vs. DR_t for UHMW–PP films. An EDR 6 extrudate prepared at 130°C was further drawn at 165°C by tensile force. (●); as–drawn, and (▲); annealed at 165°C for 30 min.

drawability of PP SGC mats and the resultant tensile properties of drawn films are not so great as in SGC mats of PE. However, a higher Mv sample gives a higher draw and higher mechanical performance for both polymers, with this effect more pronounced for PE.

The melting behavior, measured as a function of DR_t, is also different in these two polymers. For UHMW-PE, the DSC melting temperature and heat of fusion increased with DR_t and approached constant values of 146°C (at a H.R. of 3°C/min.) and 64–67 cal/g, respectively, at a $DR_t \geq 150$. In UHMW-PP, however, the heat of fusion increased with DR_t from 25 cal/g at the initial mat to 34–35.5 cal/g at a DR_t of 70, whereas the melting temperature changed complexly with DR_t. Figure 7 shows DSC melting peak temperature versus DR_t for UHMW-PP ($Mv=25 \times 10^5$) SGC mats drawn first by coextrusion at 130°C followed by tensile drawing at a higher temperature of 165°C. The DSC heating rate was 3 °C/min. The sample was allowed to shrink freely during heating. A single and sharp melting endotherm was observed at $DR_t \geq 17$. In Figure 7, the melting temperature increases with DR_t from 165°C for the SGC mat to 170°C at a DR_t of 17. Then it decreases slightly to 169°C at around a DR_t of 45 and again increases to 170°C at higher DR_t. The initial increase in the melting temperature is primarily ascribable to annealing at the higher drawing temperature (165°C) and a partial chain extention and orientation on draw. The subsequent decrease, up to a DR_t of 45, is likely due to the decrystallization of microfibrils and the successive incorporation, during drawing, of the configurational chain defects which were initially segragated to the amorphous region. At a higher DR_t of 70, the chains are almost fully extended and thus the apparent melting temperature increasesd slightly. Apart from the details, the melting temperatures of these drawn samples are significantly lower than that (176°C) for a lower EDR 6 extrudate annealed at 165°C (draw temperature) for 30 min. This is consistent with the X-ray results that the crystal disorders, both along and perpendicular to the chain axis, are significant in drawn UHMW-PP films (2–4%) as compared to those in UHMW-PE samples (0.2–0.7% and 1–2%, respectively).

Drawing of UHMW-P4M1P SGC Mats
The SGC mats of UHMW-P4M1P was also found to be ultradrawable up to a DR of 80 by a two-stage drawing technique. Although the mechanical proper-ties of this helical polymer are substantially poor compared to the

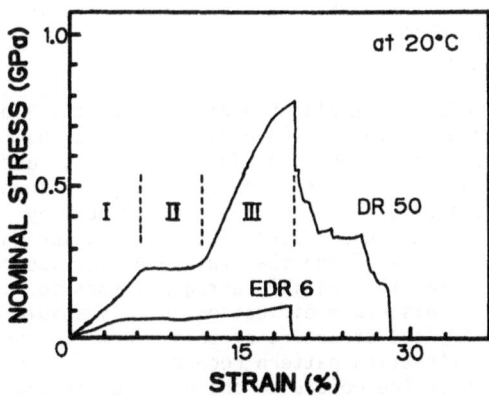

Figure 8. Stress-strain curves of UHMW-P4M1P films with a low EDR of 6 and a high DR_t of 50.

superdrawn UHMW–PE and UHMW–PP, the highly drawn samples exhibit a unique stress–strain behavior on extension at room temperature. Figure 8 shows stress–strain curves characteristic of a low and a high DR samples of UHMW–P4M1P. The extrudate with an EDR of 6 shows usual stress–strain behavior commonly observed for many polymers with low DR. However, the highly drawn samples exhibited unusual stress–strain curves as shown for DR_t 50 in Figure 8. The curve is characterized by the three distinct regions, including an initial elastic region (0–6% strain) where stress increased linearly with strain (Region I), a yield region (6–12% strain) where the stress stayed almost constant (Region II), and finally, a post-yield region (12–20% strain) where the stress again increased rapidly (Region III). Figure 9 shows tensile modulus, determined in Regions I and III as a function of DR_t for UHMW–P4M1P drawn films. The maximum initial

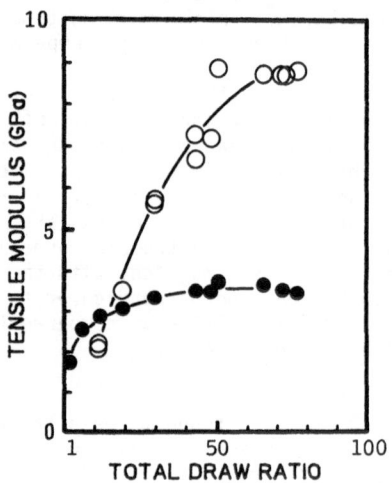

Figure 9. Tensile modulus vs. DR_t for UHMW–P4M1P films determined at Regions I (●) and III (○).

modulus of 3.8 GPa (Region I) is slightly higher than the lowest reported crystal modulus of 2.9 GPa (17) for the ordinally tetragonal form of P4M1P. Further, the modulus of 7.5–9.1 GPa in Region III is about twice the initial modulus (Region I) and even markedly higher than the highest reported crystal modulus of 6.7 GPa (15), determined by X-ray crystal strain measurements. Such unusual stress–strain behavior has been studied by X-ray diffraction. It was found that this behavior was related to the crystal transformation from the initial tetragonal form to a new crystal form during extension. This new modification is only stable under tension. Although the detailed crystal structure of this form is not clear, the characteristic X-ray diffraction pattern recorded under tension indicates a nematic structure in which the molecular chains are irregularly arranged both along and perpendicular to the chain axis.

CONCLUSION

SGC mats of PE, PP, and P4M1P formed from a wide range of Mv's (2-80 x 10^5) have been drawn by three techniques: solid-state coextrusion, conventional tensile drawing, and their sequential combination (two-stage drawing). The drawability and the resultant tensile and physical properties of drawn films were found to be sensitive to draw technique and sample Mv. Thus, the mats prepared from UHMW samples of these polymers have been drawn, by two-stage draw, to the limits in terms of tensile and physical properties approaching theoretical limits. The limiting properties and morphologies of superdrawn films of these three polymers have also been studied.

REFERENCES

1. Ciferri, A. and Ward, I.M., Eds., Ultra-High Moldulus Polymers, Applied Science London, 1978.

2. Zachariades, A.E. and Porter, R.S., Eds., High-Modulus Polymers, Plastic Engineering Series, Marcel Dekker, New York and Basel, 1987, Vol. 17.

3. Lemstra, P.J. and Kirschbaum, R., Polymer, 1985, 26, 1372.

4. Zwijnemberg, A. and Pennings, A.J., Colloid Polym. Sci., 1976, 254, 868.

5. Smith, P., Lemstra, P.J. and Booij, H.C., Colloid Polym. Sci., 1980, 258, 891.

6. Peguy, A. and Manley, R.St.J., Polymer, 1984, 25 (Commun.), 39.

7. Matsuo, M. and Sawatari, C., Macromolecules, 1986, 19, 2036.

8. Kanamoto, T., Sherman, E.S. and Porter, R.S., Polymer J., 1979, 11, 497.

9. Statton, W.O., J. Appl. Phys., 1967, 38, 4149, and Ishikawa, K., Miyasaka, K. and Maeda, M., J. Poly. Sci., Polym. Phys., 1970, 7, 2029.

10. Kanamoto, T., Tsuruta, A., Tanaka, K., Takeda, M. and Porter, R.S., Polymer J., 1983, 15, 327.

11. Kanamoto, T., Tsuruta, A., Tanaka, K. and Takeda, M., Polymer J., 1984, 16, 75.

12. Kanamoto, T. and Ohtsu, O., Polymer J., 1987, 20, 179.

13. Griswold, P.D., Zachariades, A.E. and Porter, R.S., Polym. Eng. Sci., 1978, 18, 861.

14. Kanamoto, T., Tsuruta, A., Tanaka, K., Takeda, M. and Porter, R.S., Macromolecules, 1988, 21, 470.

15. Sakurada, I., Ito, T. and Nakamae, K., J. Polym. Sci., Part C, 1966, 15, 75.

16. Obser, G. and Blasenberg, S., Colloid Polym. Sci., 1970, 241, 985.

17. Kaji, K., Sakurada, I., Nakamae, T. and Shikata, E., Bull. Inst. Chem. Res. Kyoto Univ., 1974, 52, 308.

Advances in gel-spinning technology and Dyneema fiber applications

R. Kirschbaum, DSM Research BV, J.L.J. van Dingenen, Dyneema VoF
P.O. Box 599, 6130 AN Sittard, The Netherlands

Summary

Gelspinning has proven to be a major breakthrough in processing high molecular weight polymers. For polyethylene current production routes, product properties and various applications are discussed.

1. Introduction

The Dyneema fiber was first introduced at the "International Man-made Fibers Congress" in Dornbirn (Austria) on 24-26 September 1986. Since then research has continued resulting in:

– a better understanding of the mechanism of the formation of ultra strong fibers from flexible molecules
– the optimisation of the production process of high performance polyethylene (HPPE) fibers.
– more information on the properties of Dyneema and other HPPE fibers
– the development of applications and products using the Dyneema fibers.
– new opportunities for gel technology.

2. The mechanism of formation of ultra-strong fibers

Improvement of mechanical properties of polymers by orientation and chain extension has become a major research item during the last decades. High density polyethylene was the best candidate for these investigations because in retrospect we can see:
A. That polyethylene has the biggest gap between theoretical and practical or commercial properties.
B. Polyethylene has the simplest molecular structure.
C. The only real breakthrough apart from aramids, (where chemists did the work of chain alignment by synthesizing rigid rodlike molecules) has been realised using polyethylene.
The first statement is illustrated in the table. Regular PE fibers have theoretically the same modulus and strength as HPPE high performance fibers, but the commercial values are completely different. Even HPPE still has a wide margin for improvement.

Mechanical properties of polymers

Polymer	Theoretical		Commercial	
	Modulus (N/tex)	Strength (N/tex)	Modulus (N/tex)	Strength (N/tex)
Regular fibers				
PE (polyethylene)	250 - 300	27 -33	10	0.8
PP (polypropylene)	35 - 50	~ 20	12	0.8
PVA (polyvinylalc.)	180 - 200	~ 21	27	1.0
PAN (polyacrylonitr.)	150	~ 19	17	0.8
PAM-6 (nylon-6)	~200	~ 28	5	0.9
PET (polyester)	80 - 100	~ 23	11	0.9
High performance fibers				
HPPE	250 - 300	27 - 33	90	2.7
Aramids	130	~ 33	100	2.3

How to distinguish between ordinary meltspun PE and HPPE is illustrated by the artist view below. The key words for making the high performance PE fiber or Dyneema are: "Chain alignment" and "Structural perfection".

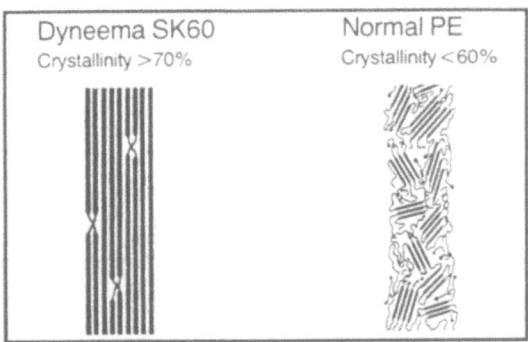

Chain alignment can be reached by unfolding PE crystals. Structural perfection is possible by using long molecules with a low number of side groups. However, these long molecules cannot be used in meltspinning.

Chain folded lamellar crystals were described early 1950 by Prof. Andrew Keller who recognised that in flexible macromolecules the thermodynamically most favourable conformation is a chain folded lamellar crystal. The figure gives the SEM photo and a drawing of his model.

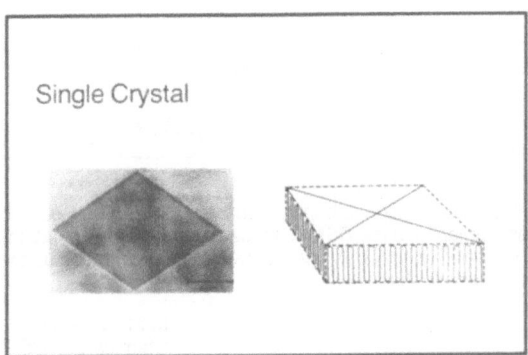

In the mid-sixties, Dr. Pennings from the DSM Laboratories found that it was possible to retain complicated structures from a sheared dilute solution of ultra high molecular weight PE (UHMWPE). These structures proved to be a combination of chain folded and chain extended molecules. In a later stage, Prof. Pennings proved that perfection of this "shishkebab " structures could lead to the extreme mechanical properties we now know for Dyneema.

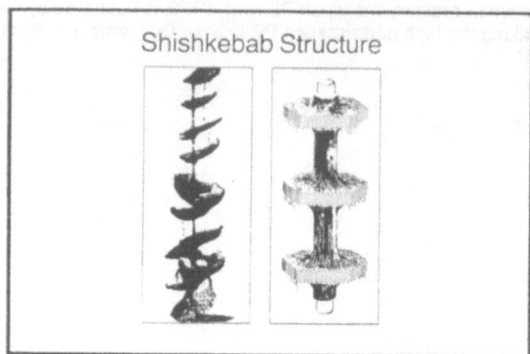

Shishkebab Structure

However, the proposed process could not lead to an economically viable product and it took another 10 years before again the DSM laboratories discovered a process that could be used on a commercial scale. The basis was the discovery of the ultra drawability of gels with a high content of chain folded lamellar crystals in the solid state. The process is generally called "gelspinning" because of the gellike appearance of the filaments after spinning and cooling.

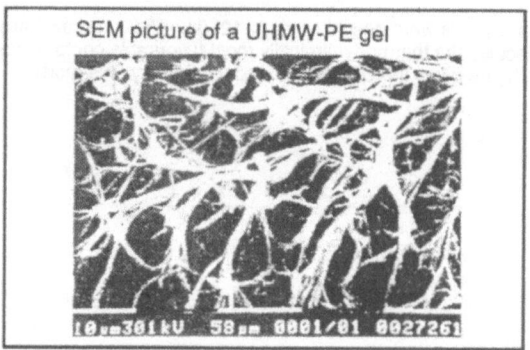

SEM picture of a UHMW-PE gel

Ultradrawing of this gel results in a high strength, high modulus yarn now called Dyneema. The figure gives the first results as published in the DSM Stamicarbon patent USP4,422,993 from 1979. Both modulus and strength increase enormously upon drawing and draw ratios up to 50 were possible, now being optimised to much higher values.

Modulus/Strength v.s. Draw ratio
Stamicarbon USP 4, 422,993

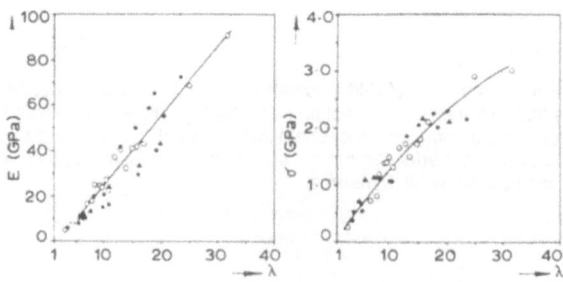

A recent study by van Aerle and Braam states that the process of ultradrawing seems to be a combination of tilting/breaking up/slipping of the chain folded crystals and converting them into chain extended micro fibrils as shown in the picture of the broken fiber. For polymer scientists, there is still an exciting area to study the enormous molecular rearrangements upon deformation.

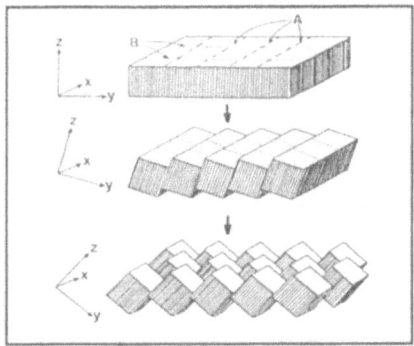

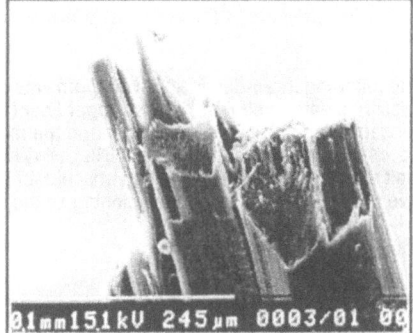

A: cleavage planes, giving rise to
   intra-microfibrillar connections

B: cleavage planes, giving rise to
   inter-microfibrillar connections

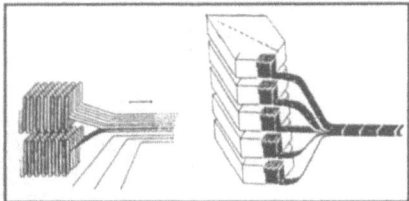

3. The gelspinning process

The figure gives a diagram of a process to produce a high performance polyethylene fiber.

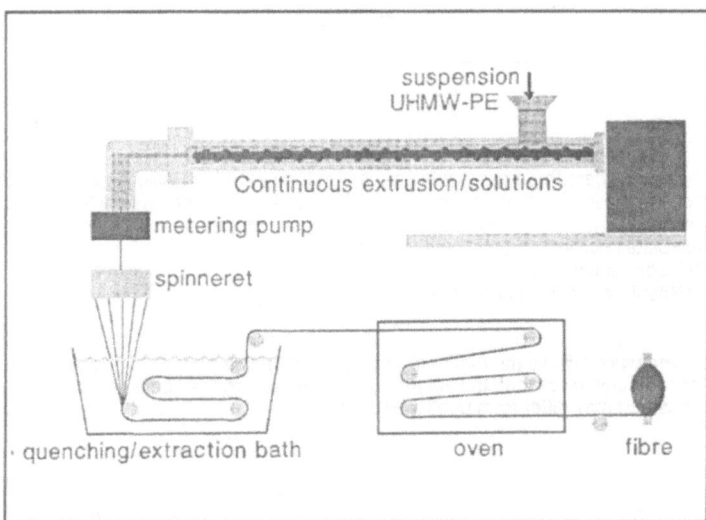

A short explanation to the main points in the process:
- The continuous extrusion of the solution of ultra high molecular weight polyethylene (UHMWPE) with a huge jump in viscosity when the solution is formed.
- Spinning of the solution and gelation/crystallisation of the UHMWPE. This can be done either by cooling and extraction or by evaporation of the solvent.
- Ultradrawing and removal of remaining solvent. Ultradrawing is of course the "heart" of the process but the other steps are essential in the production of a fiber with good characteristics.

The following three-dimensional diagram was compiled a few years ago from both DSM data and published data. It shows that a stronger fiber (higher tenacity) can be produced when the concentration of UHMWPE is lower and the molecular weight is higher (corresponds to the intrinsic viscosity). The concentration of 100% UHMWPE with the lowest molecular weight corresponds to the melt spinning technology that can only obtain (relative) low strength figures. The diagram does not give the actual situation for gelspinning as the recent results are far better than is shown here.

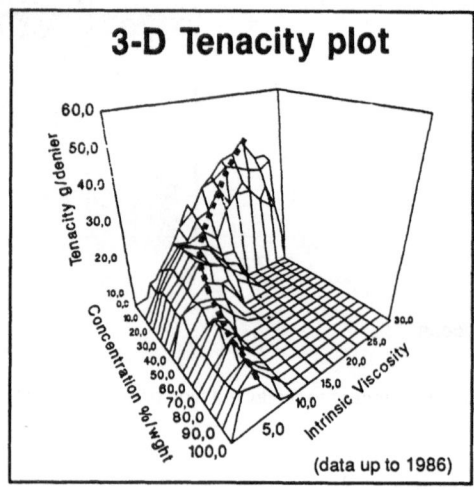

In gelspinning of PE-fibers three companies are active today of which Allied is working under a DSM license. Dyneema is a joint venture company of DSM and Toyobo. The table gives the main differences in the processes and the resulting differences in properties of the fibers:
- Dyneema: volatile solvent and the lowest creep figures
- Spectra (Allied): high modulus
- Tekmilon (Mitsui Petrochemical): lowest molecular weight.

It is difficult to measure exactly the mechanical properties of high performance fibers in absolute values, so for comparison reasons the values were measured on the same tensile tester and under the same conditions and may differ from published data.

High Performance Polyethylene Fibers

Company	Type	Mol weight (g/mol)	Cap. (tons/y)	Solvent	Tenacity (50% per minute) (g/den)	Modulus (50% per minute) (g/den)	Creep	Filament denier
DYNEEMA	Dyneema SK60	2.0×10^6			≥ 30	≥ 1000		1
			~ 60	volatile			low	
	Test grade	2.5×10^6			≥ 35	≥ 1200		1
Allied	Spectra 900	2.5×10^6			30	1000		9
			>150	paraffin oil			medium	
	Spectra 1000	2.5×10^6			35	1400		5
Mitsui Petrochem	Tekmilon	1.0×10^6	~ 30	paraffin wax	30	1000	high	10

Apart from the effect on the fiber, each solvent has its own advantages, but generally speaking the more loops there are in the plant, the more complex the plant is and the more complex the recycling processes are. The figure gives an impression of three different processes of producing high performance polyethylene fibers as derived from patent analysis and published literature.

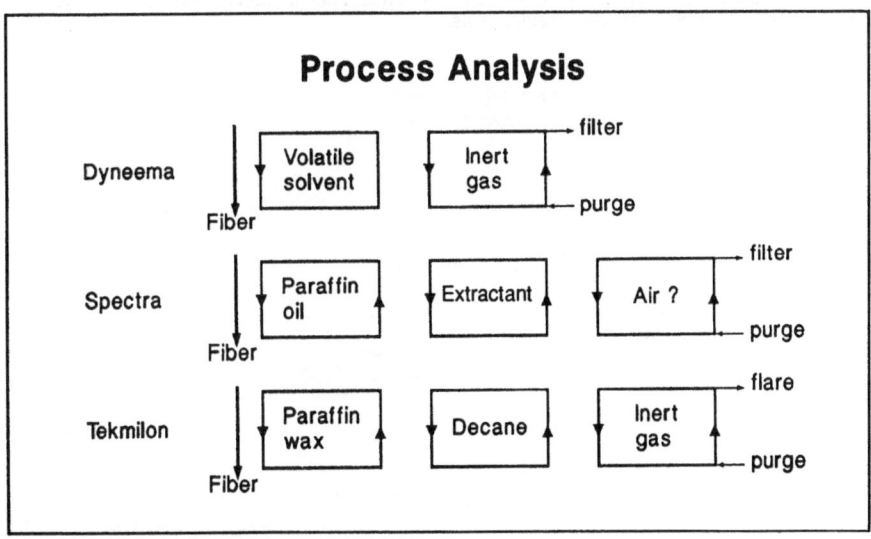

4. Dyneema SK60-fiber properties

The properties of Dyneema fibers in comparison to other high performance fibers have given rise to a real new class of materials. The first diagram gives a plot of specific strength vs specific modulus. Especially in specific strength Dyneema has unique properties and it is clear that there still remain possibilities for further improvement. In specific modulus Dyneema has higher values than aramids and almost equals certain types of carbon fiber.

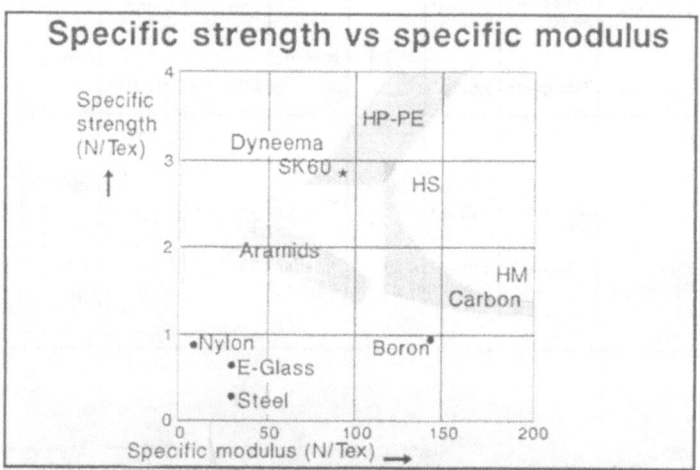

The low density of the Dyneema fibers together with high strength form of course the basis for the very high specific strength or tenacity. In the figure fibers are compared that combine high strength to a low weight (e.g. Dyneema and aramids) to fibers that try to combine high strength to a low volume (e.g. carbon fiber and boron fiber). It is clear that Dyneema has a unique combination of properties being the strongest man-made fiber on a weight basis.

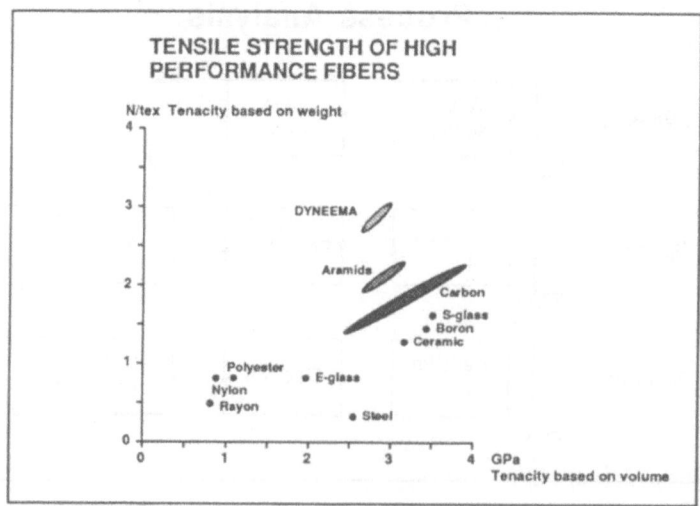

The stress/strain diagram gives another view on the Dyneema SK60 properties and of the wide range of opportunities that remain in tailoring the properties by variations of concentration, molecular weight and draw ratio.

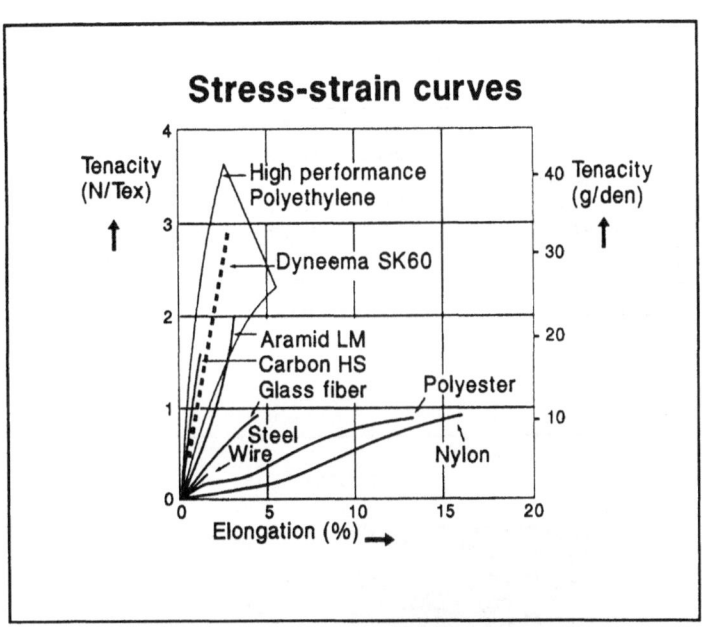

The combination of strength and elongation to break determines the impact strength or the potential energy absorption. This is shown in the next figures. Nylon performs relatively good, not because of its high strength but due to its very high elongation to break (over 15%) which is not acceptable in many applications.

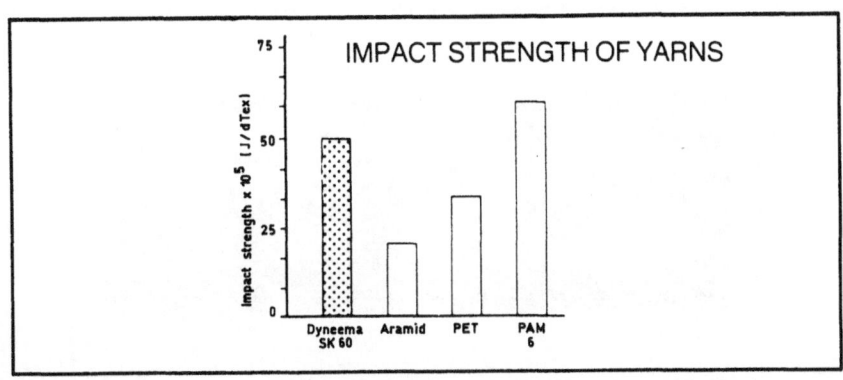

The flexible chain of polyethylene results in a flexible yarn Dyneema SK60 and it is not surprising that this yarn has a high loopstrenght and also a high knotstrength.

Loopstrength/Knotstrength

Material	Loopstrength N/tex	Loopstrength %	Knotstrength N/tex	Knotstrength %
Dyneema	1.3-2.0	40-65	1.1-1.7	35-55
Aramid	0.9-1.5	40-75	0.6-0.8	30-40
Carbon	0.01	~ 1	0	0
PET	0.6-0.7	70-75	0.4-0.5	50-60
PAM-6	0.6-0.7	70-75	0.5-0.6	60-65
PP	0.6-0.7	85-95	0.4-0.5	60-70

Loopstrength

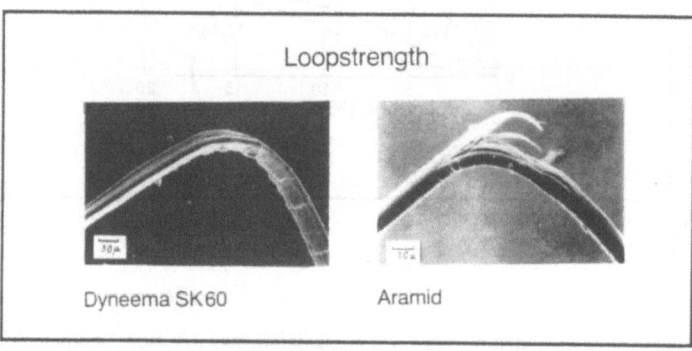

Dyneema SK60   Aramid

Knotstrength

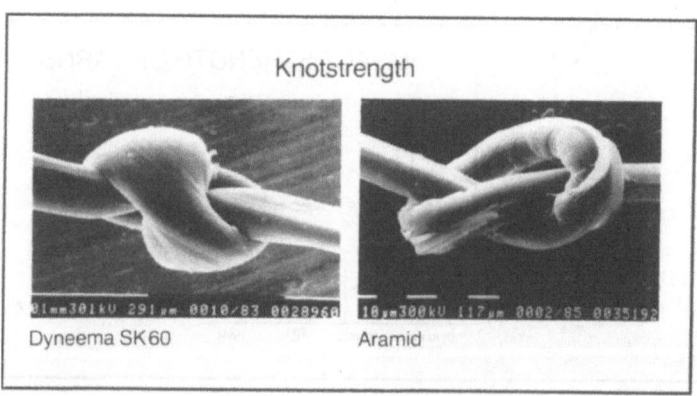

Dyneema SK60   Aramid

A good flexlife is also quite normal for a flexible chain material. The chemical nature of the ultra high molecular weight polyethylene feedstock material used for Dyneema SK60, guarantees a good abrasion resistance because that is exactly the most important property in using UHMWPE as an engineering plastic. Chemical resistance is also "built-in" in UHMWPE and so is the UV resistance: polyethylene is a very inert material.

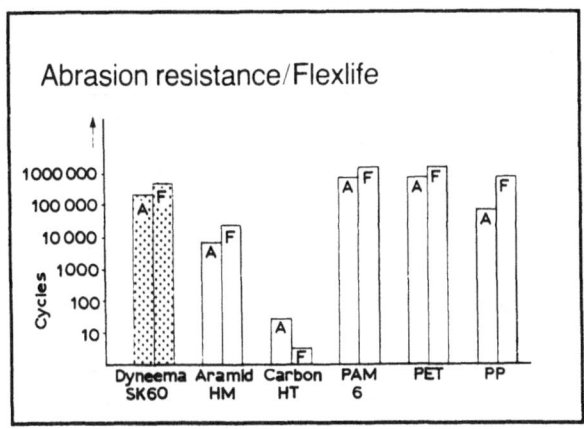

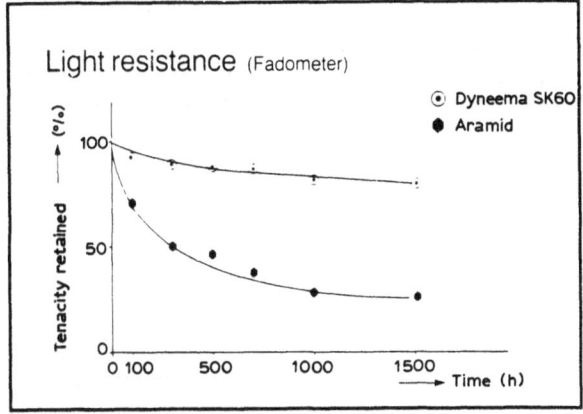

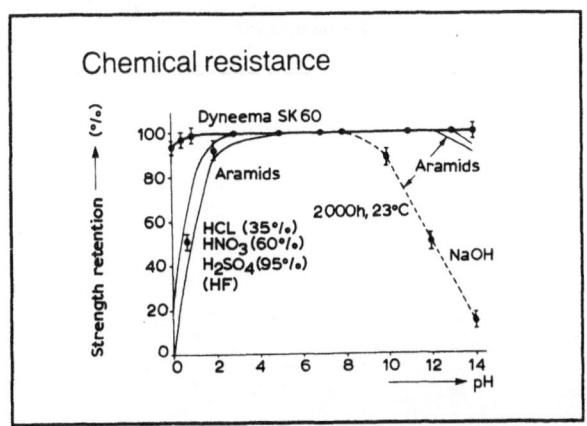

Thermal resistance may be limiting in certain applications. However, in processing the yarn it can stand relatively high temperatures for a short period of time without losing much of its properties when back to room temperature. Around the melting point, an interesting phenomena can be observed as the constrained and unconstrained Dyneema fibers show a different behaviour and have a different melting point. In practice any complex pattern can be measured between these two extremes depending on the conditions, only proving non-equilibrium conditions of the experiments.

Short term thermal exposure

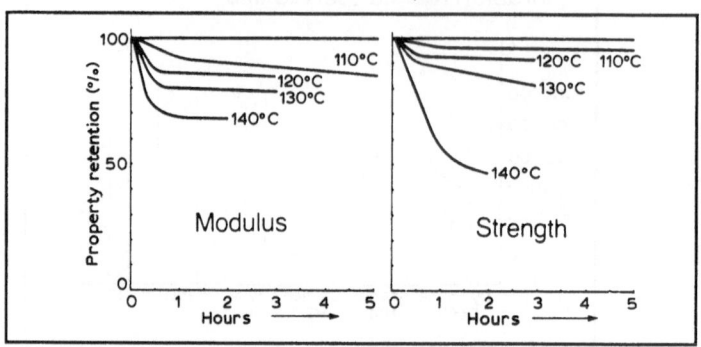

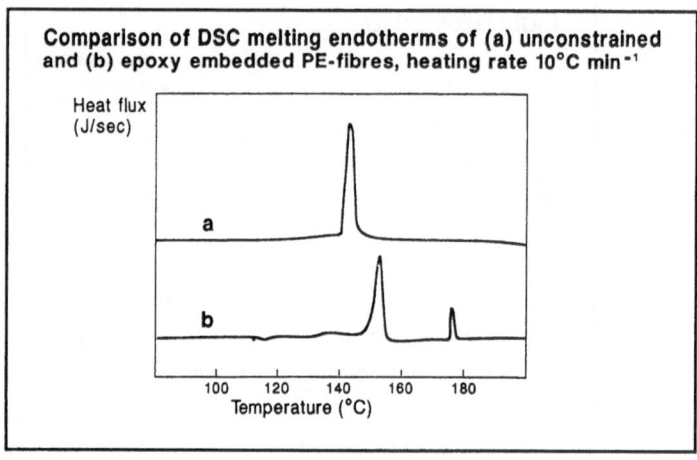

Comparison of DSC melting endotherms of (a) unconstrained and (b) epoxy embedded PE-fibres, heating rate 10°C min^{-1}

The properties of the Dyneema fiber get better when the temperature is lower. This holds especially with strength and creep. The tenacity rises fast and a brittle point has not yet been measured. Creep is also sensitive to temperature (Arrhenius type of dependency), reaching lower figures at lower temperatures. The absolute figure for creep also strongly depends on the process as mentioned before and on the grade of UHMWPE that is used to produce the fibers giving rise to differences in the plateau creep rate of more than one decade.

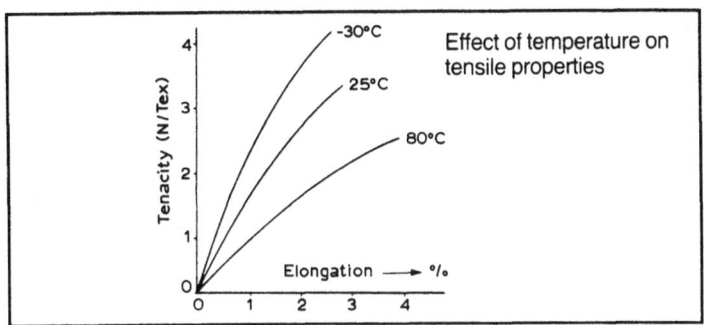

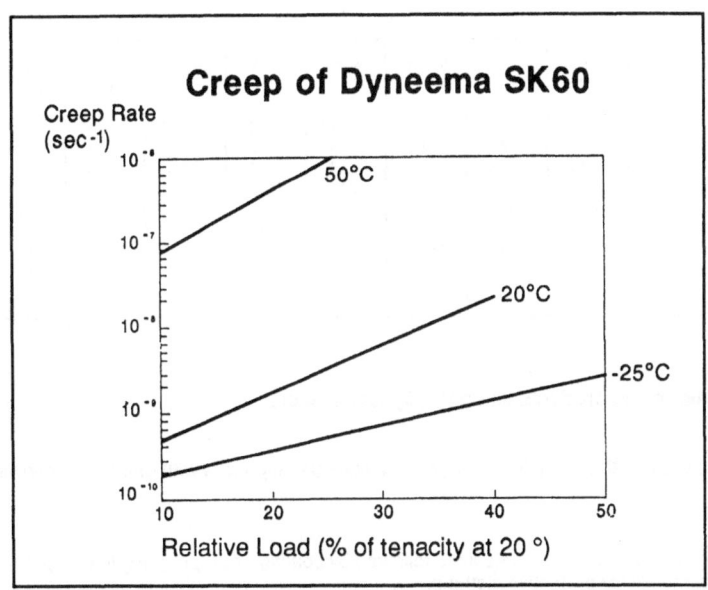

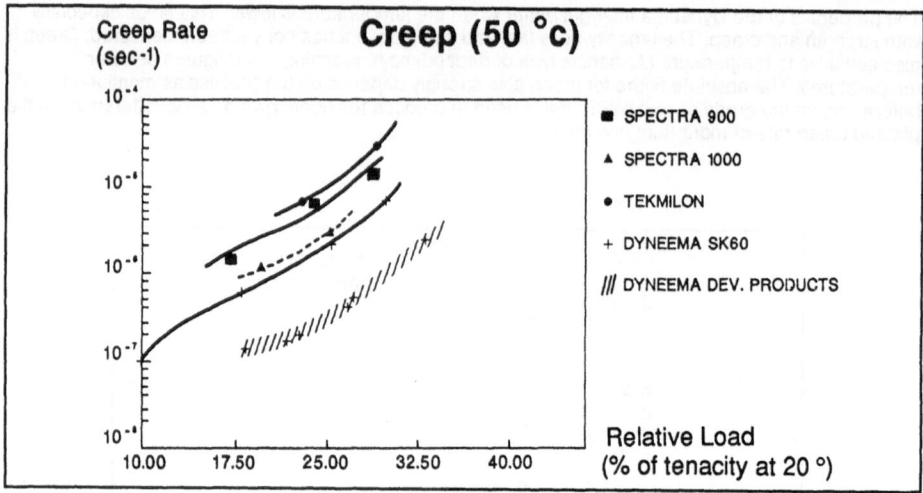

The table below compares Dyneema to other high performance fibers on the main properties and quantifies the statements made before on the first commericial yarn Dyneema SK60.

	Dyneema SK60	Aramid LM	Aramid HM	Carbon HS	Carbon HM	E-Glass	Polypro-pylene	Polyester HT fil	Polyamide 6,6 HT fil
density (g/cm3)	0.97	1.44	1.45	1.78	1.85	2.55	0.90	1.38	1.14
tensile strength (GPa)	2.7	2.7	2.7	3.4	2.3	2.0	0.6	1.1	0.9
tenacity (N/tex)	2.65	1.9	1.9	1.9	1.2	0.8	0.6	0.8	0.8
tenacity (g/den)	30	22	22	22	14	9	7	9	9
modulus (GPa)	87	58	120	240	390	73	6	14	6
specific modulus (N/tex)	90	40	83	134	210	28	6	10	5
specific modulus (g/den)	1000	450	940	1500	2400	310	70	110	56
elongation at break (%)	3.5	3.7	1.9	1.4	0.5	2.0	20	13	20

5. Applications and products using the Dyneema fibers

Producing a product that contains the Dyneema fiber, usually starts with one of the following processes.

- **Weaving or UD technics**, sometimes followed by coating or prepregging to composites
- **Knitting**, mainly for protective clothing.
- **Braiding or twisting**, for ropes and cables
- **Non-woven, staple spinning** or other processes using non-filament fibers
- **Composite forming processes** based on woven, braided or UD fabric or by pultrusion, filament winding etcetera.

All these technics can use the Dyneema SK60 yarn and keeping in mind the properties of the fiber, a long list of products is tested, under development or already on the market.

Summary of properties Dyneema SK 60		
Excellent	Good	Limiting
Specific strength	Specific modulus	Creep
Light resistance	Processability	Maximum. use temp.
Abrasion	Work to break	Adhesion
Flexlife	(Low speed)	Compression
Chemical resistance	X-Ray transparancy	Flammability
Knotstrength	Low temperature	
Floats on water	Mechanical properties	
Moisture sensitivity		
Electrical/insulation		
Work to break		
(high speed)		

A few examples from these applications can illustrate the properties of Dyneema SK60 in practice.

A rope, 17 tons breaking strength:

Maybe even more informative is the comparison of free breaking length, that is the length where a rope breaks under its own weight. Theoretically Dyneema could reach over 300 km! In water the value for Dyneema would even be infinite because the material floats and has no weight when immersed.

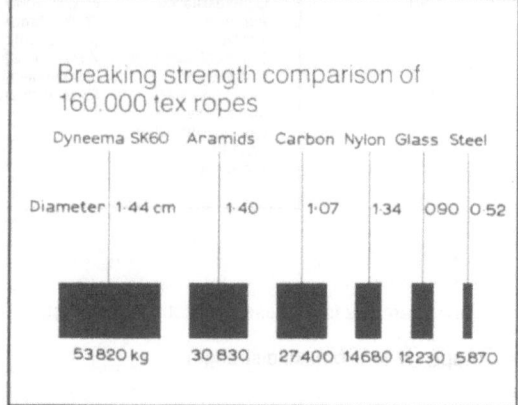

Knitted fencing suits, giving a protection of 1000 N all over the body without using extra in-lay that hinder the players.

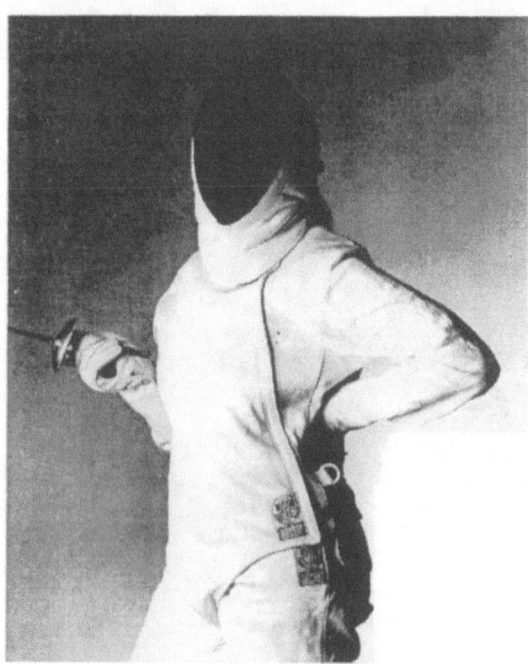

Ballistic protection can be obtained with fabric and with composite material. Both in ballistic vests and in helmets, a very high protection at a relatively low weight can be achieved.

Energy absorption is also the main mechanism in the low velocity impact absorption mechanism. The figures show the results from Dyneema composites in crushing tubes and in impacting flat plates both giving results that are very promising for applications like motorriders helmets or crash absorbing parts in e.g. cars

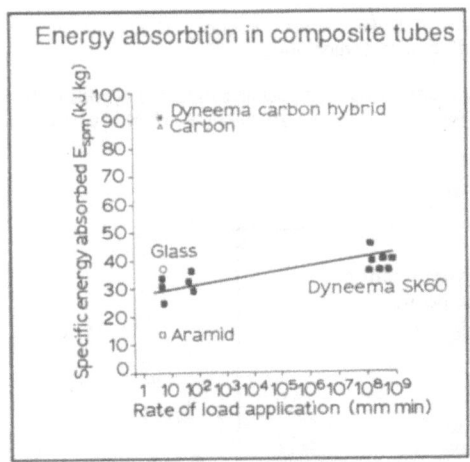

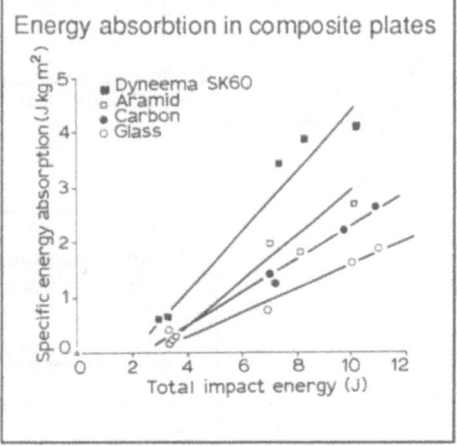

Hybrid composites using Dyneema can be made with zero thermal expansion coefficients or with a defined expansion coefficient. Especially in the tension member of optical cables this may be an important aid in the design.

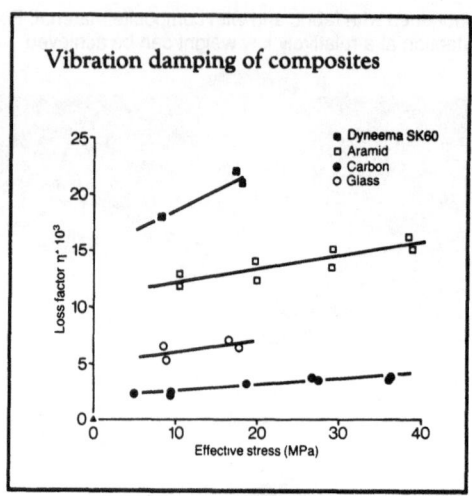

Vibration damping

Vibration damping can be important in composites under dynamic load. Especiallly in sporting goods a high damping factor can be essential to protect the sportsmen or to make a higher performance possible.

The figure gives the results of testing composite reeds of different fibers and epoxy. The figure clearly shows the superior vibration damping properties of Dyneema composites.

Dyneema can also be used to improve the damping properties of composites by making hybrids with the original fiber. The figure illustrates the better vibration damping of Dyneema/glass hybrid compared to the original composite.

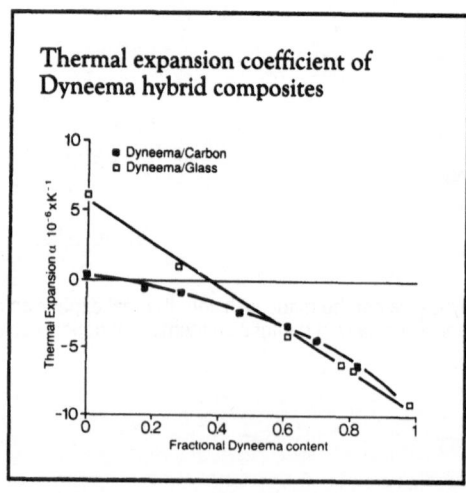

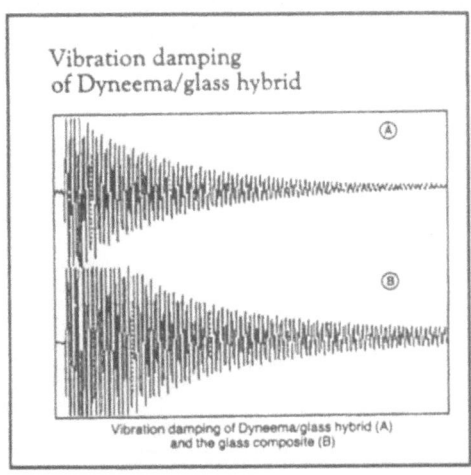

Vibration damping of Dyneema/glass hybrid (A) and the glass composite (B)

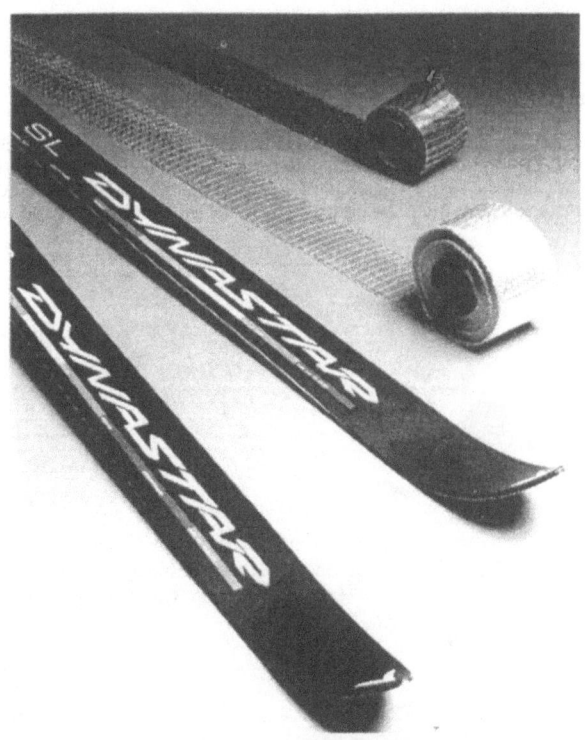

The sonic velocity in composites made of Dyneema SK60, makes the fiber very suitable for the production of speaker cones, as can be seen from the figure.

Composites

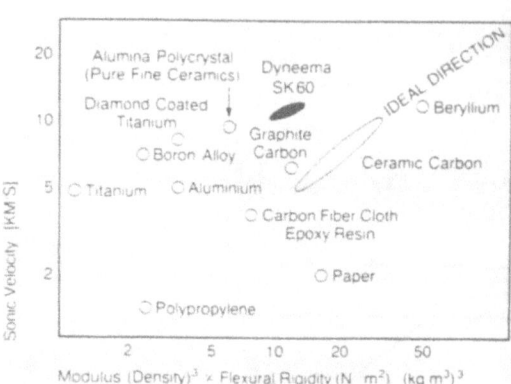

New opportunities from gelspinning

Geltechnology can be extended to other fibers and to films. In films, it is clear that biaxially stretching cannot reach the same strength as fiber in both directions. But the first experiments bring a special surface topology in the film and result in such ultra thin films that for making the TEM pictures samples could be directly transferred from the drawing machine to the electron microscope without microtoming.

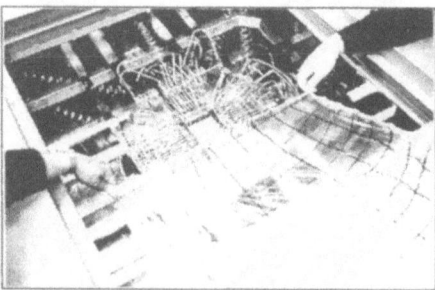

SEM and TEM photos of biaxially stretched PE-film

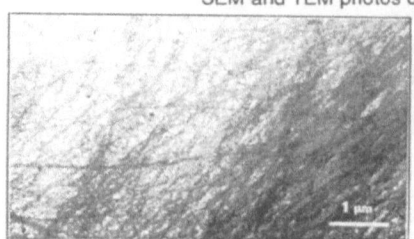
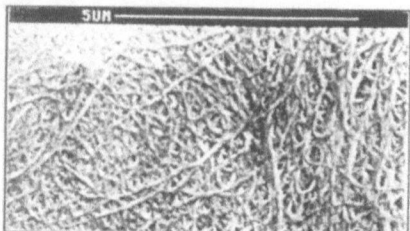

Next to DSM several companies are active in research and gelspinning of other polymers than polyethylene. The most promising seem to be PAN (polyacrylonitril) and PVA (polyvinylalcohol). In the figure the state of the art as derived from patents is plotted in the specific modulus/specific strength diagram.

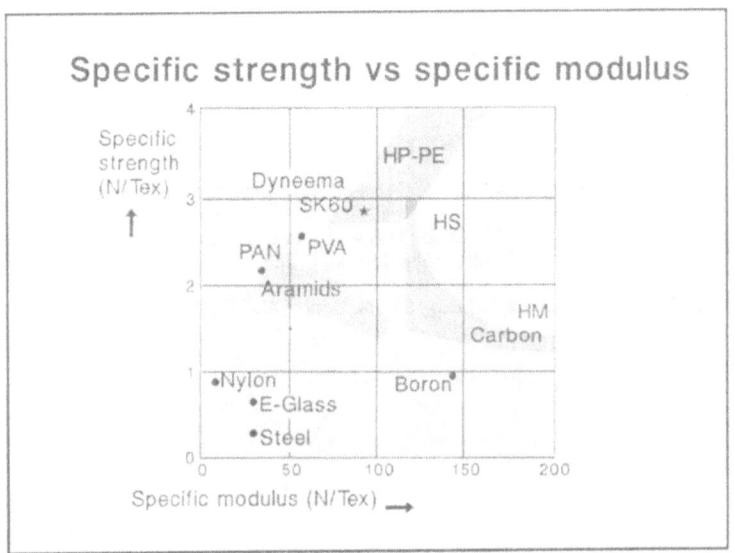

Finally a histogram that shows the exponential increase of both scientific and industrial interest in high strength/high modulus fibers over the last decades.
The real boom of patent applications only took off after the publications of the basis invention in the early '80-ies, and a continuing increase in the number of **application**-patents is foreseen.

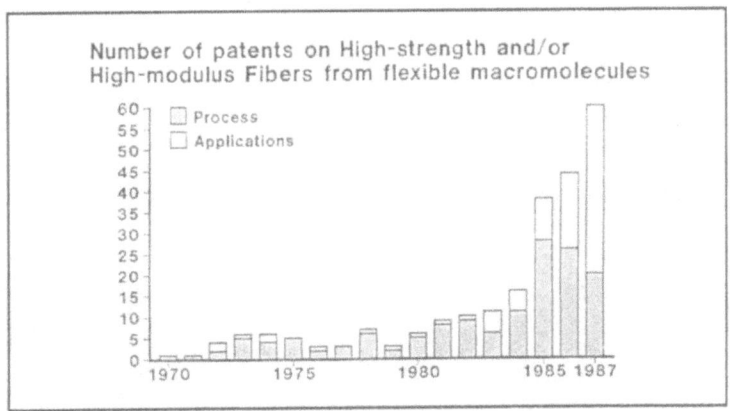

References

1. Kirschbaum R., Yasuda H. and van Gorp E.H.M., 25th Int. Chemical fibres congress, 24-26 September, 1986, Dornbirn, Austria, p. 229.

2. Lemstra P. and Kirschbaum R., Polymer, 1985, vol 26, p. 1372.

3. Patens polyethylene gelspinning:
UK Patent: GB 2042414
UK Patent: GB 2051667
US Patent: USP 4436689

4. Patents other fibers:
Japan Patent: 61-289112
US Patent   : USP 4535027

5. Van Aerle N.A.J.U. and Braam A.W.M, J. of Materials Science, to be published, 1988.

6. Scholle K.F.M.G.J., Jacobs M.J.N., Omloo H.A.A., Loete G.J.H. and Janse G.H.A., 9th Int. Conf. SAMPE European Chapter, 14-16 June, 1988, Milano, p. 253.

7. Kindervater C.M. and Scholle K.F.U.G.J., 9th Int. Conf. SAMPE European Chapter, 14-16 June, 1988, Milano, p. 277.

KINKBAND FORMATION IN ARAMID FILAMENTS

S. van der Zwaag, S.J. Picken and C.P. van Sluijs

Akzo Corporate Research Laboratories

PO Box 9300, 6800 SB Arnhem, the Netherlands

ABSTRACT

The formation of kinkbands in aramid filaments under axial compression is studied. A model for the kinkband density as a function of the strain is presented.

INTRODUCTION

Axial compression of high modulus polymeric filaments results in the formation of so-called kinkbands. These are regions of sub-micron thickness where the compressive deformation is concentrated. This mode of failure is characteristic for anisotropic materials. The processes involved in kinkband formation are not yet well understood. In this work the kinkband formation in single aramid filaments is measured as a function of the applied compressive strain. Above a critical compressive strain the kinkband density initially increases very rapidly until eventually a maximum density is obtained. The experimental results are compared with an elastic stability model. The kinkbands form before elastic instability occurs and are therefore attributed to a plastic deformation process. A model is developed to describe the kinkband density as a function of the applied strain.

EXPERIMENTAL

Single aramid filaments with a well-defined tensile pre-strain were glued onto the surface of small epoxy bars (3x3x10 mm). Araldite XB 3052 A/B was used for both the bars and the glue covering the filament. The cured specimens containing the filaments were compressed axially in a miniature compression device mounted on a Leitz optical microscope. The compressive strain was increased in small steps. After each deformation step the positions of the kinkbands were recorded. The axial compressive strain was determined by measuring the distance between two markers on the surface of the specimen.

OBSERVATIONS

Figure 1 shows an optical micrograph of a set of kinkbands in an aramid filament under 1.0 % axial compression. The micrograph shows that the deformation results in a localized failure mode, the kinkband formation. The kinkbands are at an angle of about 60° with respect to the filament axis. TEM studies [1] have shown that there is extensive shearing deformation in these kinkbands, causing delamination between fibrils. This fibrilization hardly affects the axial tensile strength of the filament (reduction < 10 %) [2].

The results of a compression experiment are shown schematically in figure 2. The horizontal bars represent a particular segment of a filament during a complete experiment. The vertical lines indicate the positions of the kinkbands. The figure illustrates that above a certain compressive strain value the number of kinkbands increases rapidly with increasing compressive strain. Here, as in most experiments, new kinkbands mainly form in-between existing kinkbands. An increase in the compressive strain not only creates new kinkbands : existing kinkbands become more visible. This suggests that the amount of deformation in a kinkband increases. The width of the kinkband does not seem to change.

In figure 3 the results of figure 2 are replotted as the kinkband density, ρ, (i.e. the number of kinkbands per mm filament length) versus the applied strain, ε. At a strain of - 0.54 % the first kinkband is formed. From - 0.55 % to - 0.75 % the kinkband density is proportional to the applied strain. At larger compressive strains the kinkband density goes to a saturation value, ρ_{max}. It is found that the pre-(tensile) strain applied on the filament during fabrication of the specimen only effects the magnitude of ε_{o}. The pre-strain has no effect of the shape of the ρ-ε curve.

The angle of the kinkband with respect to the filament axis was measured as a function of the tensile modulus. The results in figure 4 show that the angle increases with increasing filament modulus.

DISCUSSION

ELASTIC STABILITY CRITERIA

The kinkband formation process appears to be an instability problem, either of an elastic or a plastic nature. To establish the nature of the instability process, we have examined the conditions for **elastic** stability of a cylindrical rod embedded in a flexible matrix. The materials behave linearly elastic and their properties are isotropic. However, in order to account for the anisotropy of the filaments the effects of shearing deformation on the elastic stability are taken into account. The critical load for elastic instability, P_{cr}, is given by

$$P_{cr} = P_e \ / \ [\ 1 + 1.2 \ P_e \ / \ A \ G \ ] \tag{1}$$

where A is the cross-sectional area of the rod, G the shear modulus and P_e the Euler buckling load which is given by

$$P_e \ = 2 \ m^2 \ \pi^2 \ E \ I \ / \ L^2 \tag{2}$$

where m is the wave number, E the Young's modulus in the axial direction, I the moment of inertia of the bar, and L the (embedded) length. The minimum wave number, m, depends primarily on the mechanical properties of the surrounding matrix material : For an unsupported bar m = 1, for a matrix material with E_m is 10 MPa m = 70 and for E_m is 1 GPa m = 250. Inserting appropriate values for the various parameters a critical stress for elastic buckling of an aramid fibre in an epoxy matrix of 1300 to 1500 MPa is calculated. This value is significantly higher than the average stress required to form kinkbands in aramid filaments : $\sigma_{kinkband}$ **is 500 ± 50 MPa**. From this we concluded that kinkbands are not due to elastic instability of the fibre as a geometrical unit, but are due to a **plastic instability** of the filament material. This means that the compressive properties of aramid containing composites are limited by the compressive strength of the fibres and not by the properties of the matrix material or the structure [3]. Only for very weak matrix materials (E_m < 6 MPa) is the compressive strength limited by elastic stability criteria. Similar observations have been made for HM carbon fibres, which also fail due to kinkband formation rather than elastic instability [4].

RATE OF KINKBAND GROWTH

Figure 3 shows that the kinkband density increases as a function of the compressive strain. The observation is described by the following model: The formation of a single kinkband leads to an 'annihilation' of a length ΔL_{kb}. At low compressive strain values, just above the critical strain required to nucleate the first kinkband, ε_o, the additional compressive strain is used to generate more kinkbands: $\varepsilon - \varepsilon_o = \rho.\Delta L_{kb}$. At high compressive strains the rate of kinkband formation is limited by the maximum kinkband density. We assume $d\rho/d(\varepsilon-\varepsilon_o) \propto (1-\rho/\rho_{max})$. Combination of these two equations leads to the following equation for the kinkband density as a function of the applied strain

$$\rho = \rho_{max} \{ 1 - \exp(- (\varepsilon - \varepsilon_o)/(\rho_{max} . \Delta L_{kb})) \} \qquad (3)$$

Fitting the experimental data in figure 3 to equation (3) yields $\rho_{max} =$ 150/mm and $\Delta L_{kb} = 35$ nm. The curve in figure 3 is calculated using (3) and describes the data well. The calculated value for ΔL_{kb} of 35 nm is in qualitative agreement with the contraction calculated from TEM micrographs of kinkbands [1]. The calculated value for ρ_{max} seems to be a reasonable value since it yields a mimimum distance between kinkbands of 6 μm. This is the projected length of a 60° kinkband in a 12 μm diameter filament.

THE KINKBAND ANGLE VERSUS THE FIBRE MODULUS

The kinkband angle in aramid fibres increases linearly with the fibre modulus. Extrapolation of the results to the modulus for unoriented aramid fibres, E ≈ 5 GPa, yields a predicted kinkband angle of 45°. This supports the notion of kinkband formation due to plastic deformation. The increase in kinkband angle is larger than the decrease in the average molecular desorientation angle $<\Phi>$: in an aramid filament with E = 50 GNm^{-2} $<\Phi> = 14°$, while for E = 160 GNm^{-2} $<\Phi> = 5°$. The corresponding change in kinkband angle is 15°. This suggests that the kinkband angle is not directly related to the average molecular desorientation angle. A similar variation of kinkband angle with degree of orientation has also been observed in polyethylene [5]. No detailed models have been found in the literature to calculate the kinkband angle in terms of the filament anisotropy.

REFERENCES

[1] Takahashi, T. et al.; J. Appl. Pol. Sci. 28 (1983) 579
[2] Dobb, M.G. et al; Polymer 22 (1981) 960
[3] Greenwood, J.H. and Rose, P.G.; J. Mater. Sci. 9 (1974) 1809
[4] Hawthorne, H.M. and Teghtsoonian, E.; J. Mater. Sci. 10 (1975) 41
[5] Seto, T. and Tajima, Y.; Jap. J. Appl. Phys. 5 (1966) 534

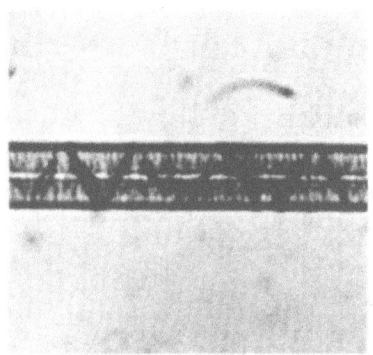

Figure 1: optical micrograph of an
aramid filament under -1.0% axial
strain.

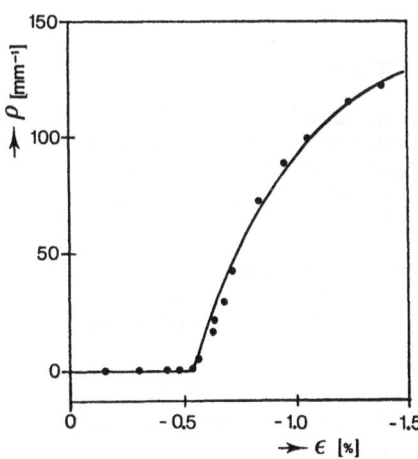

Figure 2: positions of kinkbands
in a particular segment of an
aramid filament during a
compression experiment(see text).

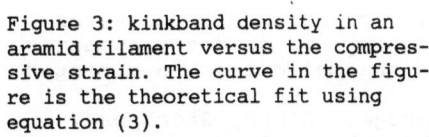

Figure 3: kinkband density in an
aramid filament versus the compressive strain. The curve in the figure is the theoretical fit using
equation (3).

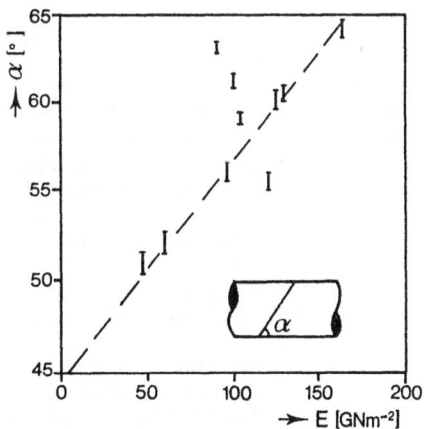

Figure 4: kinkband angle versus
the tensile modulus of some
aramid filaments.

MECHANICS OF FIBER ASSEMBLIES

JOHN W.S. HEARLE
Department of Mechanical Engineering*
University of Delaware
Newark, DE 19716, USA

ABSTRACT

The present industrial position of composites, is reviewed.
There is competition, polymeric and inorganic, in fibres and
matrices, but optimisation depends on the interaction with
developing methods of composite manufacturing. Interlaced
3-D textile structures have advantages, but their mechanics
cannot be treated well by laminate theory. Relevant advances
in textie mechanics are summarised.

INTRODUCTION

Last year at Rolduc, I presented an academic analysis of a
problem in polymer physics: a structural model for the
prediction of the mechanical properties of nylon and
polyester fibres. There are some interesting differences in
this year's presentation:

(1)   The geometrical structure has moved up over
      four orders of magnitude from the arrangement
      of linear polymer molecules to the arrangement
      of linear fibers.

(2)   At the request of the organisers, the lecture
      is more general, and directed to applications,
      particularly composites.

(3)   Although the production of nylon and polyester
      fibres is a major industry, we do not know after
      50 years of study how the molecules are arranged,

---

* Present address: The Old Vicarage, MELLOR, Stockport,
                   Cheshire, SK6 5LX, England.

because we do not have the techniques; and only
limited changes in structure are possible with
the highly efficient process of melt-spinning.
In composites, there is no difficulty in finding
how the fibres are arranged, but experimentation
with many engineering processes is currently
leading to diverse structures, some of which are
closely controlled in geometry. The problems are
that we do not know which of the many possible
structures will prove to be the most useful,
and the manufacturing processes are still primitive
and inefficient.

When methods and markets have been established, the
experience in the development of new processes for
conventional textiles shows that the expertise of the
textile machinery industry will lead to remarkable increases
in speed and efficiency. For example, yarn texturing, which
was introduced as a long sequence of batch processes 40
years ago, had been converted within 10 years to a
continuous process at 10 m/min, and with current machinery
approaches 1000 m/min.

The issues which will need to be addressed by
scientific research are not entirely clear, but they will
involve the characterisation of the topology and geometry of
structures and the understanding of the mechanics of
production processes and product performance. In order to
interact with industrial technology, the fundamental studies
will need to be linked to computer-aided design methods.

Composites bring a major change in design philosophy to
mechanical engineering. Traditionally, the materials
engineer, or the chemical engineer, is responsible for the
processes which determine the internal structure of any
selected material, and he supplies a set of property data to
the design engineer, who uses stress analysis to decide the
external shape of a component or a total structure. But, in
composites, the geometry of local fiber placement and the
overall shape of the piece are both controlled in
manufacture, and a single materials/design engineer must
optimise the internal and external structure together.

The Progress of Composites

In the opening plenary lecture at the international
composites conference in London last year, Aime Jardon, Vice
President of Renault, commented:

.....up to now, composites were rather related to
'high performance/low volume industries like
aerospace or surgery ....
....I will examine the possibility of composites
entering the mass-production industry .....

This reflects a myth, which is commonly believed, that,
apart from the Israelites making bricks with straw and
similar historical fancies, composites engineering started
with carbon fibres in 1964. The reality is that the
combination of natural (polymeric) fibres with a natural
(polymeric) adhesive to form a composite resembling a
leather was suggested in Thomas Hancock's patent of 1824;
and, following the commercialisation of flexible
cotton/rubber composites in tires 100 years ago, rigid
composites of asbestos fibers in a phenolic polymer matrix
were being produced in 1907. Paper/phenolic, cotton/phenolic
and the decorative laminates were well established 50 years
ago, and glass/polyester 40 years ago.

Commercially, the composites industry is now in three
main groups:

(1) major companies, with a large technical input,
     making high-performance, high-price individual
     items;

(2) large and small industry, with limited new
     technical input, making cheap, low-performance,
     reinforced plastic products in large quantities;

(3) the specialist hand lay-up business, typefied
     by boat-builders.

But the major industrial interest for the future must
be the mass-production of components with moderate
performance requirements at at economic cost, supplied to
the automobile, construction, and general engineering
industries, in competition with metal.

The Relevance to Polymers

Why is such a general discussion of composites relevant to a
polymer conference? It is important because of the
competition both between polymeric and inorganic materials
as fibers and matrices, and within each type. Interaction
between polymer science and industrial technology is needed
to optimise the situation.

For matrices, thermoplastic polymers are joining
thermosets, and metal and ceramic matrices are being widely
studied.

More related to this paper, there is a new generation of high-performance, man-made fibers. The main types are: (a) oriented linear polymers, such as para-aramid and HMPE; (b) carbon (graphite); (c) ceramics, such as alumina and silicon carbide; (d) single crystals. And the possibilities for glass fibers, which have been well explored, metal fibers, and the old synthetics, like PET, which have been neglected, must also be taken into account.

In addition to the detailed differences of costs and properties, there are major differences between the molecularly one-dimensional polymer fibers and the molecularly three-dimensional inorganic fibers, with the interconnected two-dimensional form in carbon showing some anisotropy, but generally being closer in behaviour to the inorganic fibers.

The high anisotropy of fibers, such as Kevlar, Twaron, Technora, Spectra and Dyneema, makes it doubly important to get the fiber orientation right in the composite; and the weak transverse bonding means that there is easy yielding in compression and relatively poor temperature/time performance. Design decisions are thus very important. The ceramic fibers are less critical in these respects.

But the counterpart to this is that the ceramic fibers are extremely brittle in bending, and will break if they are bent to a radius of curvature less than several hundred times the fiber radius. Because of the yield on the compression side, the polymer fibres can be bent completely back on themselves. They are weakened but not broken by maltreatment in manufacture and use. Their mode of fracture by multiple axial splitting absorbs much energy, and so they give greater damage tolerance in composites. It is also easier to co-mingle reinforcing fibers and matrix fibres to make a good composite.

In order to develop the best polymer fibers and matrices, it is necessary to understand the system of composite manufacture and composite structure in relation to the requirements in use. But there is a difficulty - or an opportunity - here, because the system is not settled. There are many manufacturing methods giving different structures, and the choice depends on the fiber and matrix properties. Optimisation involves the interaction of properties with the system.

Composite Manufacturing

The main methods of composite manufacturing can be summarised in the following table:

	short fibre	long fibre	continuous filament
dispersed	injection molding	press-forming	
laminated		nonwoven lay-up	pre-preg tape filament winding woven or knit fabric
integrated			3-D weaving, braiding

In going from top left to bottom right, there is an increased control of fiber arrangement within the composite. Usually this implies increased cost; but the ease of handling, and the possibilities of automated manufacture, of the integrated textile structures may well lead to these methods proving to be more economical than tape-laying or filament winding.

Three-dimensional textile preforms are of two types. Firstly, there are modifications of single-layer weaving or knitting to give honeycomb or shaped forms. Secondly, there are solid forms, with the yarns following interlaced three-dimensional paths within the structure, which may be either a thick sheet or a complex shape. 3-D weaving is a development from known multi-layer fabrics, but 3-D braiding has involved new inventions such as Cartesian braiding and two-step braiding.

An important research need is to systematise ways of characterising the topology and geometry of these complex fibre assemblies. This is a prerequisite for a scientifically based engineering approach to the control of manufacture of the preform, the flow of matrix during consolidation, the design procedures and visualisation of the structure, and the prediction of mechanical performance.

THE MECHANICS OF COMPOSITES

During the last 25 years, the applied mechanics of composites has developed in a major way. The starting point was mixture laws, which were then applied to short fiber systems by considering aspect ratios and orientation distributions, and to long fiber systems by laminate theory. The total component behaviour is then handled by finite element methods. However this approach is not right for an interlaced assembly, and it is appropriate to look at the development of textile mechanics, which has occurred in the last 40 years.

The difficulty here is that textiles are optimised as soft, flexible, porous fabrics with low fiber-volume content and high fashion aesthetics - a total contrast to rigid 3-D engineering solids. Consequently the advanced work has concentrated on large deformation, large strain, non-linear treatments of simple geometries. The methods must be modified in order to deal with the small strain problems of complex geometries.

The subject divides itself into: (a) micromechanics, which predicts the constitutive equations for the material in terms of the arrangement of fibers with the yarns and fabrics; (b) macromechanics, which determines the total deformation of the material under particular forces. A hierarchical approach is often needed in dealing with fiber, yarn, and fabric deformation.

Micromechanics

The micromechanics can be treated by the equilibrium of forces and moments [1,2] but, in my view, energy methods are the most effective.

One approach [3] is applicable to relatively dense assemblies of variously oriented fibers, in which the local fiber strain can be reasonably predicted from the global material strain by an affine deformation assumption. For nonwoven fabrics and twisted yarns, which are planar or quasi-planar (cylindrical) assemblies, the basic relations are:

$$(1 + e_f)^2 = (1 + e_1^2) \, C_{f,1}^2 + (1 + e_2^2) \, C_{f,2}^2$$

$$f_1 = \frac{\Sigma \, m_f \, f_f \, (\partial e_f / \partial e_1)}{\Sigma \, m_f} \quad , \text{ similarly for } f_2$$

where (a) m_f is the mass of a fiber element, with a strain e_f giving a specific tensile stress, f_f;

(b) e_1 and e_2 are the axial and transverse strains applied to the material, and f_1 and f_2 are the corresponding specific stresses.

(c) $C_{f,1}$ and $C_{f,2}$ are the direction cosines of the fiber orientation.

Thus the material properties can be predicted from a knowledge of the fiber orientation distribution and the fiber stress-strain properties. Extensions of the theory allow for straightening of curved fiber sections and for slippage from fiber ends.

Another approach [4] is applicable where a repetitive unit cell can be identified in the structure. In application to woven fabrics, the basic relations have the following form:

(a) a set of equations, relating the external dimensions of the unit cell, x_1 and x_2, to the yarn lengths in the cell, ℓ_1, and ℓ_2, expressed in functional form as

$$f(x, y, \ell_1, \ell_2) = 0$$

(b) partial differential equations for the forces, F_1 and F_2, involving the energy of yarn extension, V_1 and V_2,

$$F_1/F_2 = (\partial x_2/\partial x_1)_{\ell_1, \ell_2}$$

$$F_2(\partial x_2/\partial \ell_1)_{x_1, \ell_2} = dV_1/d\ell_1$$

$$F_2(\partial x_1/\partial \ell_2)_{x_1, \ell_1} = dV_2/d\ell_2$$

In order to illustrate the principles, the above examples have been kept simple, in geometry and in types of deformation, but the methods are capable of development to cover other forms of fiber deformation such as bending and twisting, and to treat three-dimensional structures, subject to three axial and three shear modes of deformation.

Macromechanics

Much of the macromechanics of textiles is related to bending, twisting, snarling, and slip in open, loose structures, and so is not relevant to composites. But the problem of drapeability is important in forming a preform sheet into a three-dimensional shape before consolidation.

Textile materials have the special property of achieving complex buckling with smoothly rounded folds of double curvature. In contrast, paper, which bends easily in single curvature, resists double curvature, and, if forced, buckles into sharp point or line discontinuities.

Recent advances in the mechanics of complex buckling [5] come from a recognition that the formation of spherical surfaces of double curvature, which is the converse of drawing a map of the world on a planar surface, demands in-plane (membrane) strain as well as bending. It has also been recognised that the simplest problem in this class is the buckling of a circular specimen pushed inwards at three equally spaced points. The material deforms into a dome of double curvature surrounded by folds of single curvature. The form of the surface, in particular the size of the central dome, has been predicted by minimising the sum of the bending energies in the inner and outer zones, the membrane energy in the inner zone, and the gravitational energy.

CONCLUSION

The industrial aim is to optimise materials, with a special bias among the Rolduc group to maximise the use of polymer composites. The scientific message must be that this is a total system problem.

In the components, we are concerned with the structure/property relations in fibers and matrices and with the interactions at the interface. In the composites, we are concerned with the manufacturing methods, and with the structural mechanics of the fiber/matrix assembly. But optimisation cannot be achieved unless both component and composite problems are treated together.

REFERENCES

1. Hearle, J.W.S., Grosberg, P., and Backer, S., Structural Mechanics of Fibers, Yarns and Fabrics, Wiley-Interscience, New York, 1969.

2. Hearle, J.W.S., Thwaites, J.J., and Amirbayat, J., Mechanics of Flexible Fibre Assemblies, Sijthoff and Noordhof, Alphen aan den Rijn, 1980.

3. Hearle, J.W.S., in [2], pages 51-86.

4. Hearle, J.W.S., and Shanahan, W.J., An Energy method for calculations in fabric mechanics, J. Textile Inst., 1978, 68, 81-91, 92-100.

5. Amirbayat, J., and Hearle, J.W.S., The complex buckling of sheet materials, Int.J.Mech.Sci., 1986, 28, 339-358, 359-370.

IMPACT PERFORMANCE OF DYNEEMA PE/EPOXY COMPOSITES IN COMPARISON WITH OTHER COMMERCIAL FIBRES

K.F.M.G.J. SCHOLLE
DSM Research
P.O. Box 18
6160 MD Geleen, The Netherlands

ABSTRACT

The impact performance of Dyneema SK 60 polyethylene fibre, Kevlar 49, E-glass and carbon T300/epoxy laminates has been determined. It was found that Dyneema has the highest ratio of absorbed energy and incident impact energy. Indicating that Dyneema can be used for energy dissipating applications. The key parameter to understand this is the low initial interlaminar fracture toughness of the material.

INTRODUCTION

Dyneema SK 60 polyethylene fibres have a high potential for use in composite applications due to the high tensile strength, modulus and low specific weight (1,2). Dyneema is produced by a gel-spinning process which provides the fibre with long and highly oriented molecules. Due to Dyneema's specific properties composite structures composed of the fibre and an appropriate matrix system are quite promising for kinetic energy dissipation applications. Applications can be envisaged e.g. in automotive and aircraft composite structures which have to be designed to dissipate kinetic energy during crash impacts (3). The use of Dyneema in protective clothing, armour and energy absorbing layers to improve the impact behaviour are also viable propositions. It has been demonstrated that polyethylene fibre/epoxy composites in UD-laminate and fabric-laminate forms perform excellently during low velocity penetration impact and high velocity impact (4,5). It was always found that a high penetration resistance inevitably correlated with a pronounced delamination damage. The aim of the present investigation was to investigate the potential mechanisms responsible for the impact behaviour. For this purpose we looked especially upon the damage formation and the absorbed energy as a function of the incident impact energy.

The advantage of such an approach is that the real (absorbed) energy needed for damage is measured and not only the penetration energy which is a sum of elastic energy and damage energy (6). A severe damage mode as caused by impact is a delamination. The key parameter to understand delamination damage is the interlaminar or delamination fracture toughness (7). It will be clear that the delamination toughness will be a means to understand the impact behaviour.

EXPERIMENTAL

Materials

UD corona treated Dyneema SK-60 Pe-, Kevlar 49, Carbon T300 and E-glass prepregs were manufactured on a drum by wet winding of fibre bundles. The epoxy system used was Schering Eurepox 730 and XE 279 as curing agent. After the winding process the Virgin prepreg was pre-cured (B-stage) in an oven. Laminates consisting of various layers were cured under pressure and vacuum control at 120 °C for two hours. The laminates were machined into testpanels by using a waterjet.

Fracture test

The interlaminar fracture toughness, in terms of strain energy release rate, has been determined by means of the Width Tapered Double Cantilever Beam (WTDCB) fracture specimen (8). In our case a thick aluminium WTDCB is adhesive-bonded to the thin laminates.
An expression for the determination of the fracture toughness from the experimental data and the procedures concerning the test are given elsewhere (8).

Impact test

Instrumented falling weight impact test were conducted on a home build machine. Details of which have already been given elsewhere (6). Test panels were circular clamped between metal plates with a circular opening of 80 mm. The mass of the falling weight was 0.9 kg. A hemispherical dart with a diameter of 10 mm was used in all cases.

Impact damage

An ultrasonic C-scanner was used to determine the impact damage area of the composite test panels. Black-white C-scan plots were used to determine the boundary of the impact damage area. The black-white criterion was set on 50 % of the height of the reflection signal of an undamaged part. An integrator was used to determine the impact damage area.

RESULTS AND DISCUSSION

Figs. 1 and 2 show the results of the non-penetrating low velocity impact tests on Dyneema SK 60, T300, E-glass and Kevlar 49/epoxy laminates with an alternating (0/90) lay up. The number of layers was different (5 layers for T300, E-glass and Kevlar; 7 layers for Dyneema) but the total thickness was kept constant 1.7 (mm); the fibre volume % was ca. 54. Fig. 1 clearly shows that Dyneema/epoxy laminates have the highest ratio of absorbed energy and total incident energy. The glass and carbon fibre composites showed elastic brittle (penetration) like behaviour with small delaminations in the case of the glass fibre - and large delaminations in the case of the carbon fibre/epoxy laminates, see fig. 2. The energy consumption of these brittle materials is low, see fig. 1. The Kevlar and PE-fibre laminates show only delaminations and splitting damage.
Fig. 3 shows the results of the fracture mechanics tests. Fracture resistance G_{1r} as a function of the cracklength of an UD PE/epoxy composite

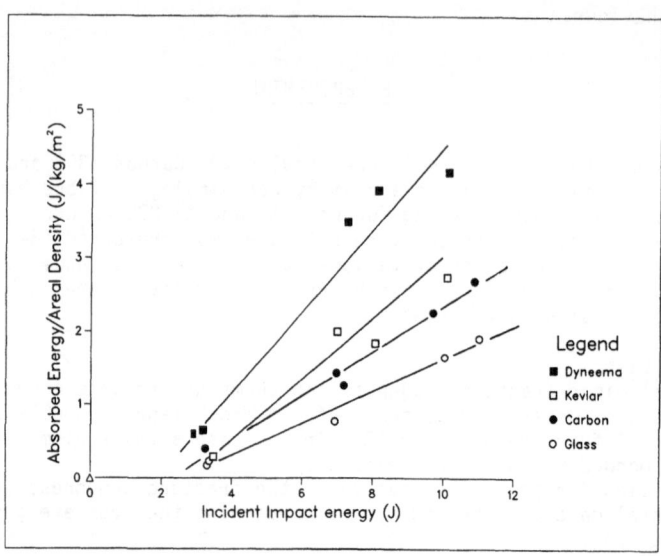

Figure 1. Low velocity impact of glass, Dyneema, Kevlar and carbon/epoxy
laminates, specific absorbed energy (related to areal density)
as a function of the incident energy.

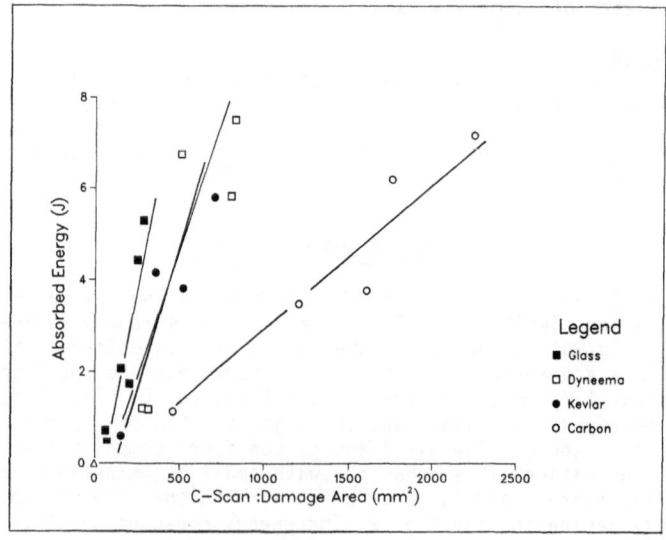

Figure 2. Low velocity impact of glass, Dyneema, Kevlar and carbon/epoxy
laminates. Absorbed energy vers. damage area as measured by
ultrasonic C-scan.

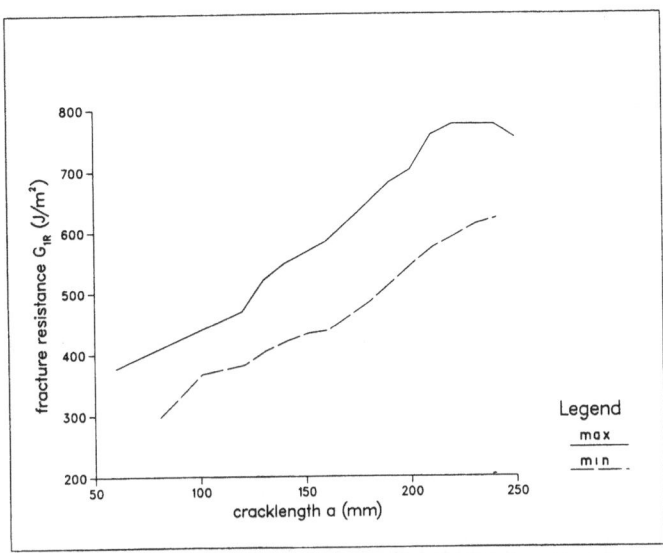

Figure 3. Fracture toughness (G_{1r}) of a (0)12 Dyneema/epoxy laminate as a function of the cracklength

consisting of twelve layers, (0)12. The figure shows the minimum-maximum range of fracture toughness values gathered for a series of tests. It clearly demonstrates that the fracture resistance increases with cracklength. The increase is due to fibre-bridging (8). The micro-mechanics explanation of the fibre-bridging phenomenon has been given elsewhere (8). In the case of the Kevlar 49 composite the fracture toughness is independent of the cracklength and is equal to ca. 1100 J/m^2 (9).
A conclusive explanation dealing with the impact performance of the four fibre systems is not possible at present; especially in the case of glass and carbon where fibre breakage occurs (due to indentation) next to delamination. However, the difference between Kevlar 49 and Dyneema SK 60 can be explained on a fracture mechanics basis due to the fact that primarily delamination damage occurs during impact. As stated earlier, the advantage of a non-penetrating impact experiment is that the real (absorbed) energy, S, needed for damage is measured (6) and not the penetration energy W:

$$W = U + S + T \qquad (1)$$

which is a sum of elastic (U), damage (S) and kinetic energy (T). In the case that the absorbed energy S is only needed for delamination damage; a plot of absorbed energy vers. damage area should give a linear correlation, the slope of which is the fracture toughness G under impact conditions. Fig. 4 clarifies this graphically. It should be stressed here that US-C-scan only determines the projection of the total 3D-damage zone. The slope G_{impact} of the Dyneema and Kevlar system are equal, see fig. 2. However, the fracture toughness G_{impact} is related to the delamination fracture toughness G_{1r}, since both are dealing with delamination damage.

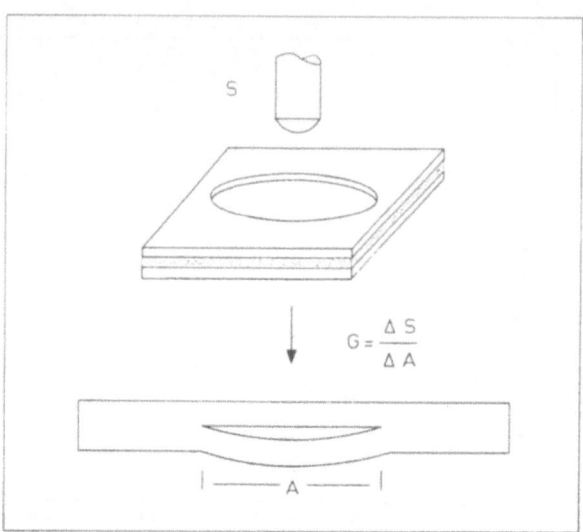

Figure 4. Graphical presentation of the impact process. Absorbed energy (S) and delamination area (A) related to interlaminar fracture toughness (G).

As mentioned above the initial fracture toughness of Pe/epoxy is lower (ca. 300 J/m^2) than for Kevlar, 1100 J/m^2. This means that multiple delaminations have to occur in the Dyneema system in order to get a similar G_{impact} slope.

Due to the low initial fracture toughness with respect to Kevlar; triggering for crack deflection during impact is easier in the Dyneema system. The triggering results in delaminations (i.e. individual laminae) and thus energy consumption. Next to this other energy sinks are present compressive yielding of the individual 'laminae' and at high impact loads the tensile failure of the PE-fibre; in order to have the highest ratio of absorbed energy and incident energy, see fig. 1.

Both mechanisms count for the excellent impact performance (i.e. energy consumption) of Dyneema SK 60.

ACKNOWLEDGEMENTS

The author wishes to thank the management of DSM Research for their permission to publish this work.

REFERENCES

1. Kirschbaum R., Yasuda H., van Gorp E.H.M., 25th Int. Chemical fibres congress, 24-26 September 1986, Dornbirn, Austria, p. 229.

2. Scholle K.F.M.G.J., Jacobs M.J.N., Omloo H.A.A., Loete G.J.H., Janse G.H.A., 9th Int. SAMPE, European Chapter, Milano, Italy, 1988, June 14-16.

3. Kindervater Ch., Scholle K.F.M.G.J., ibid.

4. Lin L.C., Bhatnagar A., Lang D.C., Chang H.W., 33rd int. SAMPE symposium, March 7-10, Anaheim, USA, p. 883.

5. Adams D.F., Zimmerman R.S., 31st Int. SAMPE symposium, Las Vegas, April 7-10, 1986, p. 1456.

6. Scholle K.F.M.G.J., Omloo H.A.A., Dorrestijn A., 3rd Int. Conf. on fibre reinforced composites, March 23-25, Liverpool, U.K. 1988.

7. Leach D., Proc. Damage Development and Failure Processes in composite Materials, Leuven (Belgium), 1987, p. 136.

8. Scholle K.F.M.G.J., 33rd Int. SAMPE symposium, March 7-10, Anaheim, USA, 1988, p. 136.

9. Verbruggen M.L.C.E., Proc. ICCM-VI & ECCM-2, London, UK, 1987, Vol. 5, p. 458.

HYBRID COMPOSITES BASED ON POLYETHYLENE/CARBON FIBRES
Compressive and Impact Behaviour

A.A.J.M. Peijs and P.J. Lemstra
Dept. of Polymer Technology
Eindhoven University of Technology
P.O. Box 513, 5600 MB Eindhoven
The Netherlands

ABSTRACT

The mechanical properties of high-performance polyethylene (HP-PE) and carbon fibres are to some extent complimentary with respect to toughness, compressive strength and long term properties.
Combining both fibres as reinforcing elements in so-called hybride composite structures can result in a unique class of structural materials in terms of strength, stiffness and impact performance. Preliminary results concerning the use of HP-PE and carbon fibres in hybride composite structures, in an epoxy matrix, will be presented emphasizing compressive and impact properties.

INTRODUCTION

Currently high-performance polyethylene fibres (HP-PE) are produced based on solution (gel)-spinning of UHMW-PE, possessing unique mechanical properties in terms of high specific values for respectively strength and modulus (1). Moreover these HP-PE fibres possess a relatively high work to break, i.e. good impact properties, in comparison with carbon, aramid and glass fibres.

However, due to the microstructure of HP-PE fibres, viz. extended
polyethylene chains with relatively weak intermolecular Van der Waals
interactions, their resistance to static loadings is relatively poor
(creep). For similar reasons the compressive strength is not very
impressive, due to shear induced failure. Consequently the new generation
of HP-PE fibres seem less suitable as reinforcing elements for structural
composites.

Comparing, however, the intrinsic mechanical properties of HP-PE fibres
with those of the modern generation of carbon fibres, one could obtain
unique materials. Carbon fibres are well-known to be rather brittle but
possess a high compressive strength/modulus in contrast to HP-PE fibres
which possess opposite characteristics. Combining both types of fibres
into so-called hybrid composite structures, one could obtain in principle
unique structured materials in terms of impact-strength-stiffness.

Since the mid-seventies various reports have been presented concerning
impact performance improvement of carbon-reinforced composites via
hybridizing by glass and aramid fibres (2,3).

In this paper we wish to report preliminary results concerning hybrid
composite structures based on carbon/HP-PE.

EXPERIMENTAL

1. Materials

In this study we used a HP-PE fibre (Spectra 1000) kindly supplied by
Allied Corp. and a high strength carbon fibre (XAS/3K) of Courtaulds PLC.
The matrix used was an epoxy system of Ciba Geigy (Araldite LY556/HY917/
DY070) based on Bisphenol A with an anhydride curing agent. To study the
effect of improved adhesion of HP-PE fibres on mechanical properties,
composites incorporating untreated and treated HP-PE fibres were
manufactured. Improved adhesion was obtained by immersing HP-PE fibres in
chromic acid for 15 min (4). This resulted in an increase of interlaminar
shear strength from 15 MPa to 29 MPa.

2. Specimen preparation

Two kinds of hybrids were prepared and tested:

1. UD-pultruded rods with intermingled yarns

   Specimens used for compressive strength measurements were prepared by
   a modified pultrusion process. Bundles containing carbon and HP-PE
   yarns in known proportions were immersed in a bath of epoxy resin and
   pulled into tubes with PTFE-inserts (6 mm internal diameter). Carbon
   and HP-PE yarns were intermingled as homogeneous as possible. The
   samples were cured for 4 hours at 80°C and then post-cured at 110°C
   for 12 hours. After curing specimens were removed from the mold and
   cut into 12 mm lengths using a low speed diamond saw. Volume fraction
   of fibres was held constant at 50% and the proportions of HP-PE fibre
   in the composite was varied from 0 to 100% with steps of 20%.

2. Laminated Plates

   Impact performance was measured on layered structures which were
   prepared by stacking pre-impregnated plies of carbon and HP-PE fibre
   woven fabric (Satin weave 8 H) together and curing them for 2 hours
   under combined vacuum- and pressure conditions in a hot press at
   110°C. After curing samples were postcured for 12 hours at 110°C. All
   laminates were 4 plies thick (nominal thickness 1,5 mm) and within
   these 4 plies stacking sequences and HP-PE/carbon ratios were varied.
   The total volume fraction of fibre in the laminate was 70%. Plane
   samples of 60 mm x 60 mm were cut from these laminates with a bandsaw.

3. Testing

Compressive strength was measured on specimens with circular cross
sections, which were placed in close fitting end loaded test fixtures. An
elastic compressive polyamide pad was used at each end of the specimen to
ensure homogeneous load transfer. This assembly was placed into a Zwick
1474 testing machine and compressed at a rate of 0,5 mm/min.

Impact performance was measured on laminated plates by an instrumented
dart-test at a velocity of 4,5 m/s. A hemispherical dart with a diameter
of 10 mm was used throughout. Laminates were clamped between two plates

with an internal diameter of 20 mm. Impact energies were obtained by recording the load-time curve during penetration. From this load time curve initiation- and total energy could be determined.

RESULTS AND DISCUSSION

1. Compressive behaviour

The compressive strength of HP-PE/carbon hybrids with different compositions is shown in Figure 1 and is in good agreement with the "rule of mixtures". At least 5 experiments were carried out for each dataprint. The error bars in the graph represent one standard deviation on either side of the mean value.

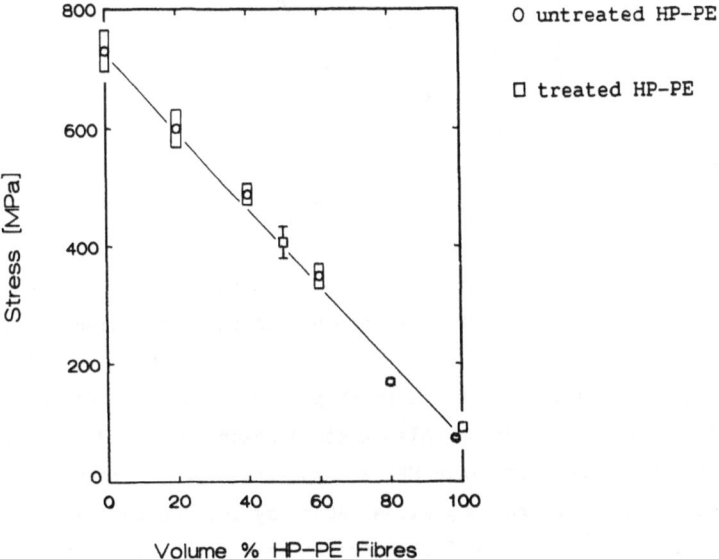

O untreated HP-PE

□ treated HP-PE

Figure 1.   Compressive strength of hybrid HP-PE/carbon composites

Compressive properties of polyethylene fibre composites are already reported to be low (5). The compressive strength of our composites consisting of 100% untreated HP-PE fibre (σ_C = 73 MPa) is more or less

equal to the compressive strength of the epoxy matrix based on 50 volume
percentage of matrix. Consequently the compressive strength of the hybrid
system follows the "rule of mixtures" of two constituents carbon and
epoxy.

The effect of improved adhesion of the HP-PE fibre on compressive
strength of the hybrid is negligible. For plain HP-PE composites an
improvement of more or less 25% was obtained (σ_c = 91 MPa).
Figure 2 shows that composites with untreated HP-PE fibres fail in a more
brittle way. On the other hand composites incorporating acid treated
HP-PE fibres show a more ductile character with a clear yield point.

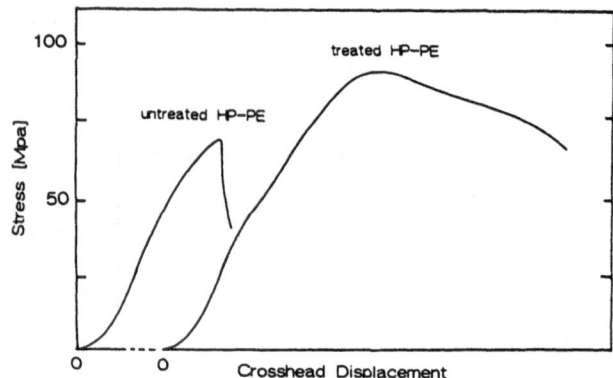

Figure 2.   Typical load vs. deformation curves for unidirectional
HP-PE/epoxy composites in compression.

Longitudial sections of tested samples were polished for optical
microscopic examination. Also these sections revealed characteristic
differences between plain HP-PE composites with untreated and treated
fibres. HP-PE fibre composites deform by the kinking mechanism. However,
samples with untreated fibres showed only little kinking (Fig. 3a),
whereas with improved adhesion the failure mode changes to more severe
kinking with highly localized shear kinking band (Fig. 3b). The good
adhesion between fibre and matrix prevents widespread catastrophic damage
which is also reflected by the stress-strain curve of plain HP-PE
composites with treated fibres.

It appears that the increase in compressive strength of plain HP-PE composites incorporating treated fibres is caused by the change in failure mode. It seems likely that compressive failure processes in the fibre itself are strongly determined by the low shear modulus and shear strength. These failure processes are comparable with those found in more detailed studies for aramid fibres (6).

Carbon fibre reinforced composites, on the other hand, failed in a brittle manner with multiple transverse and 25° shear cracking (Fig. 3c). The failure mode for HP-PE/carbon hybrids was similar to that of each of its components. Shear cracking of the carbon part was here combined with kinking of the HP-PE components (Fig. 3d).

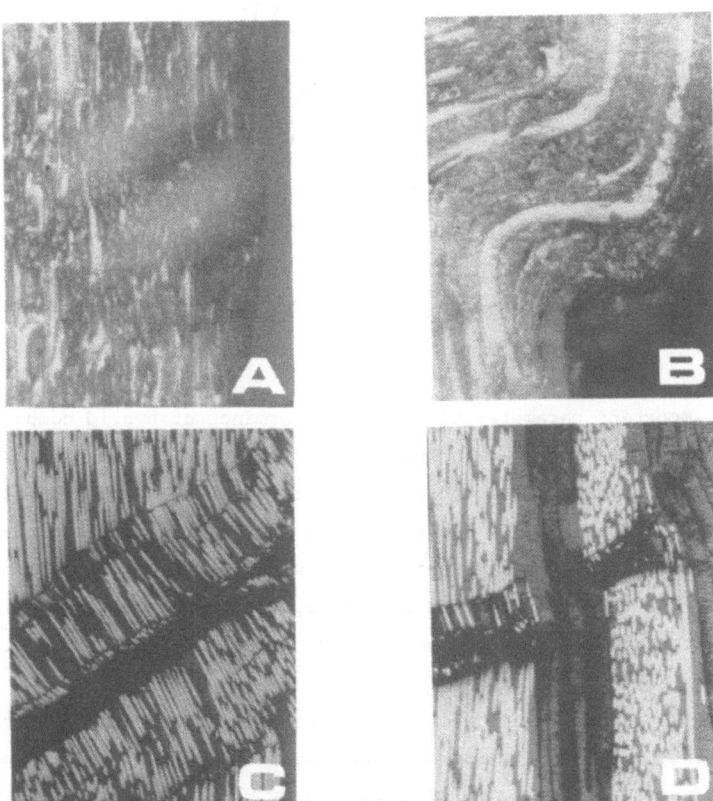

Figure 3.   Microstructure of compression-tested samples: (a) untreated HP-PE/epoxy composite (40x); (b) treated HP-PE/epoxy composite (80x); (c) plain carbon/epoxy composite (80x); (d) HP-PE/carbon hybrid (80x)

2. Impact behaviour

Figure 4 shows some typical load-time curves recorded during penetration of HP-PE, carbon and HP-PE/carbon hybrid laminates.

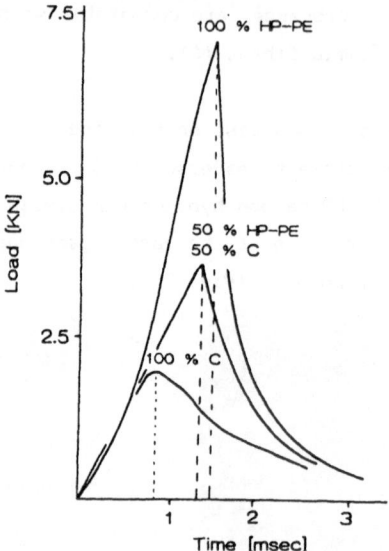

Figure 4.   Load-time curves for HP-PE, carbon and HP-PE/carbon hybrid composites

An important parameter on impact performance of hybrid composite materials is the stacking sequence. Figure 5 shows the impact performance of 4-ply HP-PE/carbon hybrids with different stacking sequences. Each laminate consisted of 2 plies carbon and 2 plies untreated HP-PE woven fabric. The highest impact energies were obtained by stacking the HP-PE plies at the opposite side of the impacted surface. By stacking HP-PE in the tensile zone of the laminate the fibre can absorb more elastic energy.

Figure 6 shows the impact performance of 4 ply PE/carbon hybrids with various HP-PE/carbon ratios. The amount of HP-PE fibre within the hybrid was varied by replacing carbon plies by HP-PE plies. To obtain an optimum effect on impact improvement HP-PE plies were positioned at the opposite side of the impacted surface. Also the influence of improved adhesion of HP-PE on impact performance was investigated. As expected we see with increasing amount of HP-PE fibre an increase in impact performance.

225

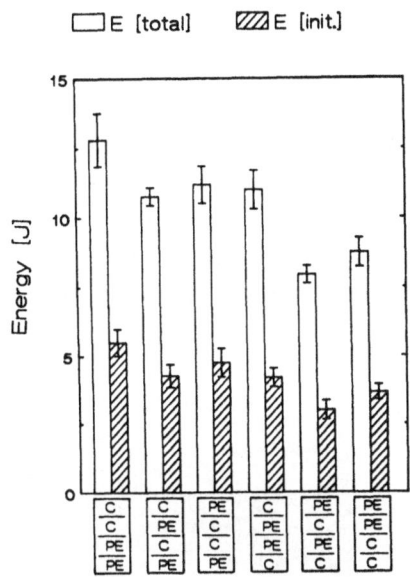

Figure 5. Effect of stacking sequence
on impact energy of hybrid
HP-PE carbon laminates

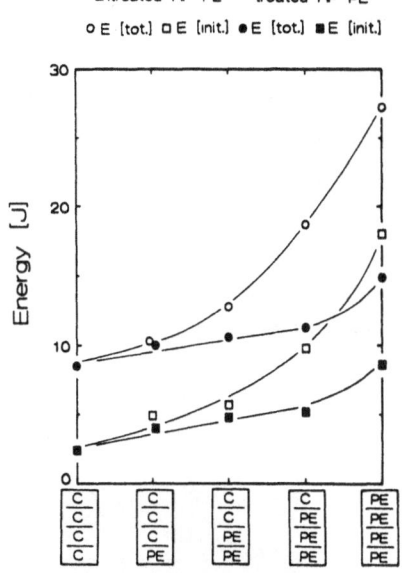

Figure 6. Dart impact energy
of hybrid HP-PE
carbon laminates

However, this increase is at a lower rate than predicted by rule of
mixtures. Important to notice is that with increasing amount of HP-PE
fibres energy absorption changes from propagation- to initiation
dominated. With other words, the hybrid shows a more ductile character.
The increase in absorbed energy can be attached to initiation or elastic
energy. Energy that is absorbed by propagation or fracture is more or
less constant. The negative influence of improved adhesion of HP-PE
fibres on impact performance is clear. Composites with untreated fibres
absorb intrinsic more elastic energy because of extended free stretching
of the fibre due to low interfacial bonding. With increasing level of
adhesion this possibility diminishes and the composite fails in a more
brittle way.

The effect of increasing ductility with hybridization opens the
possibility to use HP-PE as an elastic energy absorber for non-
penetrating impacts. By absorbing elastic energy in the HP-PE fibres less
energy is available for damage and in particular damage in the

carbon part of the hybrid. This reduction of damage in the more structural part viz. carbon, could lead in principle to improved post-impact properties.

Figure 7 shows an example of 4-ply laminates which were submitted to non-penetrating impact of 10 Joule. Impacts were performed on a falling weight impact tester kindly placed at our disposal by the composite materials group of the KU Leuven. By hybridizing carbon with one single HP-PE ply we accomplished a striking reduction in visible damage of the laminate. The indentation depth was reduced from 2,8 mm to 1,4 mm after hybridization.

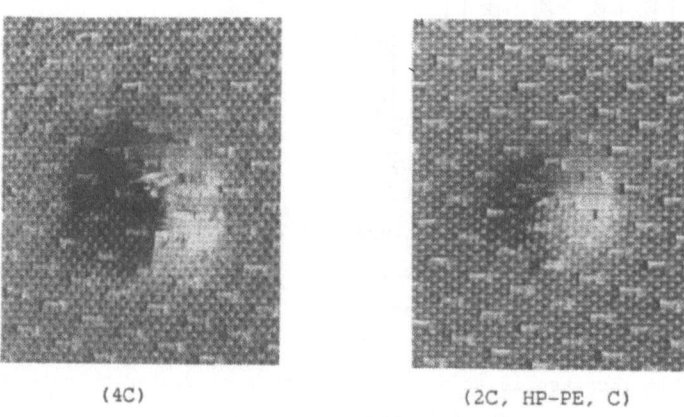

(4C)   (2C, HP-PE, C)

Figure 7.   Effect of hybridization on visible damage after falling
weight impacts of 10 Joule.

CONCLUSIONS

The present study shows that impact performance of carbon fibre reinforced composites can be increased by hybridizing with HP-PE plies. The amount of improvement depends strongly on the adhesion level and the position of these plies in the hybrid. The highest impact energy values were obtained by positioning the HP-PE plies at the opposite side of the impacted surface and a low adhesion level. Preliminary results showed another interesting feature of HP-PE/carbon hybrids viz. the possibility to use HP-PE as an elastic energy absorber under impact conditions to reduce damage. This could lead in principle to improved residual properties after impact.

The reverse of using HP-PE fibres is the reduction in static mechanical properties. In compression HP-PE/carbon hybrids follow rule of mixtures behaviour. Still, this opens the possibility to use HP-PE fibres in structural composites with high impact resistance. No significant effect of improved adhesion of HP-PE fibres on compressive strength of the hybrids was found. For plain HP-PE composites the failure mode changes drastically with improved adhesion from little kinking to highly localized kinking bands which gave a more ductile character to the composite.

REFERENCES

1. Lemstra, P.J., Kirschbaum, R., Ohta, T., Yasuda, H, in: Developments in Oriented Polymers-2, Elsevier London, 39-77 (1987), Ed. I.M. Ward
2. Mallick, P.K. and Browtman, L.J., J. Test. Eval., 5, 190-200 (1977)
3. Perry, J.L. and Adams, D.F., Composites, 4, 166-172 (1975)
4. Mercx, F.P.M. and Lemstra, P.J., to be published
5. Adams, D.F., Zimmerman, R.S. and Chang, H.W., SAMPE J., 21 (6) 44-48 (1985)
6. Van der Zwaag, S. and Van der Sluijs, C.P., this conference

TESTING OF FIBRE-MATRIX ADHESION

F. Elkink and J.H.M. Quaijtaal
Akzo Corporate Research
P.O. Box 9300, 6800 SB ARNHEM
The Netherlands

ABSTRACT

The effect of surface treatment of aramid fibers (Twaron®) on the adhesion between aramid and epoxy was measured with a bundle pull-out and the commonly used 3-point short-beam shear test. The adhesion of optimally treated aramid was compared to that of untreated aramid, glass and carbon fibres in epoxy using a 3- and 4-point short-beam shear test and a transverse bending tensile test. The adhesion treatment leads to a 40 % increase in shear strength and to a 60-80 % increase in transverse tensile strength of aramid-epoxy composites. Further improvement is unlikely because the properties of the composite are limited by the shear and transverse strengths of the aramid fibre itself. Despite the considerable adhesion improvement, aramid-epoxy composites still have a lower shear and transverse strength than glass and carbon fibre composites. However, the corresponding strains are approximately equal because of the lower shear and transverse moduli of the aramid-epoxy composites.

INTRODUCTION

The fibre-matrix interface strength is usually determined indirectly by measuring the apparent interlaminar shear strength of a small uni-directional (u.d.) sample with the short-beam 3-point bending test. In developing an aramid (Twaron®) yarn with improved adhesion to epoxies, a quick adhesion screening test was necessary. Moreover the test should demand very little material. Hence a bundle pull-out test was optimized. Basically the idea was to imitate the more difficult mono-filament pull-out test on a more convenient scale and to achieve less scatter in the results. At a later stage of development the usual short-beam test was employed. Finally, the optimally treated aramid (A-T) was compared to untreated aramid (A-U), glass (G) and carbon (C) fibres in epoxy composites with tests that can be carried out with simple u.d. beams.

MATERIALS

Carbon and aramid fibres were obtained from the Akzo fibres and polymers division. Glass fibres were supplied by Silenka B.V.. Some fibre properties are shown in Table 1. The glass and carbon fibres were provided with epoxy compatible sizing agents. Two Ciba-Geigy DGEBA epoxy systems were used. The results obtained with the anhydride (HY917)-cured system were essentially equal to those obtained with the amine (HT972)- cured system provided that the aramid yarns are dried very thoroughly. Therefore, in this paper the matrix is just referred to as epoxy. Both epoxies were cured for 6 hours at 80°C plus 6 hours at 140°C.

TABLE 1
Fibers, epoxies and their moduli in GPa

Material	type	E_L	E_T	G_{LT}
Aramid	Twaron 1610 dtex f1000	125	6.7	2.6
Glass	Silenka 084M19600EC15 f2000	73	73	30
Carbon	Tenax HTA-7 2000 dtex f3000	235	14	28
Epoxy/anhydride	LY556/HY917/DY070 (100/90/1)	3.5	3.5	1.3
Epoxy/amine	LY556/HT972 (100/27)	3.5	3.5	1.3

EXPERIMENTAL DETAILS

The bundle pull-out test: A twisted bundle of aramid yarn (1610 dtex, 90 turns/meter) is threaded through a hole (8 mm in diameter) in a metal plate provided with a 2 mm thick silicon rubber sheet at its bottom. A pretension of ~ 2 N is applied to the bundle. The combination of twist and pretension shapes the bundle into a tightly packed nearly circular strand with a diameter (d) of 0.45 mm. A 3-3.5 mm thick layer (l) of epoxy is cast into the hole and the protruding bundle below the rubber is impregnated with epoxy. After curing the bundle is cut off just above the resin and the impregnated strand is pulled out with a tensile tester. An average bundle pull-out shear strength (BPS) is calculated as the pull-out force F devided by the interfacial area (l*π*d).

The specimens for the short-beam shear and transverse bending tensile test are made by winding impregnated yarns in a U-shaped mould. Excess resin is squeezed out during screwing the lid on the mould. The resulting cross-section of the specimens is 12.7 * 2 mm. Short-beam 3-point bending tests were carried out according to DIN 29971 with 6 mm rollers, a span (l)-to-thickness (h) ratio of 5 and a width (w) of 12.7 mm. The short-beam (interlaminar shear) strength (3p-SBS) was calculated from the failure load F as 3p-SBS = 3/4 * F/(w*h).

The 4-point short-beam test yields the 4p-SBS. The test is identical to the 3-point test except that the central roller is replaced by two rollers 8 mm apart and that the span is increased from 10 to 18 mm.

The transverse tensile strength (TTS) was measured with the 3-point bending fixture at l/h=5, but now with the fibres parallel to the rollers. Fracture initiates at the point of maximum tensile stress in the bottom face of the beam below the central roller. Samples with a length of 10 mm (this is the width in the test) were cut between perspex plates with a diamond saw in order to obtain smooth cuts. The transverse tensile strength was calculated as TTS = 3/2 * F/(w*h) * (l/h).

INTERLAMINAR AND INTERFACIAL SHEAR STRENGTH

References [1], [2] and [3] give short-beam strengths (3p-SBS) of 60 % fibre volume fraction (V_f) composites and interfacial shear strengths (τ_i, average shear stress) as measured with the monofilament pull-out-test. The relations obtained from these data are in MPa:

$$3p\text{-}SBS \approx 20 + 0.93 * \tau_i \quad \text{and} \quad 20 + 1.42 * \tau_i \qquad (1)$$

for [1] + [2] and [3] respectively. The different slopes are not surprising since the stress concentrations in the pull-out tests have not been taken into account. Both equations suggest that a sample without fibre-matrix adhesion has an interlaminar shear strength (ILSS) of 20 MPa. This must be the contribution of the resin bridges between the fibres.

The ILSS of a u.d. composite may be calculated by two extreme approaches. The most optimistic estimate for ILSS is obtained by neglecting stress concentrations. Consider a triangular array of fibres that is subject to a longitudinal shear stress equal to the interlaminar shear strength in a horizontal plane. On the assumption that at fracture the horizontal resin bridges between the fibres bear the matrix yield stress τ_m and that an effective fraction α of the fibre surface bears the interface strength τ_i, it can be derived that:

$$ILSS = (1 - k\sqrt{V_f}) * \tau_m + (\alpha k\sqrt{V_f}) * \tau_i \qquad (2)$$

where $k = \sqrt{(2\sqrt{3}/\pi)} = 1.05$. The factor α is $\pi/2$ when τ_i acts on the whole fibre surface. In practice the effective fibre area that is subject to τ_i will be less than the fibre circumference (e.g. α may be 1 or less) as a result of stress concentration in the oblique resin bridges between the fibres. The stress concentration factors (SCFs) arise from the different shear moduli of fibre (G_{Lm}) and matrix (G_m), see Table 1 and [5]. For $V_f = 60$ % and $\tau_m = 100$ MPa [4] we obtain in MPa:

$$ILSS = 18.7 + 0.81 * \tau_i \quad \text{or} \quad 18.7 + 1.28 * \tau_i \qquad (3)$$

for $\alpha = 1$ and $\pi/2$ respectively. These linear rules of mixtures agree satisfactorily with eq. (1). When the fibre-to-matrix shear modulus ratio is very high (SCF is very large) the stress is mainly passed through the fibres by the oblique resin bridges and the contribution of the horizontal bridges to the shear strength will be less than indicated by equations (2) and (3).

Stress concentrations for linear elastic matrices and fibres have been calculated in [5] and [6]. The stresses are maximum at or near the interface. The composite interlaminar shear strength (ILSS) is SCF times lower than the interface strength τ_i when fracture initiates at the interface. The composite strength predicted by such calculations decreases with increasing fibre volume fraction. The decrease is very strong near the maximum packing density. These predictions may be too pessimistic because any yield behaviour will reduce the stress concentrations.

RESULTS AND DISCUSSION

Short-beam and bundle pull-out
Figure 1 shows the result of our experiments on aramid yarns with

different surface treatments. An approximately linear relation is
observed:

$$3p\text{-SBS} \approx 32 + 0.9 * BPS \qquad (4)$$

A similar plot of the 4-point short-beam shear strength versus the bundle
pull-out strength yields the relation:

$$4p\text{-SBS} \approx 26 + 1.28 * BPS \qquad (5)$$

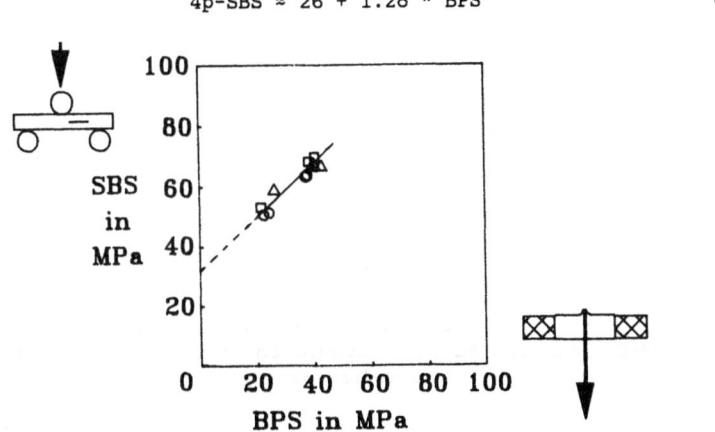

Figure 1. 3-point short-beam shear strength of aramid composites
(V_f = 60 %) versus bundle pull-out strength

Equations (4) and (5) have a close resemblance to equations (1), (2) and
(3). This suggests that the bundle pull-out strength is a direct measure
of the interface strength.

The difference between eqs. (4) and (5) is due to the damage of the
composite by the large contact and stresses near the central loading
roller(s). Especially in composites with adhesion-treated aramid narrow
kinkbands can be seen to originate at the sides of the central roller
contact area and to emanate obliquely in a downward direction. These
kinkbands initiate the final shear fracture. The damage increases with the
load and therefore with the shear strength of the composite. The damage
caused by the single roller of the 3-point test is larger than the damage
caused by each of the rollers of the 4-point test. Indeed the difference
between the 4- and 3-point SBS values is larger for treated (9 MPa) than
for untreated aramid (2.5 MPa), resulting in a larger slope and a lower
intercept in eq. (5) than in eq. (4). The 4-point test also yields larger
SBS values than the 3-point test for glass (5 MPa) and carbon fibre (3
MPa) composites. As such the 4-point SBS is a more linear and less
conservative measure of the interlaminar shear strength than the 3-point
short beam strength.

Usually aramid composites yield somewhat in compression in a bending
test before they finally fracture in shear. This also has a more
pronounced negative effect on the SBS values of composites with treated
than on those with untreated fibres. Application of the compression
correction as mentioned in [7] reduces the intercept of eq. (5) further
and leads to an even closer agreement with eqs. (1) and (3). Compressive
yielding, and hence the correction for this effect is less for beams with

more aramid fibres as the compression strength increases linearly with V_f.

The effect of the fibre volume fraction

The validity of the models was further checked by measuring the 3-point SBS of beams with different fibre volume fractions (V_f), see Figure 2.

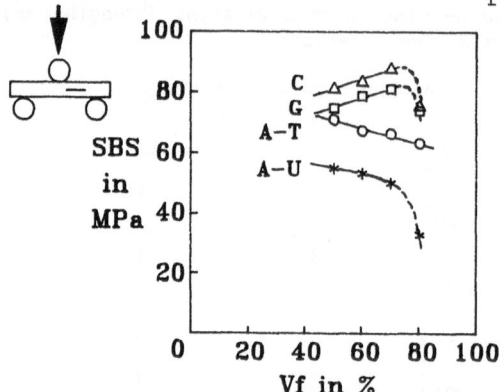

Figure 2. 3-point short-beam shear strength for carbon (C), glass (G), untreated (A-U) and treated (A-T) aramid in epoxy composites at different fibre volume fractions

The short-beam strength of carbon and glass composites increases with V_f as expected from the rule of mixtures assuming that the interface is very strong ($\tau_i \approx \tau_m$) and $\alpha > 1$. The linear elastic stress concentration factors (SCF's) of [5] predict a strong decrease of ILSS with V_f. The non-linearity of the short-beam force-deflection curves provides evidence of matrix yield behaviour since this is not due to compressive yielding of the fibres as in aramid composites. Yielding reduces the stress-concentrations and as such the rule of mixtures effect dominates. At the highest V_f the stress concentrations dominate and cause the strong reduction in SBS.

The 3p-SBS of untreated aramid composites decreases with increasing V_f. The absolute shear strength is much lower than for carbon or glass. This must be caused by the weak aramid-epoxy interface, since the SCF is very low for aramid in epoxy. On the assumption of a weak interface this decrease may also be explained by the rule of mixtures, with the exception of the strong fall at $V_f = 80$ %. This sharp decrease is a good indication that fracture occurs at the interface.

The 3p-SBS of treated aramid also decreases but does not show the sharp drop at $V_f = 80$ %. The only explanation for the apparent absence of stress concentrations is that not the interface, but the aramid itself fails in shear. This conjecture was confirmed by optical microscopy.

Transverse tensile strength

The transverse tensile strength TTS of aramid composites shows a similar dependence on V_f as the bundle pull-out shear strength BPS. The nearly linear relation between TTS and (3-point) SBS for treated and untreated aramid composites in figure 3 is described by:

$$3p\text{-}SBS \approx 35 + 0.83 * TTS \qquad (6)$$

Comparison with (4) shows that there is a general 1:1 relation between TTS and BPS for aramid. The strong sensitivity of the transverse strength to the interface or fibre off-axis strength is probably caused by the low fracture toughness (ability to yield and relieve stress concentrations) of the matrix in mode I (crack opening, as occurs in the TTS test) in comparison to the toughness in modes II and III (forward and sideward shear, which dominate in the short-beam shear test). Therefore a single debonded filament will result more easily in complete failure in a transverse tension than in a shear test.

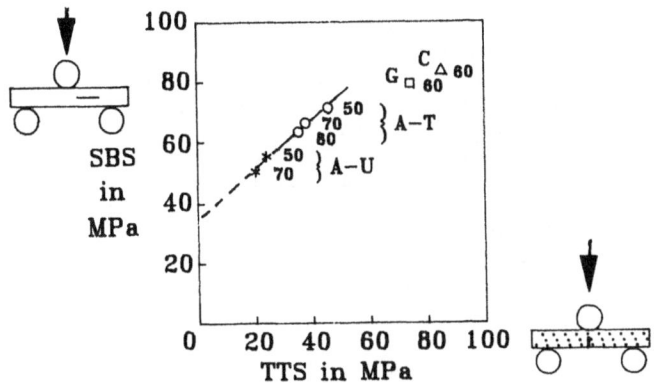

Fig. 3: Short-beam versus transverse tensile strength, V_f as indicated

Fracture surface analysis shows that the transverse crack runs straight through the adhesion treated filaments while it circumvents the untreated filaments. Further improvement in the off-axis strength of aramid-epoxy composites seems unlikely since it is limited by the shear and transverse tensile strength of the aramid fibre itself.

The transverse strength of glass fibre composites (75 MPa for $V_f = 60$ %) is slightly higher than expected (65 ± 8 MPa) on the basis of the linear elastic SCF (table 1 and [6]) and the pure epoxy tensile strength (120 MPa ± 15 MPa) as measured in the same bending test. For carbon the measured TTS (85 MPa) is only slightly lower than the calculated value (92 ± 12 MPa). This is another indication that the glass- and carbon-epoxy interfaces are nearly as strong as the epoxy itself and that the matrix has little opportunity to yield in a transverse tensile test. The scatter (15 %) in the TTS of carbon and glass composites was too large to detect a relation with V_f.

SUMMARY AND CONCLUSIONS

The 3-point short-beam shear strength is a conservative measure of the aramid-epoxy interlaminar shear strength (ILSS). This is caused by the relatively large amount of damage under the central roller and by compressive yielding in beams with higher shear strengths. These negative effects can partly be suppressed by using a 4-point short-beam test instead of the usual 3-point test and by using beams with more than the usual amount of fibres (70 instead of 60 %). A disadvantage of the 4-point test is that it demands more material.

The transverse bending tensile strength is a more direct measure of the interfacial strength than the short-beam test. The transverse tensile fracture surface is directly accessible, but the test is more time consuming than the short-beam test because the edges must be sawn very carefully.

The scatter in the results of the different tests depends upon the level of adhesion and the type of fibre in the composite. Generally the coefficients of variation are 10, 3, 3 and 10% for the bundle pull-out, 3-point short-beam, 4-point short-beam and transverse bending test respectively. The bundle pull-out test demands the least material. The 4-point short-beam test has the best balance of linearity, resolution and simplicity of the tests for u.d. composites.

The measured increase due to the aramid adhesion treatment was different for each test: $\approx$ 60 % in bundle pull-out, $\approx$ 28 % in the 3-point, $\approx$ 40 % in the 4-point short-beam and, $\approx$ 60-80 % in transverse bending strength. Further improvement of the off-axis strength of aramid-epoxy composites is limited by the shear and transverse strength of the aramid fibres. In spite of the considerable adhesion improvement aramid-epoxy composites still have a lower shear and transverse strength than glass and carbon fibre composites. However, the corresponding strains are approximately equal because of the lower aramid-epoxy shear and transverse moduli.

REFERENCES

1. Ladizesky, N.H. and Ward, I.M., Ultra-High-Modulus Polyethylene Fibre Composites: I--The Preparation and Properties of Conventional Epoxy Resin Composites. Comp. Sci. and Techn., 1986, 26, 129-164.

2. Penn, L., Bystry, F., Karp, W. and Lee, S., Aramid/Epoxy vs. Graphite/Epoxy: Origin of the difference in strength at the interface. In Molecular Characterization of Composite Interfaces,ed., H. Ishida and G. Kumar, Plenum Press, New York, 1985, pp. 93-109.

3. Favre J.P. and Merienne M.C., Characterization of fibre/resin bonding in composites using a pull-out test. Int. J. Adhesion and Adhesives, October 1981, 311-316.

4. Fisher, M. and Schmid, R., Matrix properties and composite failure. Colloid and Polymer Science, 1986, 264, 387-398.

5. Adams, D.F. and Doner, D.G., Longitudinal shear loading of an unidirectional composite. J. Comp. Mat., 1967, 1, 4-17.

6. Adams, D.F. and Doner, D.G., Transverse normal loading of an unidirectional composite. J. Comp. Mat., 1967, 1, 152-164.

7. Fisher, S. and Marom, G., A complete elastic-plastic analysis of aramid fibre composites. Fibre. Sci. and Techn., 1984, 20, 91-98.

8. Whitney, J.M. and Browning, C.E., On short-beam shear tests for composite materials. Exp. Mech., 1985, 25(3), 294-300.

MATRIX EFFECTS ON INTERLAMINAR FATIGUE CRACK GROWTH IN COMPOSITES

V. ALTSTÄDT and R.W. LANG
BASF AG, D-6700 Ludwigshafen
Federal Republic of Germany

ABSTRACT

An experimental investigation was undertaken to study the effects of neat resin properties on interlaminar crack growth of composite laminates under fatigue loading conditions. To obtain a more fundamental understanding of the parameters involved, cyclic crack growth rates were characterized over a wide velocity range for three different resin systems and corresponding laminates using fracture mechanics techniques. Of special importance, the results seem to suggest that the threshold strain energy release rate range, ΔG_{th}, in composites under Mode I conditions is controlled by the quasi-static fracture energy, G_{IC}, of the neat matrix resin.

INTRODUCTION

From the many different damage mechnisms known to occur in composite laminates, delaminations or interlaminar cracks are often regarded as the most critical failure modes. In many cases, especially in fatigue loaded structures, the total fatigue life may be controlled by the kinetics of delamination propagation. Since there is an increasing awareness that the polymeric matrix plays a dominant - if not the dominant - role in fatigue delamination propagation, it was decided to study the influence of matrix properties on delamination fatigue crack growth in more detail.

Most of the investigations on interlaminar fatigue crack growth reported so far have concentrated on using principles and techniques of classical fracture mechanics [1-3]. In analogy to the stress intensity factor approach originally proposed by Paris [4] to describe fatigue crack growth behavior in metals, interlaminar fatigue crack propagation rates, da/dN, in composites are commonly related to the prevailing strain energy release rate range, ΔG, according to the relationship

$$\frac{da}{dN} = A \cdot \Delta G_I{}^m \tag{1}$$

where a is the crack length, N is the number of loading cycles, and A and m depend on the material and test conditions. The applied ΔG_I-range (subscribt I stands for tensile or opening mode conditions) is defined as the difference between the maximum and minimum strain energy release rate (i.e., $\Delta G_I = G_{Imax} - G_{Imin}$), which in turn are related to the maximum and minimum load, respectively, in a fatigue cycle.

EXPERIMENTAL

Materials

Two of the resin systems selected for this study, designated XR-1 and XR-2, were prepared from standard type epoxy resins of different functionalities and diaminodiphenyl sulfone as curing agent. To cover a wider range in neat resin toughness, Rigidite 5212 - a novel 125°C-curable BASF epoxy resin system of proprietary formulation - was also included in this investigation. Along with these resins, a standard modulus carbon fiber (T 300 6000 X 50B, TORAY INDUSTRIES, Japan) was used to produce prepregs and laminates.

Fatigue crack propagation tests

To characterize the cyclic crack growth resistance of both neat resins and laminates, fatigue crack propagation (FCP) experiments under Mode I conditions were performed in laboratory air (23°C/50%r.h.) using a computer controlled servohydraulic fatigue testing machine. In the case of neat resins, compact-type (CT) specimens with a width of 50 mm, a hight-to-width ratio of 0.6, and a nominal thickness of 4 mm were used. For composite laminates double cantilever beam specimens (125 x 20 x 8 mm^3) were employed. FCP tests were conducted at 10 Hz. The applied waveform was sinusoidal with a constant load amplitude and a minimum-to-maximum load ratio, R, of 0.1.

Further details as to the materials, specimen preparation and test procedures are described elsewhere [5,6].

RESULTS AND CONCLUSIONS

FCP rates for the three resin and composite systems investigated are plotted in Figs. 1 and 2, respectively, as log da/dN versus log ΔG. In sharp contrast to the behavior of neat resins (Fig.1), distinctly separated curves without cross-overs are observed for laminates (Fig. 2). In good agreement with our previous results for quasi-static loading [7], the ranking in grack growth rates at high crack speeds is clearly identical for neat resins and corresponding composites. On the other hand, crack growth rates at low crack propagation rates can apparently not be translated directly from neat resins to laminates. Based on the quasi-static investigations mentioned above, it much rather appears that crack growth rates at low crack speeds are also controlled by the neat resin fracture energy G_{IC}. However, more studies with additional materials and covering a wider range in behavior are needed to verify these preliminary findings.

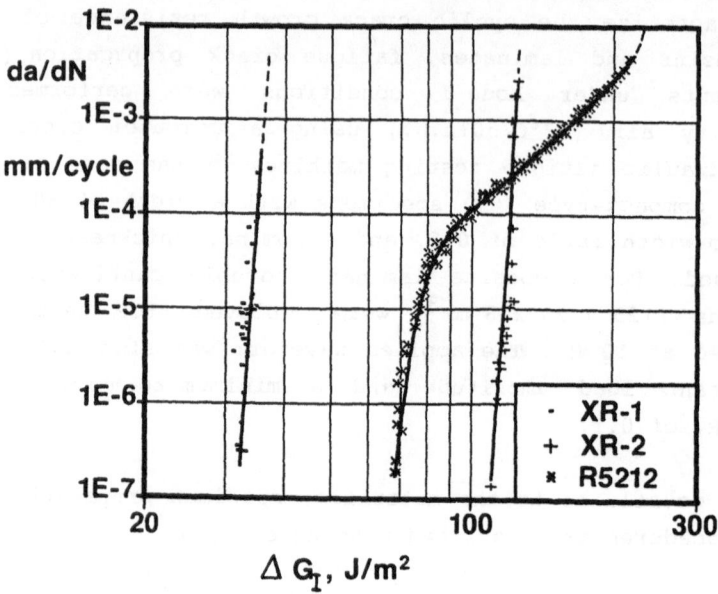

Figure 1. Fatigue crack propagation behavior of neat resins.

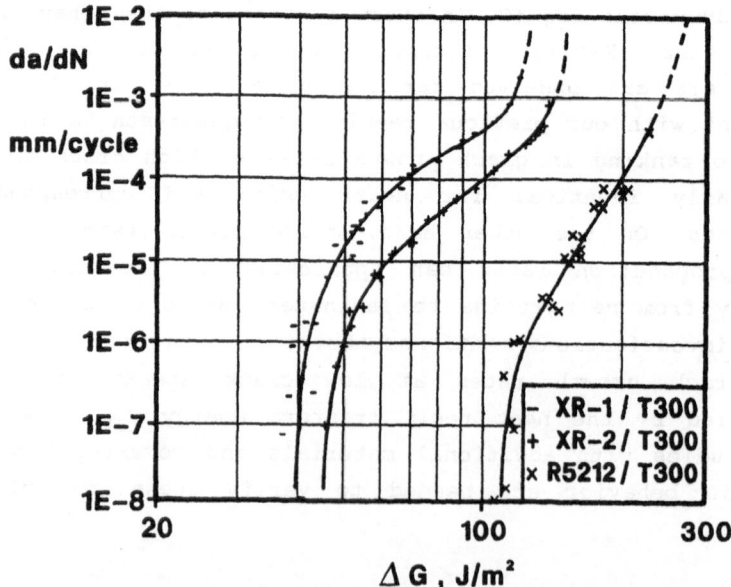

Figure 2. Fatigue crack propagation behavior of UD laminates.[1])

REFERENCES

1. C.-G. Gustafson, L. Jilken and P.A. Gradin, <u>ASTM STP 876</u>, American Society for Testing and Materials, Philadelphia, 1985, p. 200-216.

2. C.E. Bakis and W.W. Stinchcomb, <u>ASTM STP 907</u>, American Socity for Testing and Materials, Philadelphia, 1986, p.314-334.

3. M. Hojo, K. Tanaka, C.-G. Gustafson and R. Hayashi, <u>Comp. Sci. Techn.</u> 29 (1987) p. 1-20.

4. P.C. Paris, <u>Ph.D. Dissertation</u>, Lehigh University 1962.

5. V. Altstädt and R.W. Lang, <u>Proc. 9th Int.SAMPE Conf. Milano</u>, Italy, June 14-16, 1988 (to be published).

6. R.W. Lang, H. Tesch and G.H. Herrmann, <u>Proc. 9th Int.SAMPE Conf. Milano</u>, Italy, June 14-16, 1988 (to be published).

7. R.W. Lang, M. Heym, H. Tesch and H. Stutz, in "<u>High Tech the Way into the Nineties</u>", K. Brunsch, H.-D. Gölden and C.-M. Herkert (eds.), Elsevier Science Publishers B.V., Amsterdam (1986) p. 261-272.

[1] Due to a calibration error of the test equipment, Fig.2 has been published in Ref.9 with incorrect scales on the X-axis. However, this error does not alter any of the conclusions in that publication.

FIBRE DEFORMATION AND RESIDUAL THERMAL STRESSES IN CARBON FIBRE REINFORCED PEEK

R.J. YOUNG, R.J. DAY, M. ZAKIKHANI AND I.M. ROBINSON
Polymer Science and Technology Group
Manchester Materials Science Centre
UMIST, P.O. Box 88
MANCHESTER, M60 1QD.

ABSTRACT

Raman Microscopy has been used to study the deformation of carbon fibres and an experimental grade carbon-fibre/PEEK composite prepreg. It has been found that the peak position of the Raman-active bands in the fibres are sensitive to the level of applied strain. Examination of the peak positions from the carbon fibres near the surface of the prepreg shows that the fibres are subjected to a residual compressive strain. The application of a tensile stress to the composite causes the fibre strain to become tensile although significant scatter is found in the measurements. The scatter is thought to be due to variations in the local carbon fibre strain on the 1μm level. It is demonstrated that residual compressive strain is expected from differential shrinkage between the fibres and matrix on cooling the composite from the processing temperature to room temperature.

INTRODUCTION

The authors and co-workers have recently demonstrated that Raman spectroscopy is a powerful method of following the deformation of high modulus polymer fibres on the molecular level. The fibres studied have included substituted polydiacetylenes (1-3), poly(p-phenylene terephthalamide) (aramids) (4,5) and poly(p-phenylene benzobisthiazole) (PBT) (6). It is found that the peak frequencies of the Raman-active bands of such fibres are sensitive to the level of any externally applied stress or strain. In all polydiacetylene single crystal fibres shifts to lower frequencies in the order of 20 cm^{-1}/% applied strain have been found for the symmetric triple bond carbon-carbon stretching mode (1-3). For both aramid and PBT fibres the magnitude of the shift for each of the Raman bands (in terms of cm^{-1}/% applied strain) is found to be dependent upon the fibre modulus, with the higher shifts being obtained for the stiffer fibres.

In a recent publication the authors demonstrated that strain induced Raman frequency shifts can be obtained in different types of carbon fibre (7). Although there are several reports of Raman spectra being obtained from various forms of graphite (8-12) we are not aware of any reports concerning stress-induced changes. Two Raman-active bands are generally found in graphite. The 1580 cm^{-1} band, assigned to an E_{2g} mode of the infinite crystal for graphite (8) is present in all graphitic structures. In some forms of graphite such as carbon fibres another band at about 1360 cm^{-1} is obtained. This A_{1g} mode is Raman active in the crystal boundary region of the graphite and has been likened to a crystal size effect (8). The intensity of the 1360 cm^{-1} band has been observed to increase as the crystallite size decreases and the ratio of the intensities $I(A_{1g})/I(E_{2g})$ for the modes has been used to give an indirect measure of the crystallite size in graphite fibres (8). Since the Raman scattering occurs from a thin surface layer of the sample of the order of 50 nm deep, the spectra are characteristic of the surface of the fibres (10). Hence, the strain induced shifts we reported must be due to deformation of crystallites close to the surfaces of the fibres.

In addition to giving a measure of the deformation within fibres, the strain-induced band shifts in Raman spectra have been used to follow the micromechanics of fibre reinforcement in model polydiacetylene/epoxy composites (13). The critical length has been measured directly (13) and the effect of resin shrinkage has been examined in detail (14,15). It has also been demonstrated that the technique can be employed to measure fibre strain optically in a high volume fraction Kevlar 49/epoxy composite (16). In this present paper we demonstrate that Raman microscopy can also be used to measure fibre strain in carbon fibre reinforced PEEK composites (17) and to give a direct measure of residual thermal shrinkage stresses in PEEK matrix composites.

EXPERIMENTAL

The carbon fibres used were Hercules HMS4 PAN-based fibres with diameters of the order of 6-10μm. Specially prepared, experimental grade, 100μm thick composite prepregs were employed consisting of uniaxially-aligned HMS4 fibres in PEEK (Poly(ether ether ketone)) with a fibre volume fraction, V_f of about 61%.

Raman spectra were obtained from individual fibres both in a free-standing situation and in the prepreg sample using a Raman microscope system. This is based upon a SPEX 1403 double monochromator connected to a modified Nikon optical microscope. Spectra were obtained at a resolution of the order of + 10cm^{-1} using the 488nm line of an air-cooled Argon ion laser operated at an output intensity of between 5 and 15mW giving a spot size of the order of 1-2μm diameter.

Spectra were obtained from free-standing carbon fibres in air deformed in a small straining rig which fitted directly onto the microscope stage. Individual fibres were fixed between aluminium foil tabs which were placed onto the aluminium blocks of the straining rig using a cyanoacrylate adhesive, giving a gauge length of about 10 mm which was measured accurately using the stage venier gauge of the light microscope.

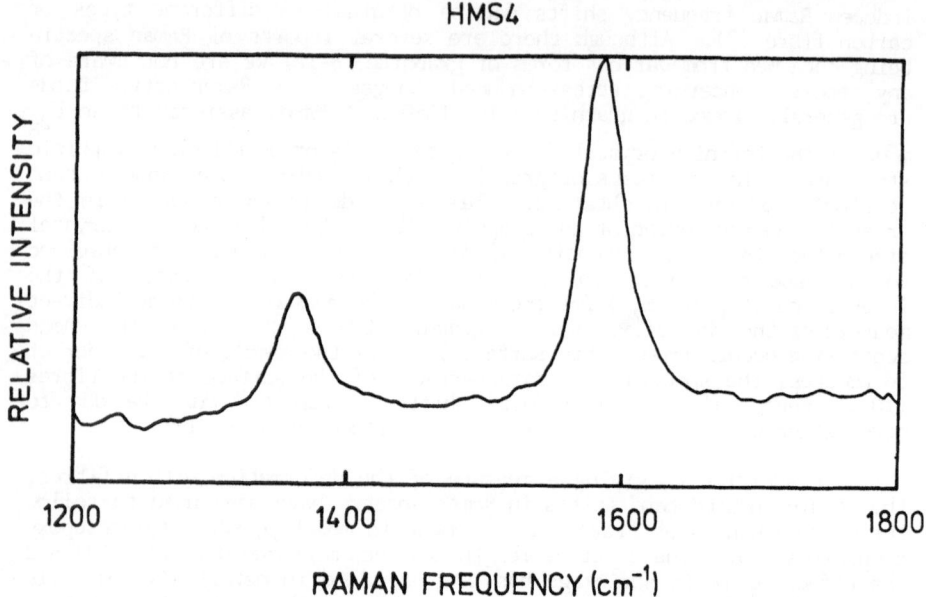

HMS4

Figure 1   Raman spectrum for a free-standing HMS4 carbon fibre in the range 1200–1800 cm^{-1}.

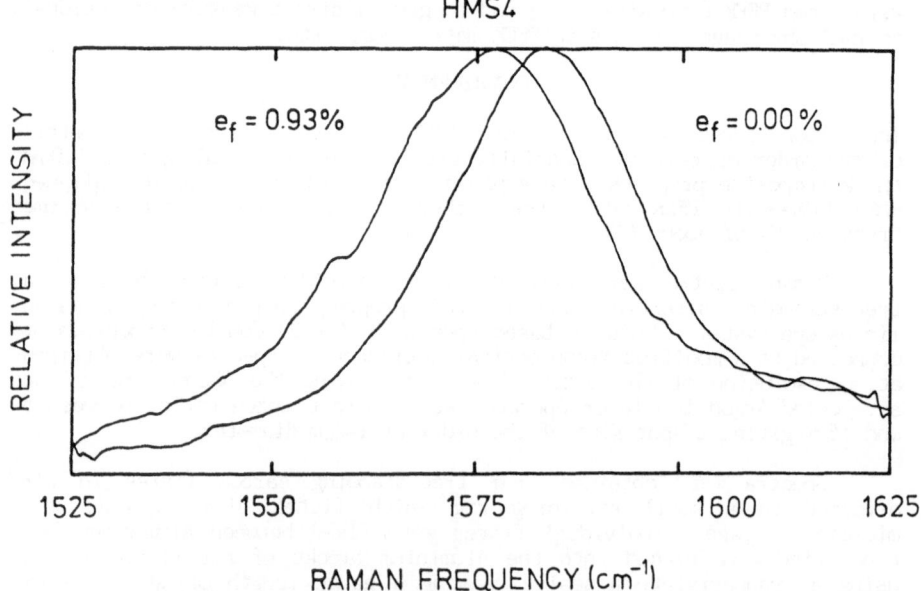

HMS4

Figure 2   Raman spectra for the 1585 cm^{-1} band of a free-standing HMS4 carbon fibre before and after straining to the levels indicated.

Spectra were obtained from fibres near the surface of prepreg samples deformed using a Polymer Laboratories "Mini-mat" mechanical testing machine which is designed specifically to fit on the stage of an optical microscope. Strips of the HMS4/PEEK prepreg of the order of 2 mm wide were deformed to fixed displacements using a gauge length of the order of 50mm. The strain was determined from the gauge length and the displacement. The load on the specimen was also monitored using a 200 N load cell to ensure that no slippage took place at the grips. The Youngs modulus of the prepreg was calculated as approximately 200 GPa which is close to that expected from its composition.

DEFORMATION OF FREE-STANDING CARBON FIBRES

Figure 1 shows the Raman spectrum obtained in the region 1200-1800 cm^{-1} from a 1 μm diameter region of a free-standing HMS4 fibre in air. The spectrum is similar to that reported for other carbon fibres (8,10,12). Two bands can be seen. There is a strong band at about 1585 cm^{-1} and a weaker one located at about 1360 cm^{-1} which correspond to the E_{2g} and A_{1g} modes of graphite described by Tuinstra and Koening (8). The effect of deformation upon the 1585 cm^{-1} band can be seen in Figure 2.

The effect of deformation upon the position of the peak of the band is shown in Figure 3. There is a linear shift in peak position up to the breaking strain which is about 1% and the rate of shift is -9.2 cm^{-1}/%. The scatter in the data is due to random error in measuring the peak position which is typically $\pm$ 1cm^{-1} for the 1585 cm^{-1} peak. It is clear that the data in Figure 3 can serve as calibration curves for the investigation of deformation micromechanics in carbon-fibre reinforced polymer composites in an analogous way to previous investigations by Young and co-workers (13-16) for polydiacetylene- and Kevlar-epoxy composites.

It is worth considering why there is a stress- or strain-induced peak frequency shift for these bands in carbon fibres. It would seem that it is another manifestation of a phenomenon also found in high modulus polymer fibres. In all these types of materials mechanical straining of the fibre is translated into direct deformation of the covalent bonds in the molecules of the fibre. The change in bond length causes a change in the vibration frequency of the particular Raman-active band. The phenomenon is related to the anharmonicity of the bond vibration and the Raman deformation technique is clearly potentially a powerful way of probing the interatomic potentials in these high modulus fibres.

CARBON-FIBRE/PEEK COMPOSITES

Figure 4 shows a scanning electron micrograph of the prepreg (HMS4/PEEK) used in this present study. Since the main aim of the present study was to establish whether or not the Raman microscope straining technique could be applied to HMS4/PEEK composite, Raman spectra were obtained first of all from a prepreg. The surface of the prepreg was examined optically in the Raman microscope and an area with a prominent fibre covered in PEEK was selected. The laser was then focussed on this area and the spectrum obtained. It was found that the spectrum for the fibre was always superimposed upon a fluorescent background from the PEEK. The

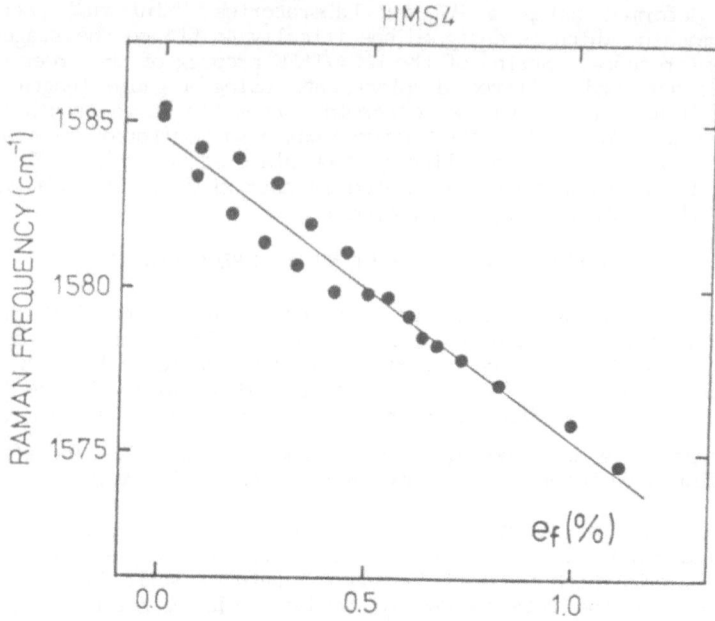

Figure 3.   Variation of the position of the 1585 cm^{-1} band with the level of fibre strain for a free-standing HMS4 fibre.

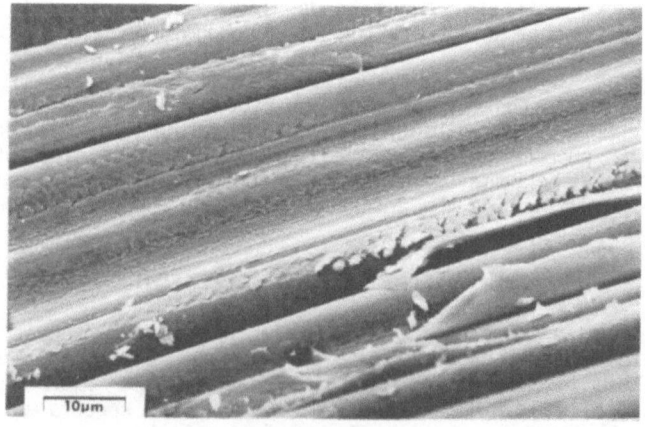

Figure 4.   Scanning electron micrographs of the specially-prepared HMS4/PEEK used in this present study.

spectrum from the region in the vicinity of the fibre is shown (after background subtraction) in Figure 5 and it can be seen that it is similar to the HMS4 spectrum in Figure 1.

Although the spectra in Figure 1 and 5 are similar, closer inspection reveals that the positions of the peaks are shifted in the HMS4/PEEK relative to their positions in the fibre in air. This is illustrated in Figure 6 where it can be seen that the Raman band of HMS4 in PEEK is shifted to a higher frequency than for the fibre in air. The important conclusion from this observation is that, since tension produces a shift to lower frequency, the HMS4 carbon fibres in this specimen are in a state of compression. It is known that the thermal expansion coefficient of PEEK is about two orders of magnitude higher than that of the fibres and so cooling the prepreg from the melt during manufacture will have the effect of putting the fibres into compression. In addition PEEK is capable of crystallising and so further contraction of the PEEK due to crystallisation will have the effect of inducing further compression in the fibres.

The effect of strain upon a fibre in the HMS4/PEEK composite prepreg is shown in Figure 7. It can be seen that on the appliction of tensile stress the position of the 1585 cm^{-1} Raman band moves to a lower frequency. The noise is due to a relatively weak signal from the fibre inside the PEEK. Even with the noise, however, the shift is significant and it shows that, as would be expected, the application of a tensile stress to the composite causes a tensile strain within the fibres. The results of a large number of similar measurements are shown in Figure 8 where the fibre strain, e_f, determined from the shift of the 1585 cm^{-1} carbon fibre peak using Figure 3 as a calibration curve, is plotted against the strain applied to the composite, e_c. Figure 8 has a considerable amount of interesting information. There is considerable scatter in the data. Since the signals are relatively weak from the fibres in the composite rather more scatter is expected than for example from the fibre deformed in air (Figure 3). However, the scatter is too large (variations in e_f of over 0.5% at a given level of e_c) to be accounted for by random errors. It would appear that there is a significant fibre-to-fibre variation in local strain within fibres. It must be remembered that a beam size of the order of 1μm was used which is well within one fibre diameter and so the Raman technique probes a very small region near the prepreg surface. One might expect significant local differences in fibre/fibre separation and fibre deformation in surface layers and so measurements in Figure 8 reflect variations in the state of stress and strain at the surface of the composite on the 1μm level. However, it is not clear as to how far these surface measurements reflect the behaviour of the fibres in the bulk of the composite.

The solid line in Figure 8 is a least squares fit of the data. It has a slope of 0.92 and an intercept of 0.28% strain. This line reflects the average behaviour of the fibres during deformation. The dashed line in Figure 8 has a slope of unity and corresponds to $e_f = e_c$ which would be the behaviour expected from the Cox model (18) for an ideal uniaxially-aligned continuous fibre composite with no complication due to thermal stresses. The experimental line confirms that the surface fibres are on average subjected to a thermal compressive strain of 0.28% if it is assumed that the linear dependence of peak position upon e_f shown in

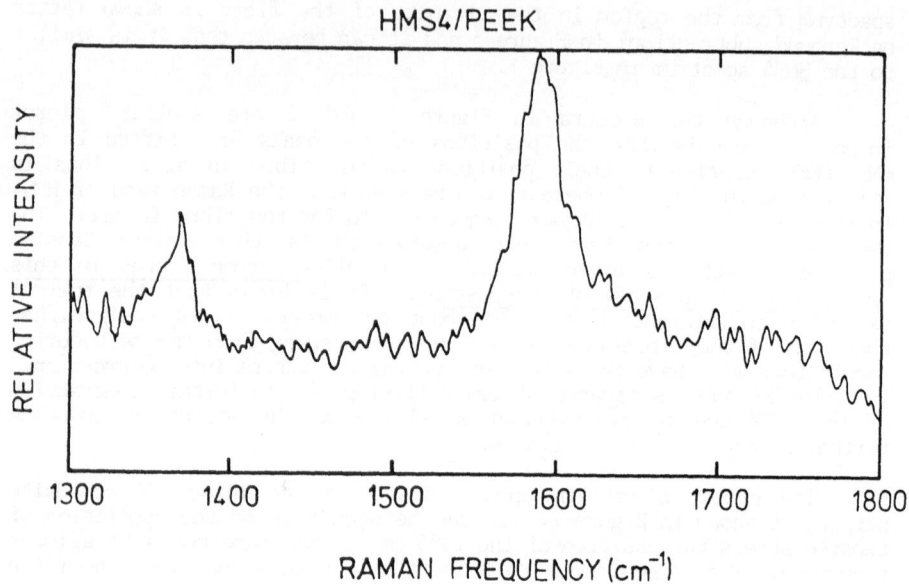

Figure 5.   Raman spectrum from the specially-prepared HMS4/PEEK prepreg in the range 1300–1800 cm^{-1}.

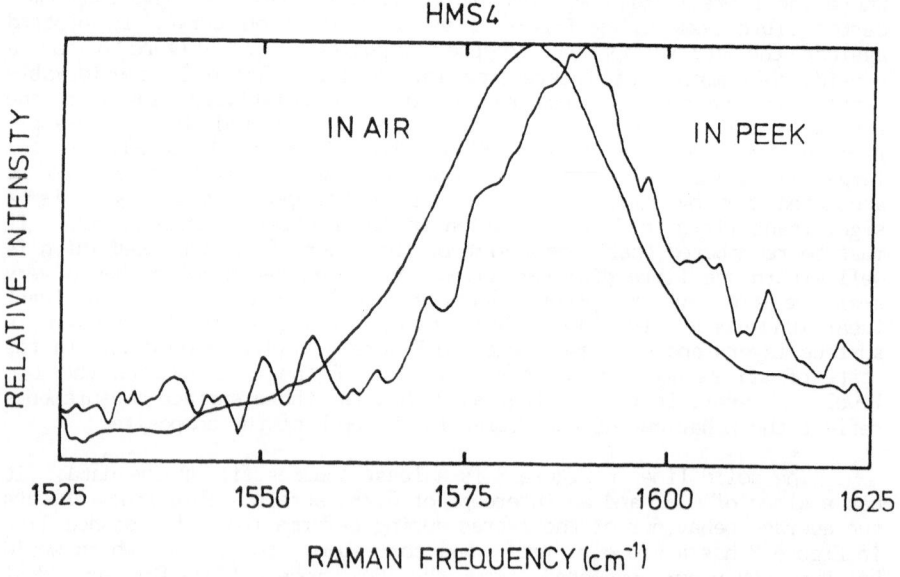

Figure 6.   Raman spectrum for the 1585 cm^{-1} band showing the variation in peak position for a free-standing fibre in air and in PEEK.

Figure 3 can be extrapolated for compressive strains. However, since the slope of the line is close to unity (0.92) it shows that when the prepreg composite is deformed the fibres are strained in tension as predicted by the Cox model. We feel that these are two important observations relevant to the deformation of both carbon-fibre/PEEK matrix composites and composites in general.

The authors have recently calculated the residual thermal stresses in HMS4/PEEK (19) and have shown that significant levels of axial compression would be expected in the fibres because of high levels of thermal contraction in PEEK on crystallisation and cooling from the processing temperature.

CONCLUSIONS

It has been shown that the peak positions of the Raman-active bands of carbon fibres are strain-sensitive and that Raman Microscopy can be used to follow the deformation of carbon fibres both in air and in a thermoplastic PEEK matrix. It has been demonstrated that the fibres near the surface in the carbon-fibre/PEEK composite examined are subject to a residual compressive strain of the order of 0.28% which is of the same order as that expected (19) from matrix shrinkage due to crystallisation and thermal contraction on cooling from the processing temperature. It is found that when the composite is subjected to an externally-applied tensile deformation then, as expected, the change in fibre strain is similar to the applied strain as expected from simple composite theory.

It is clear that this present study introduces a new method of following the micromechanics of fibre-reinforced composites. The ability to measure the strain at a point in an individual fibre should lead to a significant increase in our understanding of the deformation of these important materials and of fibre/matrix interactions.

ACKNOWLEDGEMENTS

The authors are grateful to Dr. Judith A. Peacock of ICI plc for the supply of the materials and characterisation data and for useful discussions. The work was supported by research grants from the SERC and UMIST.

REFERENCES

1. Mitra, V.K., Risen, W.M. and Baugham, R.H. J.Chem.Phys. 66 (1977) 2731.
2. Batchelder, D.N., and Bloor, D. Polym.Phys.Ed., 17 (1979) 569.
3. Galiotis, C., Young, R.J. and Batchelder, D.N. J.Poly.Sci.,Polym.Phys.Ed., 21 (1983) 2483.
4. Galiotis, C., Robinson, I.M., Young, R.J., Smith, B.J.E. and Batchelder, D.N. Polymer Comm., 26, (1985) 354.
5. van der Zwaag, S., Northold, M.G., Young, R.J., Galiotis, C., Robinson, I.M. and Batchelder, D.N. Polymer Comm., 28 (1987) 276.
6. Day, R.J., Robinson, I.M., Zakikhani, M. and Young R.J. Polymer, 28 (1987) 1833.

7. Robinson, I.M., Zakikhani, M., Day, R.J., Young, R.J. and Galiotis, C. J.Mater.Sci.Lett. 6 (1987) 1212.
8. Tuinstra, F. and Koenig, J.L. J.Chem.Phys., 53 (1970) 1126
9. Nemanich, R.J. and Solin, S.A. Phys.Rev. B, 20 (1979) 392.
10. Tuinstra, and F. Koenig, J.L. J.Comp.Mat., 4 (1970) 492.
11. Morita, K., Murata, Y., Ishitani, A., Murayama, K., Ono, T. and Nakajima, A. Pure and Appl.Chem., 58 (1986) 455.
12. Chieu, T.C., Dresselhaus, M.S. and Endo, M. Phys.Rev.B, 26 (1982) 5867.
13. Galiotis, C., Young, R.J., Yeung, P.H.J. and Batchelder, D.N. J.Mater.Sci., 19 (1984) 3640.
14. Robinson, I.M., Yeung, P.H.J., Galiotis, C., Young, R.J. and Batchelder,D.N. J.Mater.Sci. 21 (1986) 3440.
15. Young, R.J., Galiotis, C., Robinson, I.M. and Batchelder, D.J., J.Mater.Sci. 22 (1987) 3642.
16. Galiotis, C., Robinson, I.M., Batchelder, D.N. and Young, R.J. paper presented at Comp '86 Patras, Greece, 1986 (to be published by Omega, Oxford, 1988).
17. Crick, R.A., Leach, D.C., Meakin, P.J. and Moore, D.R., J.Mater.Sci. 22 (1987) 2094.
18. Cox, H.L. Brit.J.Appl.Phys. 3 (1952) 72.
19. Young, R.J., Day, R.J., Zakikhani, M. and Robinson, I.M. Comp.Sci. and Tech. in press.

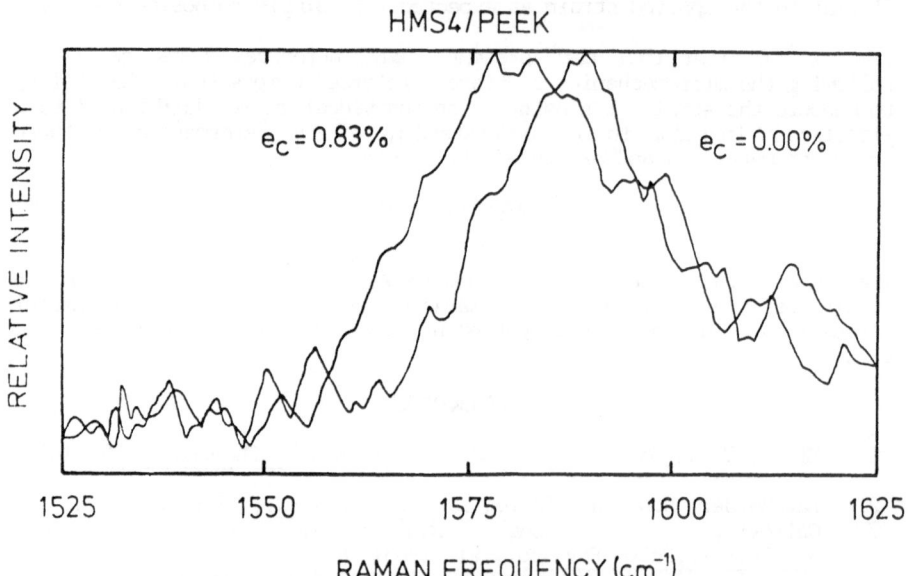

Figure 7.   Raman spectrum from carbon fibres in PEEK before and after straining the composite prepreg to the level indicated.

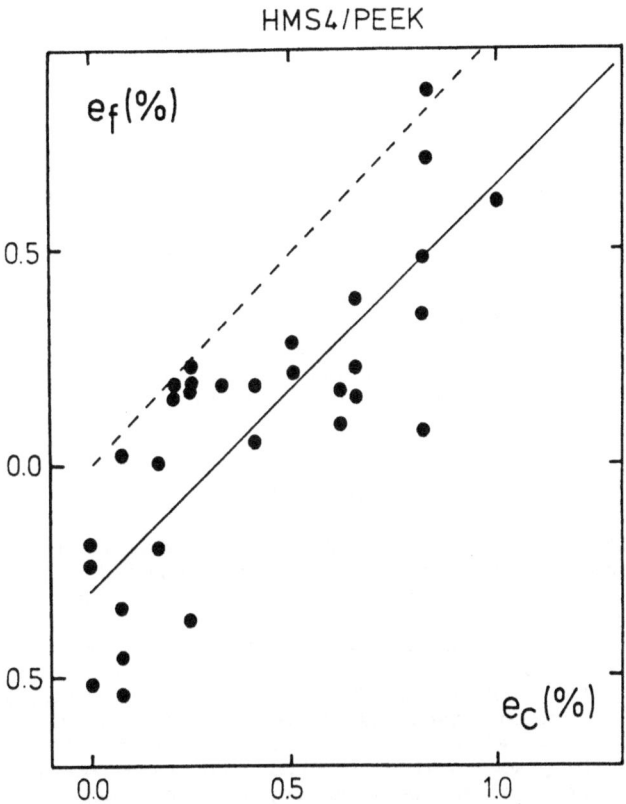

Figure 8.   Variation of fibre strain, e_f, measured using Raman spectroscopy with applied composite strain, e_c for a specially prepared HMS4/PEEK prepreg. The dashed line is for the expected behaviour of an ideal composite where $e_f = e_c$. The solid line is a least squares fit of the experimental data with a slope of 0.92 and an intercept of -0.28%.

MECHANICAL BREAKDOWN OF SINGLE FIBRES AND MICROCOMPOSITES

H. DANIEL WAGNER, L. W. STEENBAKKERS
Polymeric Composites Laboratory
Department of Materials Research
The Weizmann Institute of Science
Rehovot 76100 (Israel)

ABSTRACT

Recent results on the fracture behaviour of advanced fibers and model composite materials (microcomposites) are presented. The fiber data utilized are strength results for polydiacetylene whiskers and Kevlar 149 fibres. The probabilistic approach adopted for the effect of fibre diameter on strength is shown to be at least as appropriate as well-known empirical and LEFM-based schemes. A modified Poisson/Weibull scheme deals with a currently unsolved problem inherent to the classical Weibull distribution function, used as a model for strength. Preliminary experimental results with model microcomposites (which consist of single fibers carefully placed within a matrix film using specially developed positioning techniques) are reviewed and, in particular, the usefulness of video/microphotographic techniques in the study of failure dynamics and fracture modes in composites is emphasized.

INTRODUCTION

Tensile failure of unidirectional composites is a complex statistical process involving scattered failure of fibres at flaw sites, and local overloading and failure of neighbouring fibres by way of stress transfer through the matrix. Final failure follows via the development of a catastrophic crack of critical dimensions. The Poisson/Weibull probabilistic approach is the most popular scheme in use for describing the ultimate mechanical behaviour of solids having a linear stress-strain behaviour up to breakdown [1]. It has been used to describe extreme-value problems as varied as the propagation of cracks along a geological fault [2], the collapse of fractal trees [3], the failure of vacuum-tubes [4], the analysis of carcinogenesis test results [5], and more. The length and diameter of a fibre are known to be key factors with respect to the mechanical stress necessary to promote failure, and different methods are usually utilized to quantify both types of such size effects: linear elastic fracture mechanics [6] (lefm) and related schemes provide the theoretical basis for the effect of diameter variability upon strength whereas statistical theories, generally based upon the Weibull probability distribution [7] combined with the weakest-link theorem [8], describe length effects.

Using polydiacetylene (PDA) crystal fibre strength data produced by Galiotis et al. [9-11] and aramid Kevlar 149 (K149) fibre fracture data generated in our laboratory, we show in the first part of the present paper that simple modifications of Poisson/Weibull concepts, resulting in a new failure probability function, can be used as a modeling scheme for the study of diameter effects on strength in a way at least as satisfactory as previously used lefm-based schemes. Moreover, this new scheme is shown to deal adequately with a well-known problematic issue inherent to the Weibull/weakest link model, namely that the Weibull shape parameter value obtained from conventional Weibull plots at a given diameter is significantly different from the value obtained from the slope of a log(strength) vs. log(size) plot.

The second part of the paper demonstrates the potential usefulness of microcomposite models, as studied by (video) microphotography, in the characterisation of basic failure modes in fibre-reinforced composites, and as a probe of existing strength theories in such materials.

MATERIALS AND METHODS

The single fibre materials used were PDA single crystal fibres produced recently by Galiotis et al. [9-11], and aramid Kevlar 149 single fibres (du Pont de Nemours) tested in our laboratory using a now well established procedure [12-14]. In particular, the diameter of each Kevlar 149 fibre was measured (by optical microscopy) at three sites along the fibre length and the mean value taken as the diameter of the fibre.

Preliminary strength tests were performed with microcomposites. The manufacturing method for the microcomposites, which consist of thin epoxy (CY223/HY956, Ciba Geigy) films reinforced with Kevlar 149 single fibres, is described elsewhere [12]. Accurate control of the interfibre distance is possible, and very thin films (≈ 30 μm) may be produced in certain cases. The microcomposites were tested at room temperature using a custom-made minitensile testing machine fitted to the stage of a StereoZoom microscope (Bausch and Lomb) equipped with a video camera, so that the failure process could be easily studied.

The method of analysis of single fibre strength results is based on a new stochastic scheme as already mentioned, which involves the following failure distribution function (justified elsewhere [12, 13]):

$$F(\sigma) = 1 - \exp[-k\omega^\delta \sigma^b] \qquad (1)$$

where k is a constant, ω is the nondimensional surface or volume over which defects are distributed, σ is the strength, b is the shape parameter, and δ is a constant which, in principle, can take any real value. The main advantage of this distribution function over conventional Weibull distribution functions is that the diameter size effect translates as

$$\langle \sigma \rangle \sim d^{-\delta/b} \qquad (2)$$

where $\langle \sigma \rangle$ is the mean strength of the population, as is easily computed from the expected value of the distribution, and d is the fibre diameter. The slope of a $\log(\langle\sigma\rangle)$ vs. $\log(d)$ plot is equal to $-\delta/b$, rather than to $-1/b$ and $-2/b$ for conventional Weibull plots in which the critical defects are distributed over the external surface or throughout the bulk of the fibre [12, 13], respectively. Thus, $-\delta/b$ is now the key factor determining the size effect, rather than the shape parameter alone as in previous statistical models. A particularly attractive feature is that provided that δ takes on the appropriate numerical value, the values of b obtained from a single surface or volume Weibull plot, and from the slope of the $\log(\text{strength})$ vs. $\log(\text{diameter})$ plot, are now reconciliated.

RESULTS AND DISCUSSION

Taking the PDA strength results first (namely, set A in [15]), the lefm model ($1/\sigma \sim d^{1/2}$) yields a fairly good fit for the strength-diameter dependence, with a coefficient of correlation $r^2 = 0.72$. The Poisson/Weibull prediction, however, gives a significantly better fit with $r^2 = 0.79$. With Kevlar 149 strength results the same trend is observed: the lefm model gives a fair correlation with $r^2 = 0.36$, whereas the Poisson/Weibull model gives a much better fit with $r^2 = 0.50$. More details may be found elsewhere [16].

In Table 1 the value of $-\delta/b$ obtained from Maximum Likelihood Estimation (MLE) is compared with the value obtained from the experimental slope on the $\log(\text{strength})$ vs. $\log(\text{diameter})$ plots. Since the agreement is very good, the suggested failure distribution (equation 1) resolves the well-known problem of the discrepancy between the values of b from different sources in simpler Weibull models.

TABLE 1

Maximum likelihood estimates of the scale (a), shape (b) and diameter exponent (δ) parameters. The values of the calculated ratio $-\delta/b$ and of the experimental slope are compared.

Material	a [GPa]	b	δ	Experimental slope	r^2	$-\delta/b$	n
PDA [9–11]	0.67	7.07	3.93	−0.61	0.79	−0.56	22
K-149 [16]	2.74*	8.70	13.93	−1.94	0.496	−1.60	50

Preliminary results on the elastic and fracture properties of several types of microcomposites, including hybrids, have recently been obtained. In Figure 1 the modulus of Kevlar 149/epoxy microcomposites made of 1 single fibre, and 8 single fibres, is seen to follow quite well the Halpin-Tsai (or rule-of-mixtures, RoM) equation, even at such low volume fraction. In Figure 2 the strength is shown to increase linearly according to the RoM.

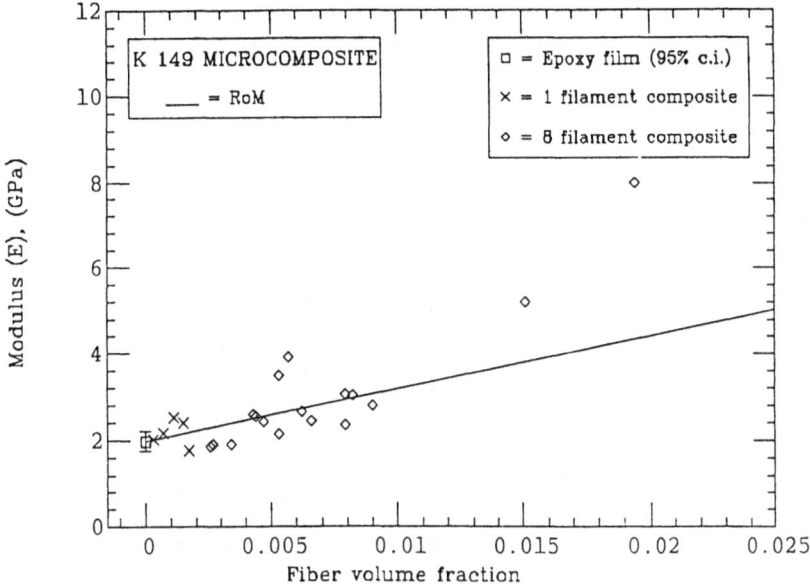

Figure 1. Young's modulus of Kevlar 149 microcomposites as a function of fibre content.

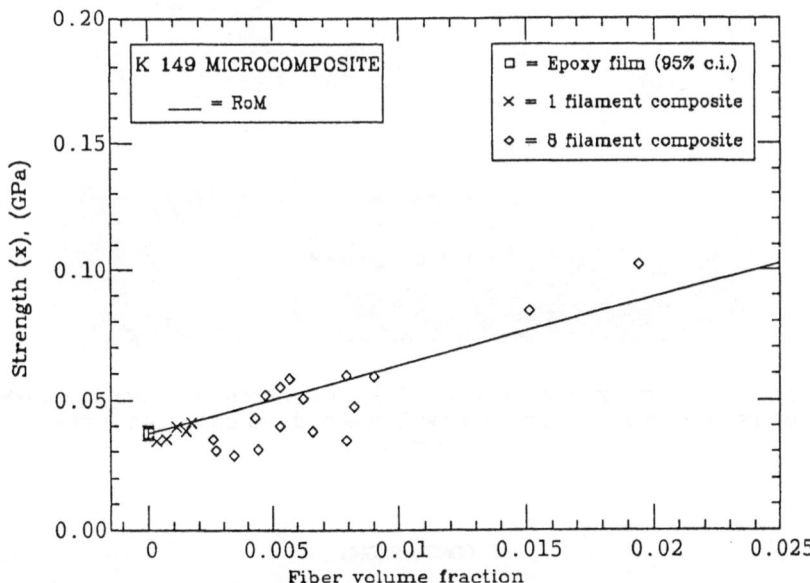

Figure 2. Tensile strength of Kevlar 149 microcomposites as a function of fibre content.

Figure 3 is a typical videomicrographic view of a microcomposite undergoing a progressive damage growth. These results are encouraging and point out that the quality of the manufactured microcomposites is quite high. The study of microcomposites via videomicrographic techniques is particularly relevant to the dynamics of mechanical fracture in such materials. Indeed, with respect to failure dynamics, the first findings of our research are as follows: (i) As predicted by several authors (see references in [12]), a critical number of neighbouring fibres must break before fast failure occurs. Based on a very limited number of preliminary tests, this number was found to be equal to 4 in E-glass/epoxy microcomposites. In Kevlar 149/epoxy microcomposites, the crack growth process is extremely fast (Figure 3), but some fibre integrity is conserved (fibre splitting rather than snapping occurs). (ii) In principle information on the type of load sharing rule in effect in a given fibre/matrix system system can be obtained. A fast-propagating damage array is observed to nucleate and grow in Kevlar 149/epoxy systems, at several places simultaneously. This may indicate that equal-load sharing is in effect in this system. (iii) In principle, the effect of changes in the chemical or physical nature of interfaces on the failure nucleation, growth, and criticality can be studied. This is now being pursued. (iv) The effect of fibre bunching (that is, the fibre-to-fibre interaction) and/or of volume fraction, as well as of fibre misalignment, on the mechanical and fracture properties can easily be assessed. This is also currently being studied.

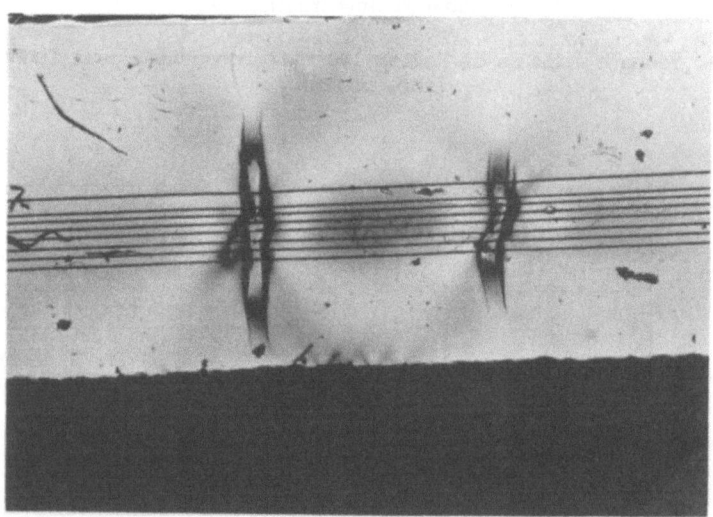

Figure 3. Fracture growth in Kevlar 149 microcomposites. Two independent sites are involved, the darker zone between these undergoes virtually no stress.

CONCLUSIONS

It has been shown that simple modifications of the classical Poisson/Weibull form for the strength of single fibres used in composites yield a new failure probability function which not only may be used to

255

adequately model strength-diameter effects, but also resolves in a satisfactory way some of the current problems inherent to the Weibull/weakest link model. A new technique was developed for the manufacture of single fibre reinforced polymeric films, or microcomposites, and preliminary strength tests were performed. The elongation and fracture process was followed by videomicrophotography. The combination of this technique with the use of simple composite models is a promising tool in the study of failure modes and mechanical efficiency in composites.

ACKNOWLEDGEMENTS

This work was supported in part by the National Council for Research and Development, Ministry of Science and Development, Israel. Thanks are due to A. S. Taylor for help with some of the experimental work. L. W. Steenbakkers is a Visiting Graduate Student from the Department of Chemical Engineering, Eindhoven University of Technology, The Netherlands. H. D. Wagner is the recipient of the J. and A. Laniado Career Development Chair.

REFERENCES

1. Phoenix, S. L., Composite Materials: Testing and Design (Third Conference), ASTM STP 546, 1974, pp. 130-151.

2. Smalley Jr., R. F., Turcotte, D. L., Solla, S. A., J. Geophys. Res. 90, B2, 1985, pp. 1894-1900.

3. Turcotte, D. L., Smalley Jr., R. F., Solla, S. A., Nature 313, 6004, 1985, pp. 671-672.

4. Kao, J. H. K., Computer Methods for Estimating Weibull Parameters in Reliability Studies, I. R. E. Trans. on Reliability and Quality Control, PGRQC No. 13, 1958, p. 15.

5. Williams, J. S., Biometrics 34 (2), 1978, pp. 209-222.

6. Hertzberg, R. W., Deformation and Fracture Mechanics of Engineering Materials, John Wiley and Sons, Inc., 1976.

7. Weibull, W., J. Appl. Mech. 73 1951, 293.

8. Mann, N. R., Schafer, R. E., Singpurwalla, N. D., Methods for Statistical Analysis of Reliability and Life Data, John Wiley and Sons, Inc., 1974, Chapter 9.

9. Galiotis, C., Young, R. J., Polymer 24, 1983, pp. 1023-1030.

10. Galiotis, C., Young, R. J., Batchelder, D. N., J. Mater. Sci. Letters 2, 1983, pp. 263-266.

11. Galiotis, C., Read, R. T., Yeung, P. H. J., Young, R. J., Chalmers, I. F., Bloor, D., J. Polym. Sci. (Polym. Phys.) 22, 1984, pp. 1589-1606.

12. Wagner, H. D., in Application of Fracture Mechanics to Composite Materials (R. B. Pipes, K. Friedrich, Eds.), Elsevier Sc. Publ. B. V., 1988 (to appear).

13. Wagner, H. D., Submitted, 1987.

14. Wagner, H. D., Phoenix, S. L., Schwartz, P., J. Comp. Mater. 18, 1984, pp. 312-338.

15. Wagner, H. D., Submitted, 1988.

16. Steenbakkers, L. W., Wagner, H. D., Submitted, 1988.

COMPOSITES WITHOUT CHEMISTRY
(Continuous fibre reinforced thermoplastics)

ir. Willem H.M. van Dreumel
Royal Nijverdal Ten Cate NV
Technical Division

THE INDUSTRY

The world of advanced fibre composites is changing rapidly. Technical and commercial battles take place in the composites arena, where big chemical companies have joined the show. The diagram shows a basic idea behind these involvements. The chemical industry thinks volume. Commodities and energy, products with a relatively low added value, have been their business over the past decades. Nowadays, however, High Tech is the message. Specialities with a high added value seem attractive. This is demonstrated by the amount of recent take overs of relatively small "high tech" composite companies by the chemical giants. What are the drivers behind this development. Why are advanced fibre composites so attractive.

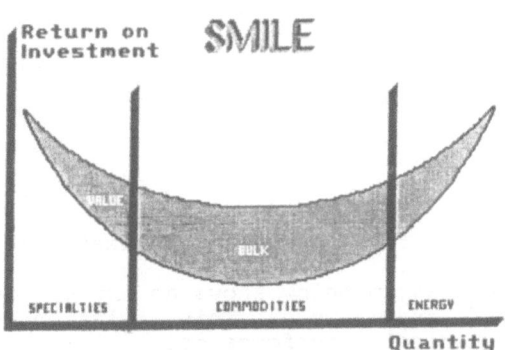

A rather common application of fibre reinforced polymers is car repair. This poor mans technology is translated to the "high tech" area. The most important difference is found in the addition of the word "advanced". For also in the aircraft industry composites is a sticky business. Prepregs are draped in moulds by hand, moreover, they are cured in expensive autoclaves.

Composites   Advanced Composites

Chemical industries are entering this small scale and highly competitive composite business. Most of them, however, realize that it is not enough to supply just resins or fibres. The pipe line to the end user is long and complex. The fibre supplier sells to the textile industry, while the resin supplier has the impregnator as customer. Weaving and impregnating is often integrated in one company. The basic knowledge of such companies is in textile technology. This textile technology is missing in the chemical industry, which explains the interest to aquire fibre processing companies. There is more, however. Another gap between raw materials and end users lies in the transformation of raw materials into parts. At the moment this step brings the highest added value.

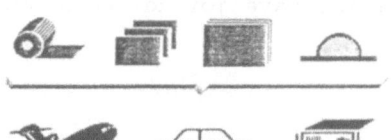

TEXTILE TECHNOLOGY

impregnating-cutting-consolidating-moulding

Therefore, part making seems to be the ultimate goal. Part making, in relation to the critical mass required for bigger companies, is only realistic if thermoset hand lay up is replaced by a more industrial manufacturing approach which has been made possible by a new generation of thermoplastics. In the volume required for successfull development of industrial composites and the price that can be paid for industrial applications, however, a contradiction is found.

The diagram below shows an indicative relation between polymer cost and volume. This relation is dictated by thermal and mechanical performance.

The low modulus cheap polypropylene is not an attractive composite matrix, because the expensive fibres will not be fully utilized. On the other hand, PEEK is not applicable in the automotive industy because of it's high price. In the long run, it seems obvious, that the chemical industry will reduce its interest in smaller specialised high tech market niches and concentrate on

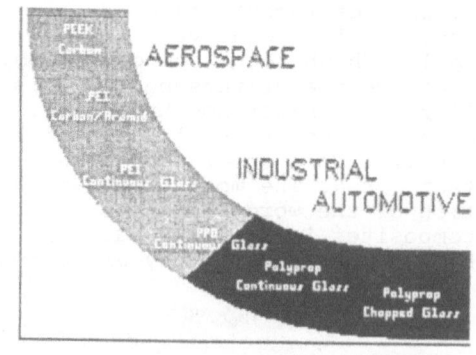

volumes that have to be created yet. One of the examples in this area is the automotive industry. The use of plastics in cars is increasing. A considerable part of a car is made of unreinforced plastic. Even lamp glasses and window elements are made of plastics nowadays. Short and random fibres are used in flow and injection moulding.

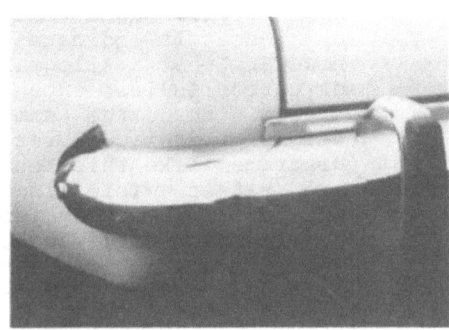 

For load carrying parts, continuous fibre reinforcement is necessary. Because of the relatively high volume of the automotive industry, this is the area that most likely will be covered by the bigger companies. Glass fibre, the less "high tech" material obviously will be the best candidate in this field because of the importance of cost.
In the next paragraph will be demonstrated, however, that for a certain type of applications carbon fibre performs better, also cost wise.

MATERIAL ASPECTS
Some present ideas about reinforcing fibres are stated below:

- Although expensive, carbon fibres are usefull for light weight structural applications.
- Aramid fibres are hardly used in load carrying composite applications (partly due to misunderstanding the nature of this attractive material)
- The intrigueing PE super fibre has no real opportunity to be used in structural composites because of its low service temperature and lack of mechanical properties in compression.
- In many applications glassfibre is a winner, also in aircraft applications, because of low cost.

The last statement is especially interesting because it reflects the idea of the majority of composite users. A look at material costs seems to prove this rule. It is interesting to analyse the development of prices over the last years. The figure on the next page shows this development.
In 1986, prices reached a stable level. If we look to 1988, however, a remarkable discontinuity is observed. The investment for a carbon fibre plant is relatively low. A certain minimum capacity, however, is required for efficiency reasons. Due to large overcapacity of the carbon fibre manufacturing facilities, prices have come down substancially. In contrast, the

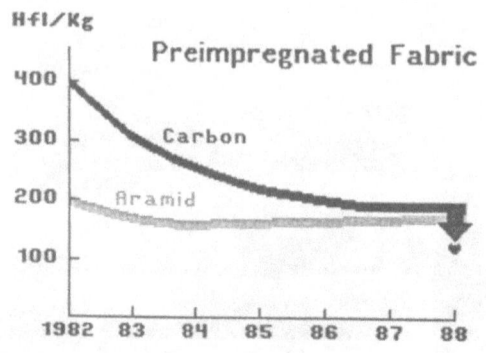

price of aramid fibres kept stable. The diagram only gives a rough idea because price dependence on Tex values has been averaged more or less. It indicates, however, that a kilogram of carbon is less expensivenow than the same weight of aramid fibres. Diagrams like this one are rather misleading, however, since they do not include performance. This phenomenon is best illustrated with a practical example.

Consider a profile made of fibre reinforced composites, which has to meet a certain bending stiffness. Glass and carbon are compared. Instead of comparing cost per unit weight, we should compare cost per unit bending stiffness. Bending stiffness is defined as Young's modulus multiplied by the moment of inertia of the profile. For high profiles, the moment of inertia is governed by laminate thickness. Due to the difference in Youngs modulus between glass and carbon, a glass laminate has to have twice the thickness of a corresponding carbon laminate. Moreover, glass is heavier because of the higher density of this fibre. Table I reflects the influence on the material cost of a product.

TABLE I

	Unidirectional composite					
	E-Modulus	Density	Price/kg (Hfl)	Thickness (equal E*t)	Weight	Price
Carbon	90 Gpa	1.5	140	100%	100%	100%
Glass	45 GPa	1.9	53	200%	250%	90%

Surprisingly, carbon is very attractive with respect to weight and price in applications dominated by bending stiffness. Add to it the value of weight reduction (which varies from 100 to 1000 Hfl/kg in the aircraft industry) and carbon is definitly less expensive.

The example above is based on unidirectional material. It does not imply, however, that UD-material gives the best result in profiles. On the contrary. The design of a profile in composites requires slightly more than just an indicative cost calculation. There are however few usefull guidelines for the design of composite parts. Practical software support for composite designers is especially not readily available. Building a practical source for designers should be a real challenge for scientists and students at technical universities.

In the development of continuous fibre reinforced composites, three generations can be distinguished.

Generation 1
Wet hand lay up is a widely used technology at the low end of the market. We saw this technology demonstrated in the first figure of this paper.

Generation 2
Thermoset prepreg materials are used in aerospace and sport article manufacturing. Prepregs have to be stored at low temperature and are cured under pressure at high temperature. Due to the chemical nature and the complex manufacturing methods, a large amount of waste (typically 30-40%) is produced. It is not surprising that the use of thermoset composites will not increase considerably in the civil aircraft structure as is indicated by the diagram.

Generation 3
The last generation of fibre reinforced composite materials was made possible by the creative chemical industry. High performance thermoplastic polymers like PEEK, PEI and PES were introduced. Polymers with excellent thermal performance.

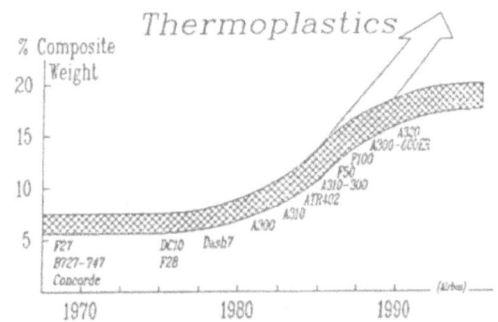

POLYETHERIMIDE
Like thermosets, continuous fibre reinforced thermoplastics are also available as preimpregnated material. In contrast to thermosets, however, they have an infinite shelf life even at higher temperatures. PEI is a thermoplastic polymer that shows very attractive thermal, mechanical and flammability behaviour. Due to its amorphous character, PEI is easy to process. Over the last years, an extensive research program has been carried out with the aircraft industry into the properties of a PEI based continuous fibre reinforced thermoplastic PEI of Ten Cate. In cooperation with MBB, Westland Helicopters and Hispano Suiza, a wide range of properties has been established.

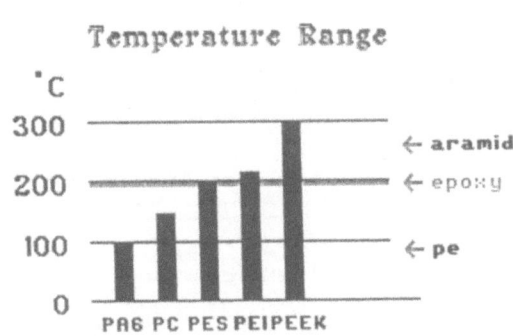

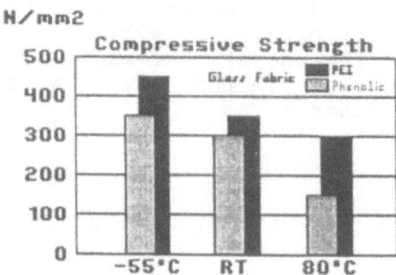

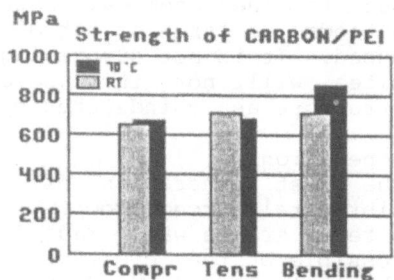

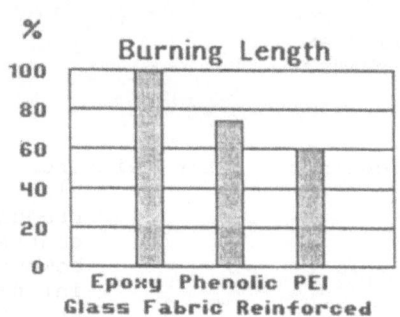

Since PEI fulfils the new requirements for aircraft interiors with respect to flammability, smoke and toxicity, a range of sandwich panel configurations made of this material has been tested. It passes the requirements for heat release according the Ohio State University test method (OSU).

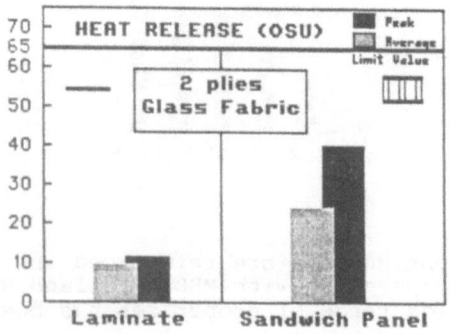

PART MANUFACTURING

In contrast to fibre reinforced thermoset systems, thermoplastics are easy to process. By consolidating a stack of prepregs, a laminate is formed which can be regarded as a sheet material like aluminium and steel. Organic sheet is a characterization which is sometimes found in the literature.

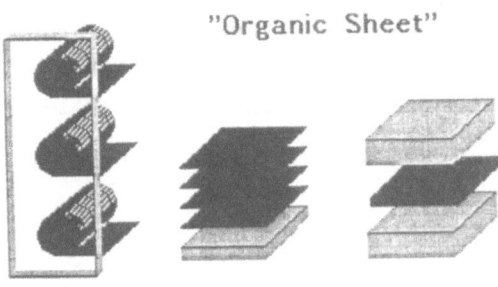

"Organic Sheet"

Prepreg   Stacking   Consolidating

A sheet is heated, transported to a press and moulded into shape. The possibility of production rates as short as three minutes has been demonstrated. Due to the absence of chemistry, the percentage of waste material is low. Moreover, the moulding cycle can be repeated several times, since recycling is a typical thermoplastic characteristic.

Filament winding is an intrigueing technology. The use of yarns
impregnated with thermoplastic makes on mould consolidation
possible with a concentrated heat source. Although the use of
impregnated fabrics and unidirectionals will cover most of the
applications, impregnated yarns can be useful if a drapeable
fabric or braiding is needed. New impregnation technologies
produce impregnated yarns that have sufficient flexibility for
a weaving or braiding process. Combined with less expensive
resin systems like polycarbonate and PPO, impregnated yarns are
highly attractive for sport articles and general applications.

Continuing development by the composites industry will lead to
thermoplastic systems that are capable of performing under
severe conditions with respect to stress, temperature, chemical
resistance and environmental exposure.

Materials for the nineties:

COMPOSITES
without
CHEMISTRY

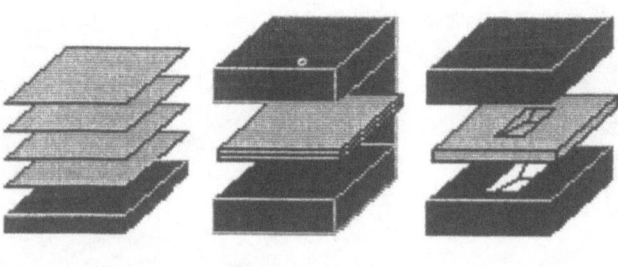

Stacking   Heating   Moulding

CURING OF EPOXY MATRICES

KAREL DUŠEK
Institute of Macromolecular Chemistry,
Czechoslovak Academy of Sciences,
162 06 Prague, Czechoslovakia

ABSTRACT

Special features of curing of epoxy resins derived from N,N-diglycidyl-amines and diffusion control of curing due to increase in T_g are discussed. The special features of amine curing of diglycidylamines include extensive formation of small rings and dependence of reactivities of epoxy groups. The relative intensity of etherification depends on curing temperature. Diffusion control of curing sets in when T_g approaches T_{cure}. The decrease of the rate constant is relatively steep. The decrease of segmental mobility and physical aging have been taken into account in the theoretical approach.

INTRODUCTION

Matrices for high-performance epoxy composites should exhibit a sufficiently high modulus, strength, temperature resistance and toughness. Also the flow properties during cure should be programmable in order to optimize the curing conditions.

The required temperature resistance depends on the application conditions. For the lower temperature regions, the classical systems based on diglycidyl ether of Bisphenol A (DGEBA) cured with aromatic diamines like 4,4'-diaminodiphenylmethane (DDM) or 4,4'-diaminodiphenylsulphone (DDS) and systems containing dicyandiamide have been used. To increase the bond density and the temperature resistance, polyfunctional epoxy resins have been introduced. (N,N,N',N'-tetraglycidyl-4,4'-diaminodiphenylmethane (TGDDM) is an example.) The glass transition temperature, T_g, of these systems may increase far above 200°C.

The changes of the flow properties during curing (chemorheology) are determined by temperature and structural changes, i.e. by the increase in molecular weight, occurence of the gel point and changes in the gel fraction and crosslinking density.

Toughness is another important property of the matrix and composite. The toughness of the matrix of the neat resin is not very good and an improvement is desirable. In commercial prepregs the toughness was improved by inclusion of a phase-separated rubber. The rubber is a liquid low-molecular-weight elastomer with reactive end groups which is soluble in the uncured epoxy resins but becomes incompatible during curing as a result of increasing molecular weight of the epoxy-curing agent reaction products. Important is the size of the separated phase [1]. Newer developments proceed along the following lines: directed formation and shaping of the (phase separated) elastomer phase (on matrix-fiber interface or as interlayers) and changing the structure of one of the components in order to make the cured one-phase or two-phase resin tougher. Modification of the epoxy monomer in such a way that the epoxy groups are separated by a tough oligomer chain (like a polysulfon) is one of the ways to achieve this goal [2,3]. Preparation of semi-interpenetrating networks using a soluble tough oligomer is another option [4]. Formation of a new elastomer or tough polymer phase is a necessary condition for increasing the toughness. Phase separation must occur at a certain conversion: i.e., not at the very beginning of cure, where it may cause a macroscopic phase separation, and not when the network structure is so dense that it either entirely prevents the formation of a new phase or limits the size of the separated phase far below the level where it is still effective. To understand and predict the phase separation, one has to learn the structure growth in these systems and the thermodynamic interactions of mixture of the formed polymer.

This demonstrates the importance of the understanding of the network formation which determines both the chemical (topological) and physical structure of the resulting matrices. In this contribution, the special features of the network build-up from N,N-diglycidylamines will be reviewed. Also, attention will be paid to diffusion control of curing caused by the increase of T_g.

SPECIAL FEATURES OF CURING OF EPOXY RESINS DERIVED FROM N,N-DIGLYCIDYLAMINES

Network Build-Up in Amine Curing of Diglycidyl Ether of Bisphenol A (DGEBA)
It has been shown (cf.e.g. [5,6] that curing of epoxy resins with diamines can be well described by statistical branching theories taking into account different reactivity of hydrogens of the primary and secondary amino group

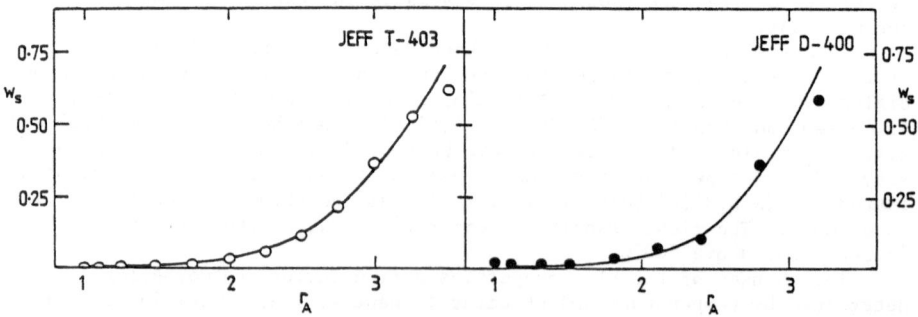

Figure 1. Solf fraction, w_s, of DGEBA-poly(oxypropylene)polyamine networks in dependence on the initial molar ratio of amine hydrogens to epoxy groups, r_A. a) Jeffamine [R] T-403, b) Jeffamine D-400; ——— calculated by branching theory (Ref. [6]).

in the addition to the epoxy group. Two important features characterize the curing: (1) the reactivity of epoxy groups in DGEBA is independent and (b) cyclization is usually very weak due to the relative stiffness of the DGEBA bridge. (However, non-negligible cyclization has been found in curing with polyoxypropylene polyamines [6]. Good agreement between theoretical and experimental gel points, molecular weights, sol fractions and concentration of elastically active network chains (EANC) has been obtained for a number of DGEBA-diamine systems. Figure 1 shows some newer data on crosslinking of DGEBA with polyoxypropylene diamines and triamines (Jeffamines®).

In many commercial systems, etherification accompanying the amine-epoxy addition is an important curing reaction. It is released (initiated) by the OH group formed in the primary addition step.

The experimental results obtained on model systems – phenylglycidyl ether and an OH-containing adduct have shown that a strong chain transfer interferes. The products are of relatively low molecular weight and their number is larger than the number of reacted OH groups. Using HPLC, the first two addition products I and II have been identified as well as an appreciable amount of the product III which can be formed by transfer reactions:

$$R_1OCH_2\underset{\underset{OH}{|}}{CH}CH_2NR_2R_3 \quad + R_1OCH_2\underset{\overset{\diagdown O \diagup}{CHCH_2}}{} \longrightarrow \quad R_1OCH_2\underset{\underset{O}{|}}{CH}CH_2NR_2R_3$$

(initiator)   (epoxide, E)

$$\underset{\underset{OH}{|}}{CH_2}CHCH_2OR_1$$

(I)

$$+ E$$

$$R_1OCH_2\underset{\underset{OCH_2CHCH_2OR_1}{|}}{CHCH_2NR_2R_3}$$
$$\underset{\underset{OH}{|}}{OCH_2CHCH_2OR_1}$$

(II)

$$R_1OCH_2\underset{\underset{OH}{|}}{CH}CH_2OR_1$$

(III)

Only small quantities of polyethers of higher molecular weight than that of II are contained in the system even at high conversions of epoxy groups. The transfer reactions are accompanied by formation of C:C double bonds [7]. The critical conversion at the gel point in systems with epoxy groups in excess is higher than that predicted for polyetherification in the absence of transfer reactions [5].

Reaction mechanism and network build-up for curing of N,N-diglycidyl-aniline (DGA) and N,N,N',N'-tetraglycidyl-4,4'-diaminodiphenylmethane (TGDDM): The main features of the reaction of N,N-diglycidylamine derivatives with amines are [10]:

a) formation of small cycles,

b) dependent reactivity of glycidyl groups.

Six- to eight-membered rings are formed by intramolecular reaction of DGA or TGDDM or their reaction products with amines. It has been found that at temperatures $\geq 150°C$ compounds with quinoline ring are formed [8,9]

(IV)

So far, these quinoline rings have been found in reaction products of N-alkyl-N-glycidylanilines. It is clear, however, that in DGA two such rings can exist and consequently in TGDDM there may be found up to four such rings.

The rings in reaction products of N,N-diglycidylamines with amines involve [10] an eight-membered ring V formed from the adduct of diglycidyl-amine with primary amine [11]

(V)

and six and seven-membered rings formed by intramolecular etherification

(VI)   (VII)

Depending on the reaction conditions, up to 60-70% of formed bonds may be ring closing. This is a very important feature for this family of epoxy resins. Thus, the crosslinking density may not be much higher than in DGEBA formulation, but the bond density is higher and so is T_g.

The second feature is the dependence of reactivities of glycidyl groups. Denoting by E_i (E_0, E_1, E_2) and A_i (A_0, A_1, A_2) the respective diepoxide and amine units with i reacted groups (amine hydrogens) one can describe the acyclic reaction steps by the scheme

$$
\begin{aligned}
E_0 + A_0 &\xrightarrow{k1} E_1 + A_1 \\
E_0 + A_1 &\xrightarrow{k2} E_1 + A_2 \\
E_1 + A_0 &\xrightarrow{k3} E_2 + A_1 \\
E_1 + A_1 &\xrightarrow{k4} E_2 + A_2
\end{aligned}
\tag{1}
$$

It is known that there exists a substitution effect within the amino group which is characterized by the ratios of rate constants k_2/k_1 and k_4/k_3. The substitution effect within the diepoxide is characterized by the ratios k_3/k_1 and k_4/k_2. If these substitution effects are mutually independent, $3k_2/4k_1 = k_4/2k_3$ and $2k_3/4k_1 = k_4/3k_2$.

So far, it has been found that the substitution effect in DGA is positive since $k_4/3k_2$ is considerably higher than unity for uncatalyzed reaction as well as for the reaction catalyzed by OH groups [11]. The observed positive substitution effect is explained by the intramolecular catalysis of the epoxy-amine addition by the OH group in the same molecule. However, the independence of the substitution effects is not operative.

It is to be remarked that the analysis of the reaction product and evaluation of the reaction kinetics is complicated by the existence of separable stereoisomers of DGA and reaction products in which the reactivity of functional groups is not the same [9,12].

Cyclization and substitution effects influence network formation. In contrast to systems with DGEBA, the gel conversions or critical molar ratios exhibit a strong dependence on dilution [13] which is typical for ring formation. Not only the gel point itself, but also network formation beyond the gel point is affected. Figure 2 shows that the dependences of the sol fraction on the molar ratio of amine hydrogens to epoxy groups, r_A, are very dissimilar for DGA and DGEBA systems. It can be well described by branching theory taking into account the substitution effect in the amine for DGEBA but not for DGA systems [14].

Although N,N-diglycidylamine systems are more inclined to etherification, at low curing temperatures cyclization is weak. Figure 3 shows the typical dependence of the sol fraction and reduced equilibrium modulus of DGA-DDM system cured at 110°C, where etherification was negligible [15]. However, the fact that the dependence is asymmetrical points again to the specificities of curing of N,N-diglycidylamines. At elevated temperatures (150°C), etherification becomes important. Figure 4 shows that the gel fraction increases with increasing conversion of excess epoxy groups into

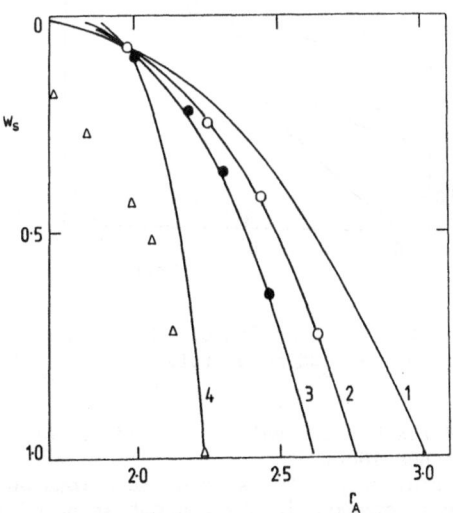

Figure 2. Dependence of the sol fraction on r_A. —— theoretical curves calculated for the substitution effect in amine given by k_2/k_1 (Scheme I): 0.5 (1), 0.32 (2), 0.22 (3), 0.05 (4); experimental data: DGEBA-4-4'-diaminodiphenylmethane (DDM), DGEBA-1,8-diaminooctane, DGEBA-DGA (Ref. [14]).

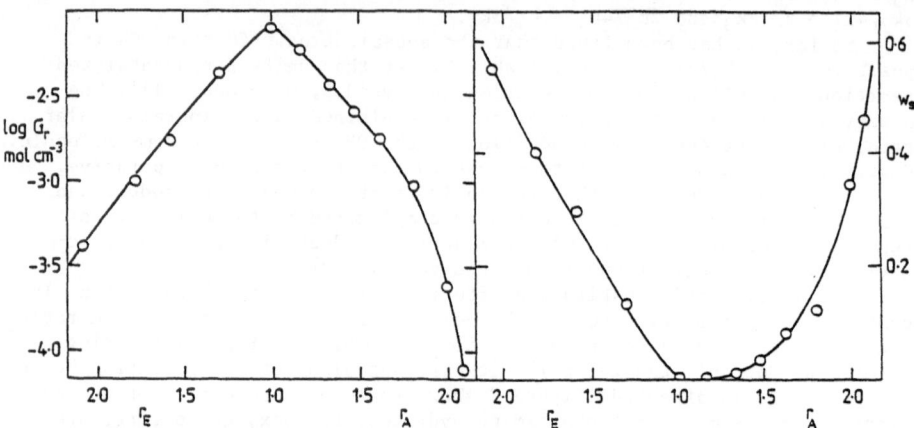

Figure 3. Dependence of the sol fraction and reduced equilibrium modulus, G_r, of DGA-DDM systems cured at 110°C on the ratios r_A and $r_E=1/r_A$ (etherification negligible, Ref. [15]).

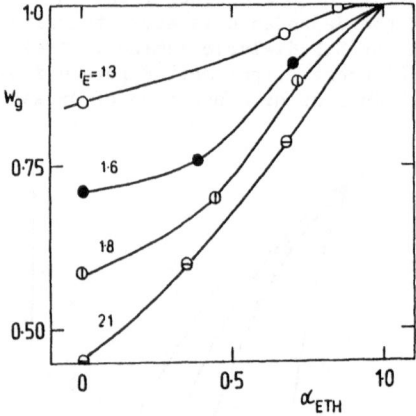

Figure 4. Dependence of the sol fraction of DGA-DDM systems on etherification conversion for various molar ratios $r_E=1/r_A$, Ref. [15].

ether bonds, α_{ETH}, and reaches the value 1 regardless of the initial molar ratio when α_{ETH} approaches 100%.

At even higher curing temperatures, side reactions become important. They are manifested by a decrease in the concentration of OH groups and increase in the concentration of secondary amine groups [15,16]. Thus, high temperature curing of N,N-diglycidylamines, e.g. TGDDM, is rather complex not only because of complicated kinetics and mechanism of the addition reaction itself but also because of a number of side reactions.

DIFFUSION CONTROL OF CURING DUE TO INCREASE IN T_g

The reaction rate and changes in rheological properties during curing are very important for processing of composite materials. The rheological properties are determined by structural changes (increase of molecular weight, gel point, changes in network structure) and a considerable increase in T_g typical for many thermosets. The problem of curing and vitrification has been addressed in terms of TTT (time-temperature trans-formation) diagrams [17].

Let us consider here the effect of segmental mobility on curing rate. The Eiring theory of reaction rates, which considers formation and decom-position of an activated complex, can deal with the simultaneously occuring chemical kinetics and diffusion control characterized, respectively, by some rate constants k_c and k_d. Under certain conditions (prevailing chemi-cal kinetics or diffusion control), the apparent reaction rate constant, k_{app}, can be expressed via the Rabinowitch equation

$$1/k_{app} = 1/k_c + 1/k_d \qquad (2)$$

The rate constant for chemically controlled reaction, k_c, is determined by the reaction mechanism (mass-action law) and k_d is proportional to the dif-fusion coefficient of the segmental motion. While the temperature depen-dence of k_c is given by the Arrhenius equation (log $k_c \propto 1/T$), the segmental diffusion is determined by the distance of the curing temperature from T_g of the reacting system. Using the free-volume or Gibbs-DiMarzio theory of glass transition, one finds that the diffusion coefficient (and conse-quently k_d) falls down to zero when $T_{cure} = T_g - T'$; experiments show that

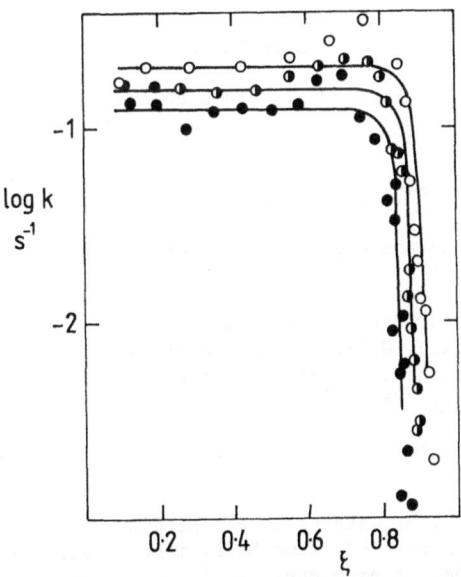

Figure 5. Calculated and experimental rate constant for curing of DGEBA with 1,3-diaminopropane as a function of conversion. Arrows indicate conversion at which T_g was reached; T_{cure}: ● 50°C, ◑ 60°C, ○ 70°C (Ref. [18]).

for majority of polymers T' is of the order of 50 K. In order to predict variations of the curing rate with conversion, one has to know also the dependence of T_g on conversion. A comparison of the theory with experimentally measured curing rate of DGEBA - 1,3-diaminopropane systems (Fig. 5) shows that a relatively steep fall of the rate constant begins in the vicinity of T_g [18].

However, in the theoretical approach described above [18] physical aging of the glassy state has not been taken into account although it plays a role when the curing rate is slow. Then, the diffusion-controlled rate constant k_d is not only a function of T and conversion, ξ, but also of time, i.e., $k_d(T, \xi, t)$, on the other hand, k_c is only a function of temperature. Using the Adam-Gibbs theory of cooperative rearrangements, one can express the diffusion coefficient as a function of configurational entropy S_c. Thus

$$k_d = k_{do} \exp(-B/TS_c) \tag{3}$$

where k_{do} and B are constants. In treating the effect of physical aging on curing rate, it has been assumed that the time change of S_c is given by [19]

$$dS_c(t)/dt = -W[S_c(t) - S_c(T,\xi)] \tag{4}$$

where $S_c(T,\xi)$ represents the equilibrium $(t \to \infty)$ value of S_c. The "rate" constant W is again proportional to the diffusion coefficient and depends on S_c in a similar way as k_d does. Consideration of physical aging in the theory of diffusion-controlled reaction rate leads to an improved agreement between the theory and experiment.

REFERENCES

1. Siebert, A.R., Elastomer-modified epoxies and their utility in composites. Makromol. Chem., Makromol. Symp., 1987, 7, 83.

2. Bauer, R.S., New, high performance epoxy resins for hot/wet environments. Polym. Preprints, 1987, 28 (1), 33.

3. Multiphase epoxy thermosets having rubber within disperse phase. US Patent 4,680,076, 1987.

4. Setton, N.S., McGrail, P.T., Peacock, J.A., Wilkinson, S.P., Crick, J.A., Davies, M. and Almen, G. Semi-interpenetrating polymer networks as a route to toughening of epoxy resin matrix composites. 19th Internat. SAMPE Techn. Conf., 1987, 700.

5. Dušek, K., Network formation in curing of epoxy resins. Adv. Polym. Sci. 1986, 78, 1.

6. Dušek, K., Ilavský, M., Štokrová, S., Matějka, L. and Luňák, S., Formation and structure of networks from poly(oxypropylene) polyamines and diglycidylether of Bisphenol A. II. Reactivity of amino groups, gelation, sol fraction and equilibrium modulus. In Cross-linked Epoxies, ed. B. Sedláček and J. Kahovec. Walter de Gruyter, W. Berlin, 1987, 279.

7. Fedtke, M.,Acceleration mechanism in curing reactions involving model systems. Makromol. Chem., Macromol. Symp. 1987, 7, 153.

8. Reyx, D. and Costes, B.,Homopolymerisation of tetraglycidyl-4,4'-diaminodiphenylmethane (TGDDM). Identification of cyclic reaction products. Polymer Commun., in press.

9. Attias, A.J., Ancelle, J., Bloch, B., Laupretre, F.,Chemical structure of networks resulting from curing of diglycidylamine-type resins with aromatic amines. 1. Detection and characterization of cyclization reactions on model compounds. Polym. Bull. 1987, 18, 217.

10. Dušek, K., and Matějka L.,Amine curing of epoxy resins derived from N,N-diglycidylamines. Proc. ACS Div. Polym. Mater. Sci. Eng. 1987, 56, 356.

11. Matějka, L., Tkaczyk, M., Pokorný, S. and Dušek, K.,Cyclization in the reaction between diglycidylamine and amine. Polym. Bull. 1986, 15, 389.

12. Doskočilová, D., Matějka, L., Pokorný, S., Březina, M., Štokr, J., Dobáš, I. and Dušek, K.,Curing of epoxy resins: Configurational structure and reactivity of stereoisomers in the model reaction of diglycidylaniline with N-methylaniline. Polym. Bull. 1985, 14, 123.

13. Matějka, L., Dušek, K. and Dobáš, I., Curing of epoxy resins with amines. Gelation of polyepoxides derived from diglycidylaniline. Polym. Bull., 1985, 14, 309.

14. Dušek, K. and Matějka L.,Advances in network formation and rheology during curing of epoxy resin systems including reactive liquid elastomers. Polym. Mat. Sci. Eng. 1987, 57, 765.

15. Zelenka, J., Structure and viscoelastic behaviour of cured nitrogen-containing epoxy resins. Ph.D. Thesis, Institute of Macromolecular Chemistry, Prague, 1988.

16. Morgan, J.R., Structure-properties relations of epoxies used as composite matrices. Adv. Polym. Sci., 1985, 72, 1.

17. Aronhime, M.T. and Gillham, J.K., The time-temperature transformation (T.T.T.) cure diagram of thermosetting polymeric systems. Adv. Polym. Sci., 1986, 78, 83.

18. Havlíček, I., Dušek, K., Štokrová, Š., Biroš, J., Kenny, J. and Nicolais, J., Theoretical modelling of the rheological behaviour of an epoxy-amine system during curing. Proc. Europ. Symp. Polym. Mat.-Chem. Aspects in Processing Operations and Use of Polymeric Mater., 1987, prepr.

19. Havlíček, I., Bulatov, V.V. and Dušek, K., The influence of diffusion control and structural relaxation on the kinetics of curing of epoxy-amine systems. Proc. Conf. Mechanics and Technology of Composite Materials. Varna 1988.

A NOVEL APPROACH TO REACTION INJECTION MOULDING TECHNOLOGY

AREND NOORDAM, Hans R. de Groot and Jan E. Stamhuis
Koninklijke/Shell-Laboratorium, Amsterdam
(Shell Research B.V.)
Badhuisweg 3, 1031 CM Amsterdam, The Netherlands

ABSTRACT

The present paper describes a novel proprietary reaction injection moulding technology, developed by Shell Research B.V.. The technology makes possible the high-rate manufacture of high-performance fibre-reinforced EPIKOTE[*] epoxy resin based composites that combine a high glass loading with a low void content. This is accomplished by means of a semi-permeable open-cell polyurethane-foam mould seal.

Experimental results show that composites with a glass-fibre content of up to 73 %(mass/mass) and a void content of only 1 %v can readily be made. The cycle time of the moulding operation is less than five minutes, which intrinsically renders this process very attractive for automotive applications.

INTRODUCTION

Reaction injection moulding (RIM) is a specialised form of injection moulding where the two highly reactive reactant streams (resin and curing agent) are brought together with high-pressure impingement mixing and then enter directly into the mould. The relatively high mould temperature in combination with the system exotherm allows very rapid cure. Therefore, the RIM process overcomes the problems of combining high reactivity with poor pot life or shelf life, and is intrinsically very attractive for the high-rate production of composite components based on epoxy resin matrices for the automotive industry.

High performance composites for this industry must have continuous reinforcement. This contrasts with the chopped fibres used in reinforced RIM (known as RRIM) [1], where short fibres are injected into the moulding via a slurry in one (or both) of the reactant streams. The use of a contin-uous reinforcement necessitates the placement of the reinforcing package (preform) in the mould prior to dispensing the epoxy resin system. This

---

[*]EPIKOTE is a Shell Trade Mark.

approach poses a large number of problems, especially with regard to fibre placement, loading, impregnation and resin flow, which have hampered the commercialization of epoxy resins in RIM so far.

An additional problem is the presence of air in the mould; this air has to be removed before the epoxy resin system gels to ensure that the resultant composite mouldings have a low void content. Various ways to overcome these difficulties have been reported such as evacuation of the mould, injection via a manifold, and the use of overflow bellows and back pressure [2,3], but none has met with commercial success.

The novel technique developed by us [4], makes use of a semi–permeable open–cell polyurethane–foam mould seal. During the moulding operation such a seal allows the air to escape from the mould initially; as soon as the resin system impregnates the foam, however, the seal becomes air–tight. Injection a small excess of resin system into the mould leads to some pressure being generated during gelation. Both the fact that the air can flow from the mould and the fact that gelation takes place under pressure contribute to a low void content.

EXPERIMENTAL

RIM machine
All composites mentioned in this paper were made using a Krauss–Maffei RIM star 16/40 dispensing unit, fitted with a MK 10–2k impingement mixing head.

During all experiments the resin and the curing agent were kept under a nitrogen pressure of 1.5 bar. The impingement mixing pressure was 50 bar for both components and the output was set at a total of 6 kg/min.

Moulds
Two types of moulds were used.

1. An aluminium, oil–heated plaque mould (cavity dimensions 50x35x0.6 cm), with the injection point in the centre of the long edge and with a full–width film gate.

2. A bowl–shaped mould, the mould cavity 40 cm in diameter and 12 cm deep. The width of the cavity was 6 mm and the mixing head was mounted centrally on the bottom female part of the mould.

Both moulds were placed in a press with a closing force of 80 tonnes. For each trial the mould was treated with Dow Corning silicone release agent.

Preforms
For the plaque mould, performs were made by dipping the woven cloth in a 5 %(m/m) solution of the resin system in an organic solvent, followed by drying in air and curing in the mould for 10 minutes at 120 °C.

The preforms for the bowl–shaped mould were made with the required number of layers of fabric or mat. The fabrics were formed manually on an auxiliary tool, identical with the male part of the mould. When the desired number of plies was reached the preform was stabilized by stitching with Aramid fibres.

Preforms made from continuous random mat were shaped by pressing the required number of plies into the hot mould and allowing to cool. Finally the edges were trimmed.

Semi-permeable foam mould seal

The open-cell polyurethane foam for the seal had a density of 25 g/l. The foam was cut to shape, positioned between the mating surfaces and then compressed by a factor of about 35.

EPIKOTE epoxy resin system

The resin system is based on two components: namely EPIKOTE epoxy resin, code name 1153/172 A, and a curing agent code name 1153/172 B.

Component A was kept at a temperature of 80 $^\circ$C (viscosity about 100 mPa.s) and component B at ambient temperature (viscosity about 30 mPa.s). The mixing ratio of A to B was 4:1 (by mass). The time to string in a hot-plate gel time experiment at 150 $^\circ$C amounted to 60 s and the time to no string 67 s.

Moulding procedure

After the mould had been loaded with reinforcement, the foam seal was put in place and the mould, which had been set at a temperature of 150 $^\circ$C, was closed.

The filling time, depending on the glass loading and the type of mould, was of the order of 8 s. A slight excess of resin system (5 %) was injected to ensure pressure (about 15 bar) during gelation. The pressure was regulated by ending the injection period when the two mould parts moved apart some 0.2 mm (indicated by a micrometer fitted to the mould).

Demoulding generally took place after 4 min.

RESULTS AND DISCUSSION

Some illustrative results of the experiments with the plaque mould and the resultant mechanical properties are given in Table 1. The properties quoted were obtained after a post-cure of 2 h at 150 $^\circ$C.

TABLE 1
RIM trials with plaque mould

Exp. No.	Reinforce-ment*	Glass content, %(m/m)	Void content, %v	Flexural modulus**		Flexural strength**		Flexural strain at break, %
				GPa	95 % CL	MPA	95 % CL	
1	14 M	42	1.2	12.9	0.2	249	22	2.1
2	16 M	50	0.9	14.8	0.3	280	13	2.2
3	18 M	53	1.2	15.4	1.3	274	7	2.1
4	20 M	58	0.8	17.8	0.3	289	24	1.8
5	14 D	53	2.6	15.8	0.3	394	33	2.8

* Number of plies and type of reinforcement. M — Marglass, type 138; D—Diagonap type 04.20 x 2; M and D are both woven cloth
** Values are mean values from 5 experiments. The 95 % confidence limits (CLs) are based on these results

Experiments 1-4 clearly show that a high glass content can be achieved whilst a relatively low void content (~ 1 %v) is maintained. This is only possible with the use of the foam seal, since comparable experiments without such a seal, or experiments in which the seal ruptured, resulted in composites with void contents ranging from 5 to as high as 12 %v.

An increase in the flexural stiffness is not reflected in an increased flexural strength.

Experiment No. 5 was carried out with Diagonap, 04.20 x 2 glass. The mechanical flexural properties, when measured in the usual way, show a relatively high strength of about 400 MPa, which is in the same order of magnitude as found for fabric-reinforced laminates obtained via compression moulding of prepregs. Standard polyester and vinyl-ester sheet-moulding compounds typically show flexural strengths of about 100 and 150 MPa, respectively. The relatively high void content (2 %v) of experiment No. 5 is probably caused by the fact that the much heavier Diagonap fabric (compared to Marglass) restricts resin flow to some extent.

Results obtained with the bowl-shaped mould are given in Table 2. The mechanical properties were determined after a post-cure of 2 h at 150 °C. Table 2 shows that composite components with a glass content as high as 73 %(m/m) have been made in combination with a low void content.

Furthermore, Table 2 shows that a direct correlation exists between the fibre content and the flexural stiffness, whereas the flexural strength is invariably about 350 MPa.

The inter-laminar shear strength is higher for the Interglas fabric and for the Vetrotex mat than for both types of Diagonap fabrics (at constant void content). Of these last two fabrics, type 04.40 x 2 shows the lower inter-laminar shear strength, although both types contain the same surface coating. This discrepancy is not fully understood. However, it

TABLE 2
RIM trials with bowl-shaped mould

Exp. No.	Reinforcement	Glass content, %(m/m)	Void content, %v	Inter-laminar shear strength, MPa	Flexural modulus, GPa	Flexural strength, MPa	Strain, %
1	Diagonap, type 04.20 x 2, UD±45°	56	2.3	33	11.8	326	3.4
2		61	0.4	35	13.7	349	3.1
3		65	1.7	31	-	-	-
4	Diagonap, type 04.40 x 2, UD±45°	73	0.7	27	15.5	302	2.7
5		73	1.5	24	15.7	346	2.1
6	Interglas, type 92125, satin weave	57	0.9	> 37*	13.7	370	3.1
7		65	2.2	-	15.5	348	2.6
8	Uniflo, type 750 continuous random	55	-	> 33*	12.2	334	3.4
9		59	-	> 41*	14.0	377	3.3

* Specimen failed in the tension mode.

should be noted that type 04.20 x 2 was moulded at 120 °C as opposed to 150 °C for the other type. Moreover type 04.40 x 2 had to be cut radially to make the manufacture of the preform possible.

It is striking that the continuous random-mat-based moulding shows essentially the same level of properties as those based on fabrics, even though these random fibres contain some 6-8 %(m/m) of a thermoplastic fibre binder. This implies that there is no reason to use fabric-based preforms in a non-oriented fibre lay-up pattern, since continuous random-based preforms are far easier to make and result in composites with comparable properties. This finding is very important for automotive applications, as it increases the likelihood of an automated process for making preforms being developed.

CONCLUSIONS

- Reaction injection moulding with EPIKOTE epoxy resin systems to produce high-performance fibre-reinforced composites at a high rate is quite feasible.

- The use of semi-permeable open-cell polyurethane-foam mould seals enables composites with a void content as low as 1 %v and a fibre loading of 73 %(m/m) to be made.

- For a high production rate of continuous fibre-reinforced RIM composites, preform manufacture must be automated. This technology is not, as yet, available.

REFERENCES

1. Liedtke, H.W., Journal of Cellular Plastics, March/April (1978) 102.

2. de la Mare, H.E., Brownscombe, T.F., Gottenberg, W.G. and Overcashier R.H., "High Modulus Epoxy RIM Systems", paper presented at the SAE Congress and Exposition, Cobo Hall, Detroit (U.S.A.), February 25-29, 1980.

3. Farris, R.D., de la Mare, H.E., Overcashier, R.H. and Gottenberg, W.G., "Structural Parts from Epoxy RIM Using Preplaced Reinforcement", paper presented at the 37th Annual Conference, Reinforced Plastics/Composites Institute (SPI), January 11-15, 1982.

4. de Groot, H.R., Wintraecken, J.J.M.H. and Noordam, A., Patent application, "Process for the preparation of fibre-reinforced composites", K 724 G.B. (1986).

279

NOVEL HOMOGENEOUS X-RAY CONTRAST POLYMER SALT COMPOSITES

YADOLLAH DELAVIZ, ZHAO XI ZHANG, ISRAEL CABASSO and JOHANNES SMID
Polymer Research Institute, Chemistry Department,
College of Environmental Science and Forestry
State University of New York, Syracuse, NY 13210, USA

ABSTRACT

Present X-ray contrast plastics are generally heterogeneous materials. This paper describes composites in which radiopacifying additives such as barium bromide dihydrate and bismuth bromide are solubilized in acrylic resins at the molecular level.

INTRODUCTION

The increased use of plastics in dentistry and medical implants calls for materials that can be detected and analyzed by rapid, nondestructive methods [1,2]. For example, ingested dentures or restorative dental materials made from radiolucent plastics cannot be easily detected· by common radiographic diagnostic techniques. The difficulty to detect plastic fireweapons by current screening devices could be cited as another reason for the need of X-ray contrast or radiopaque resins.

Most available radiopaque plastics contain certain glasses or heavy metal salts embedded in the polymer matrix [2-6]. However, most additives are incompatible with the resins, and this can adversely affect their properties. Halogenated polymers have also been used [7]. (e.g., iodated hydrogel particles in endovascular embolization [8].) In our approach we have researched polymer-salt composites where through chelation with the polymer the salt is solubilized and homogeneously distributed through the polymer matrix at the molecular level [9,10]. We briefly discuss here results obtained with barium bromide dihydrate and bismuth bromide added as radiopacifying agents to acrylic resins.

MATERIALS AND METHODS

Polymer composites were either cast from polymer-salt solutions (from ethyl acetate or acrylate for bismuth halides and methanol for barium bromide), or were made by polymerizing a monomer-salt solution with benzoyl peroxide or azobisisobutyronitrile at $70^{\circ}C$ (in a number of cases 2 wt% tetraethyleneglycol dimethacrylate was added as cross-linker). $BiBr_3$

(Alpha) could be dissolved directly into methyl methacrylate (MMA) while for $BaBr_2 2H_2O$ (Alpha) the monomer $CH_2=C(CH_3)COO(CH_2CH_2O)_{22}CH_3$ from Polysciences (abbreviated as MG22) was used.

Infrared spectra were recorded on a Nicolet 20 DX FTIR spectrometer, and glass transition temperatures were obtained on a Perkin Elmer DSC-4 differential scanning calorimeter. Radiographic measurements were carried out with a Picker X-ray unit operating at 90 kV and 6 mA using cylindrical polymer specimens of known thickness. Radiopacities were compared with that of an aluminum stepwedge with 1 mm steps. Experimental details have been published elsewhere [9].

RESULTS AND DISCUSSION

Acrylic Composites with Barium Bromide
Neither barium bromide, nor its mono- or dihydrate are soluble in MMA. However, molten MG22 ($50^{\circ}C$) will dissolve the salt, but only the dihydrate. Polyethers are known to chelate barium ions [11], but additional energy supplied by hydrogen bonding forces between barium-bound water molecules and polyether oxygen atoms is apparently needed to complete the solubilization process. The monohydrate will dissolve in MG22 only when some methanol or water is added, but removal of the methanol under vacuum causes the salt to precipitate.

The composites of poly-MG22 and $BaBr_2 2H_2O$ are only transparent for molar ratios of ethylene oxide units (EO)/barium ≥ 10. The radiopacity of a 1 mm specimen with 19-20 wt% salt (EO/Ba ≈ 22, that is, on the average one barium ion per side chain) was found to be equivalent with that of 1 mm aluminum. DSC scans reveal an endotherm at about $110^{\circ}C$, the same temperature where the dihydrate is known to decompose into the monohydrate and water (12). At the same time the transparent sample turns opaque, apparently because the barium monohydrate phase-separates from the polymer. A second endotherm is found at $220^{\circ}C$, close to the temperature where anhydrous $BaBr_2$ is formed from its monohydrate [12]. More details on this system will be published shortly.

Composites of Bismuth Bromide
$BiBr_3$, unlike the barium salt, solubilizes in MMA and other carbonyl-containing compounds. It also does not phase-separate in PMMA at elevated temperatures. As long as moisture is excluded, $BiBr_3$ gives a clear solution in MMA up to 45 wt%, and polymerization with BPO or AIBN at $70^{\circ}C$ yields transparent, hard, glassy materials. Lewis acids are known to form 1:1 and 1:2 adducts with carbonyl compounds [13]. The IR spectrum of PMMA-$BiBr_3$ (fig. 1, 40 wt.%, CO/Bi=6) clearly shows bands at 1685 (the CO-adduct) and at 1732 cm^{-1} (free carbonyl). A crystalline adduct of $BiBr_3$ and dimethylmalonate melting at $115^{\circ}C$ yielded carbonyl bands of nearly equal intensity at 1747 and 1701 cm^{-1}, the $\nu(c=o)$ for the free ester being 1735 cm^{-1}. This suggests formation of a 1:1 adduct, a result that in the PMMA-$BiBr_3$ complexes is supported by the observation that the T_g of the polymer slightly decreases on adding the salt (fig. 2). Formation of 1:2 adducts probably would have increased T_g as this results in chain stiffening. Proton and ^{13}C NMR chemical shifts in MMA-$BiBr_3$ mixtures also point to strong carbonyl-Bi interactions. No free $BiBr_3$ (mp $218^{\circ}C$) in the polymer composites could be detected in the DSC scans.

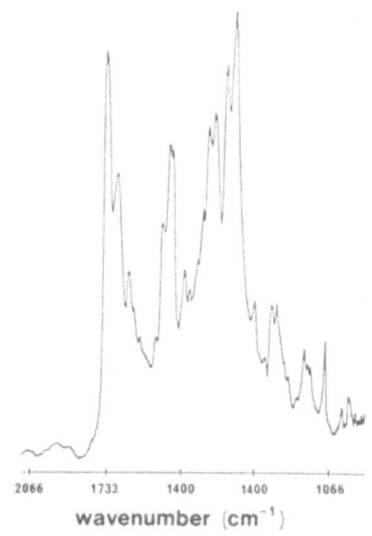

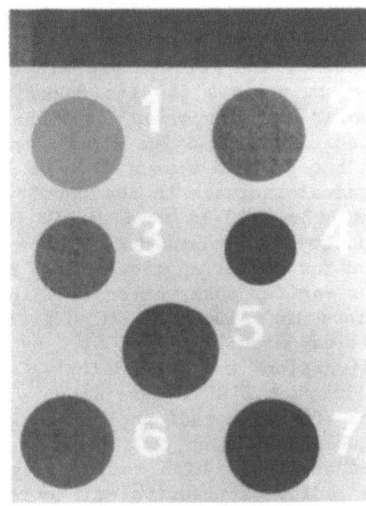

Figure 1

Figure 3

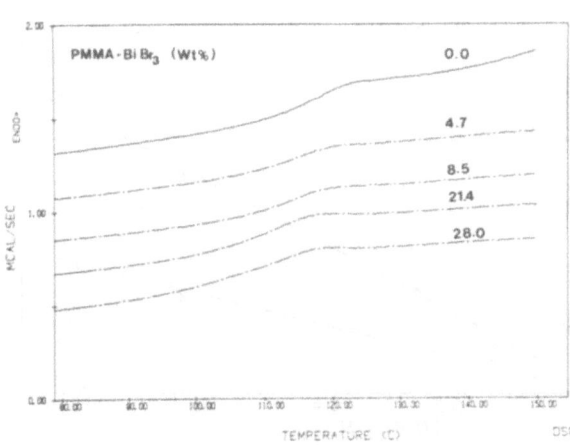

Figure 2

Figure 1. Infrared spectrum of a homogeneous composite of PMMA and 40
wt.% BiBr$_3$ (ratio C=O / Bi = 6).

Figure 2. DSC scans of PMMA/BiBr$_3$ composites.

Figure 3. Positive radiographs of PMMA/BiBr$_3$ composites. Wt.%/thickness
(in mm): (1) 14.4/1; (2) 14.4/2; (3) 25/1; (4) 25/2; (5)
20/1.15; (6) 33/1 and (7) 33/2.0.

Radiographic Measurements

Positive radiographs of 2% cross-linked, transparent, one and two mm thick cylindrical pellets of PMMA containing 14 to 33 wt.% $BiBr_3$ are depicted in fig. 3. Specimens containing from 2 to 10% salt can also be easily detected. Plots of radiopacities expressed in mm aluminum verses the thickness of the pellets are linear. A linear relationship for both one and two mm pellets is also found when the radiopacity is plotted versus the molar concentration of $BiBr_3$ (fig. 4; the intercepts represent radiopacities found for 1 and 2 mm pellets of salt-free PMMA). Similar graphs but plotted versus wt.% of $BiBr_3$ curve slightly upward [10] due to the gradual increase in the density of the sample. Our results show that 0.36 M $BiBr_3$ (13-14 wt.% salt) in a 2 mm PMMA specimen will impart a radiopacity equivalent to that of 2 mm aluminum, a radiopacity standard adopted for dental applications. $BiCl_3$ is less soluble in MMA than $BiBr_3$, but our earlier observation (10) that clear solutions can only be obtained with less than 10 wt.% $BiCl_3$ is not correct. Cloudiness is caused by impurities, most likely $BiOCl$. By removing the insoluble material through centrifugation it is found that clear solutions can be obtained up to at least 30 wt.% $BiCl_3$, consistent with earlier reported observations (6). Another effective radiopacifying agent is uranyl nitrate hexahydrate which is soluble in MMA up to at least 40 wt.%. It forms transparent, yellow-colored resins of high radiopacity, 11 wt.% of salt being sufficient to give a radiopacity equivalent to that of aluminum. These results will be discussed in more detail elsewhere.

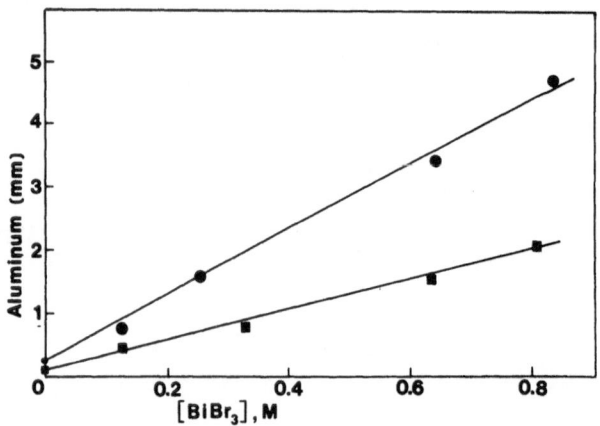

Figure 4. Plot of radiopacity expressed in mm of aluminum versus the molar concentration of $BiBr_3$ for PMMA/$BiBr_3$ composites

An advantage of $BiBr_3$ as radiopacifying agent is that it appears to be non-mutagenic and its toxicity index is below that of other dental formulations. A drawback of the compound is that ambient curing of acrylic resins with BPO and amine accelerators is hindered as the amine apparently interacts with the bismuth. Another problem is the moisture sensitivity of $BiBr_3$ which causes a transparent $PMMA-BiBr_3$ resin to become hazy with time when exposed to water. Formation of water- and resin-insoluble $BiOBr$ is the probable cause of the opaqueness [6]. The material does not decrease in radiopacity but it becomes more brittle. The water stability improves at higher cross-linking density of the acrylic resins.

Acknowledgment

The authors gratefully acknowledge the financial support of (the Department of Health and Human Services through) the National Institute of Dental Research, Grant No. R01DE06179, and the Polymers Program of the National Science Foundation, Grant No. DMR8504999.

REFERENCES

1. The desirability of using radiopaque plastics in dentistry: a status report. J. Am. Dent. Assoc., 1981, 102, 347-349.

2. McCabe, J.F. and Wilson, H.J., A radiopaque denture material, J. Dent., 1976, 4, 211-217.

3. Chandler, H., Bowen, R.L., Paffenbarger, G.C. and Mullineaux, A.L., Clinical investigation of a radiopaque composite restorative material, J. Am. Dent. Assoc., 1970, 81, 935-940.

4. Cook, W.D., An investigation of the radiopacity of composite restorative materials, Austr. Dent. J., 1981, 26, 105-112.

5. Watts, D.C., Radiopacity vs. composition of some barium and strontium glass composites, J. Dent., 1987, 15, 38-43.

6. Combe, E.C., Further studies on radio-opaque denture base materials, J. Dent., 1972, 1, 93-97.

7. Davy, K.W.M. and Causton, B.E., Radiopaque denture base: a new acrylic co-polymer, J. Dent., 1982, 10, 254-264.

8. Xia, D.W., Silberman, R., Cabasso, I. and Smid, J., Novel polymer salt complexes for X-ray diagnostics, Polym. Prepr. (Am. Chem. Soc., Div. Polym. Chem.) 1985, 26, 72-73.

9. Smid, J., Cabasso, I., Rawls, H.R., Obligin, A., Delaviz, Y., Sahni, S.K. and Zhang, Z.X., Novel homogeneous polymer-heavy metal salt complexes for X-ray imaging, Makromol. Chem., Rapid Commun. 1987, 8, 543-547.

10. Takaki, U. and Smid, J., Complexes of crown ethers and glymes with difluorenylbarium, J. Amer. Chem. Soc., 1974, 96, 2588.

11. Borchandt, H.J. and Daniels, F., Differential thermal analysis of inorganic hydrates, J. Phys. Chem., 1957, 92, 917.

12. Lappert, M.F., Coordination compounds having carboxylic esters as ligands. Part II. J. Chem. Soc., 1962, 542.

Part 4

MORPHOLOGY/STRUCTURE

MOLECULAR STRUCTURE, CRYSTALLIZATION AND MORPHOLOGY
OF VERY LOW DENSITY POLYETHYLENE (VLDPE)

V.B.F. Mathot, R.A.C. Deblieck, M.F.J. Pijpers
DSM Research
P.O. Box 18
6160 MD Geleen
Netherlands

ABSTRACT

The heterogeneous ethylene/1-octene copolymer studied crystallizes from 120 to -60 °C and melts from -60 to 130 °C, as shown by DSC measurements. At -60 °C, the enthalpy-based weight crystallinity is 37 %, and at 23 °C it amounts to 25 %. The results of fractionation experiments strongly suggest that the heterogeneity is intermolecular in nature and that VLDPE is actually a reactor blend of molecules ranging from the uncrystallizable type to the HDPE type. The morphology, as investigated by electron microscopy, confirms this conclusion. It is characterized by regions of highly different crystallinity, ranging from structureless regions to compact semi-crystalline domains, and by lamellar crystallites of varying size.

INTRODUCTION

Recently [1], the crystallization and melting behaviour of Very Low Density Polyethylenes (VLDPEs) was described on the basis of DSC experiments. It was concluded that the crystallization and melting ranges of VLDPEs are extremely wide, typically covering about 200 °C, while the DSC curves show several peaks. This is why VLDPE was classed among the so-called heterogeneous ethylene copolymers, as opposed to homogeneous copolymers. The present study concerns the nature of the heterogeneity and the morphology of the material.

MOLECULAR HETEROGENEITY

Fractionation techniques are excellently suitable for exploring the nature of the heterogeneity present. In the case of intermolecular heterogeneity, crystallization/dissolution fractionation is a convenient method of research. In the present study, a VLDPE sample was crystallized in xylene, after which four fractions, K-1 through K-4, were drained at increasing temperatures.

Figure 1 shows the specific heat capacity curves [$c_{p_{hc}}(T)$] of the fractions, as obtained from the DSC heating curves, plotted for the mole percentages [X_8 (%)] of 1-octene in question. In comparison with the curve for the whole sample [1], these curves clearly have fewer peaks. For the first fraction obtained, K-1, the final melting temperature is about 90 °C and high-melting material is absent, whereas in the last-obtained fraction, K-4, with a final melting temperature of more than 130 °C, the low temperature peak is absent.

The very fact that it is possible to split a VLDPE into portions with the characteristics mentioned strongly suggests that VLDPE is an ethylene copolymer in which the molecules are intermolecularly heterogeneous. An important finding is that the first fraction constitutes an appreciable portion of the material, viz. 66 wt %. It can be concluded, then, that VLDPE is to a large extent, if not wholly, intermolecularly heterogeneous.

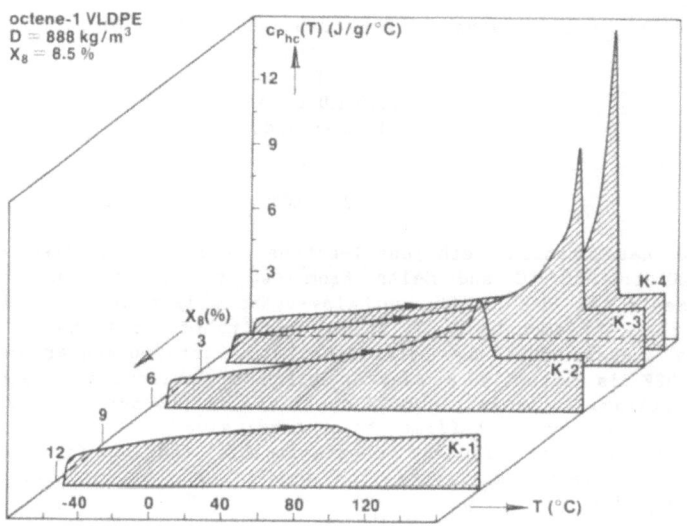

Figure 1. DSC-2 heat capacity heating curves as a function of the mole percentage of octene for fractions of a VLDPE obtained by a crystallization/dissolution method.

This means that VLDPE should be regarded as a blend of ethylene copolymer molecules with widely different comonomer incorporation percentages, the blend having been obtained by polymerization. In fact, the whole range of molecules, from uncrystallizable and poorly crystallizable material up to and including HDPE-like material, seems to be present.

The enthalpy-based weight crystallinity curves, W_{hc}^c (T), as obtained within the two-phase model via numerical integration of the heat capacity heating curves in combination with the reference enthalpy data for purely amorphous and purely crystalline polyethylene, are given in Figure 2 for the VLDPE studied and its fractions. They illustrate even more clearly that the fractions differ greatly in melting behaviour and crystallinity.

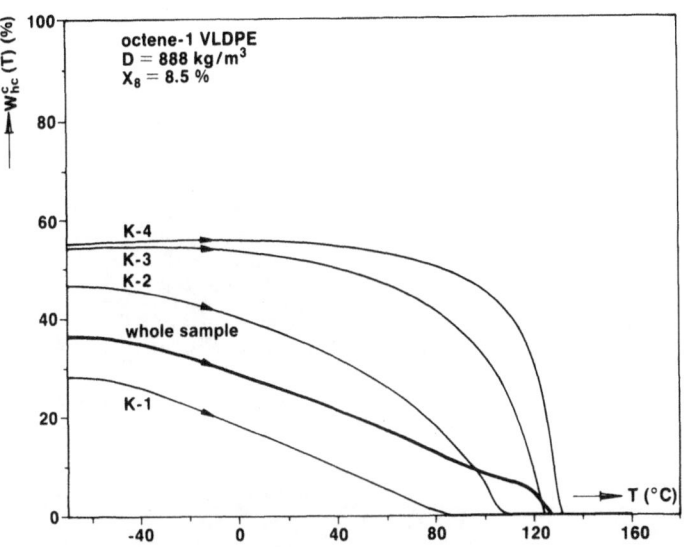

Figure 2. Weight crystallinity heating curves for a VLDPE and fractions of it obtained by a crystallization/dissolution method.

MORPHOLOGY

The blend character of the VLDPE under consideration is reflected in the morphology in a peculiar way. Figure 3 shows thick (13-14 nm) and relatively long (0.5-1 μm) lamellae crossing compact semi-crystalline domains (CSDs) and reaching out into the surrounding low-crystalline phase. They also occur as single lamellae and are thought to be related to the HDPE component in VLDPE. On these lamellae, small lamellar crystallites are observed. The complementary phase is, to all appearances, amorphous.

It should be borne in mind that at the temperature at which the fixation/staining treatment was performed (45 °C) an appreciable portion of the material was in the molten state. Moreover, crystals smaller than 3-4 nm cannot be made visible by the fixation/staining technique used. This means that the phase that seems to be amorphous consists of uncrystallizable to poorly crystallizable material.

The morphology of the VLDPE under consideration is characterized by distinct co-continuous regions of highly different crystallinity. Internally, the CSDs show a mixture of the above-mentioned long lamellae and shorter ones (about 0.15 μm), which are also slightly thinner (down to 6-7 nm). Besides segregation by crystallization, thermodynamic demixing in the melt is being studied as a possible cause of the occurrence of CSDs, on the basis of evidence for blends of molecules differing in branching [2].

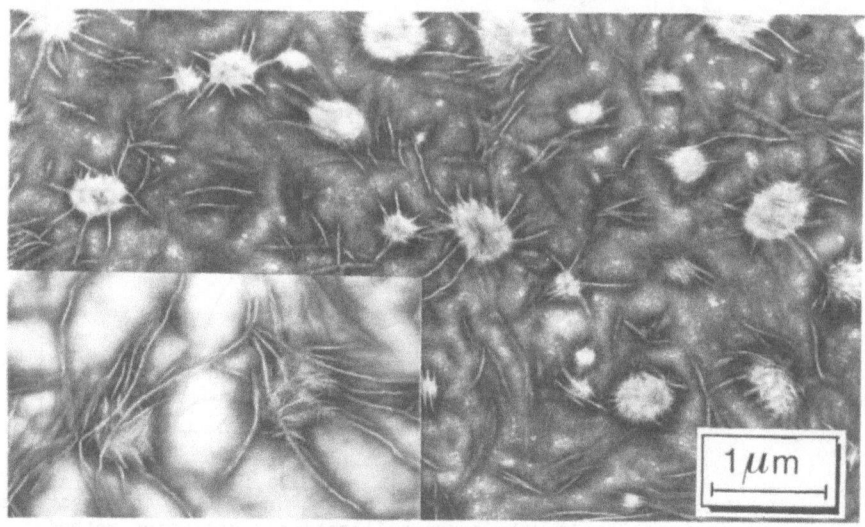

Figure 3. Morphology by transmission electron microscopy of the VLDPE of
Figs. 1 and 2, after staining/fixation by chlorosulphonation. The
inset shows a 1.6 times magnified view of a more severely treated
sample revealing the internal morphology of the CSDs.

REFERENCES

1. Mathot, V.B.F. and Pijpers, M.F.J., Structure and crystallization of
   homogeneous and heterogeneous ethylene copolymers including very low
   density polyethylenes. In Integration of Fundamental Polymer Science and
   Technology, Proceedings of International Discussion Meeting, Rolduc
   Abbey, Netherlands, April 26-30, 1987. Elsevier Applied Science
   Publishers Ltd, London, 1988, pp. 381-4.
2. Barham, P.J., Hill, M.J., Keller, A. and Rosney, C.C.A. These
   Proceedings.

A CASE FOR LIQUID-LIQUID PHASE SEGREGATION IN POLYETHYLENE MELTS

P.J. Barham, M.J. Hill, A. Keller, C.C.A. Rosney
H.H. Wills Physics Laboratory, University of Bristol,
Tyndall Avenue, Bristol BS8 1TL. U.K.

ABSTRACT

A pair of polyethylenes, one linear, one branched, have been studied after crystallization from melts of different concentration blends. Results obtained by TEM and DSC show striking differences with blend concentration and crystallization conditions. We are only able to interpret these results on the basis of liquid-liquid phase segregation.

INTRODUCTION

Linear and lightly branched polyethylenes are generally considered to be fully compatible in the melt; whereas polyethylene and polypropylene melts are generally considered to be completely incompatible. Polypropylene can be regarded as extremely branched polyethylene. Thus there should be an intermediate branched polyethylene which is partially compatible with linear polyethylene in the melt. We believe we have found such a pair and we have started to map the relevant phase diagram.

EXPERIMENTAL

The polymers used in this work were HDPE, Sclair 2907 ($M_w \simeq$ 98000, $M_n \simeq$ 28000 1 long branch/1000 carbon atoms) and LDPE BP PN 220 ($M_w \simeq$ 208,700 $M_n \simeq$ 25,300, 16 short and 10 long branches/1000 Carbons). The blends were prepared by dissolving HDPE and LDPE in xylene (at 1% $^w/v$), and quenching into acetone at -20°C. In all our experiments we have examined material after crystallisation at room temperature (i.e. not actually in the melt). The observations thus made led us to the inference that very rapid quenching preserves a structure that must have been present already in the melt.

EVIDENCE FOR LIQUID-LIQUID PHASE SEGREGATION

On reheating rapidly quenched samples in the DSC we observe either a single peak, indicative of a single crystal phase, or a double peak. Fig. 1 shows a sequence of DSC traces of blends of various concentrations quenched from 160°C into iced acetone. Note that blends with 100% - 50% (fig. 1a) HDPE show a single endotherm whose temperature decreases with increasing LDPE content. At lower HDPE concentrations a second peak is found at lower temperatures, and both peak positions remain independent of concentration (fig. 1b).

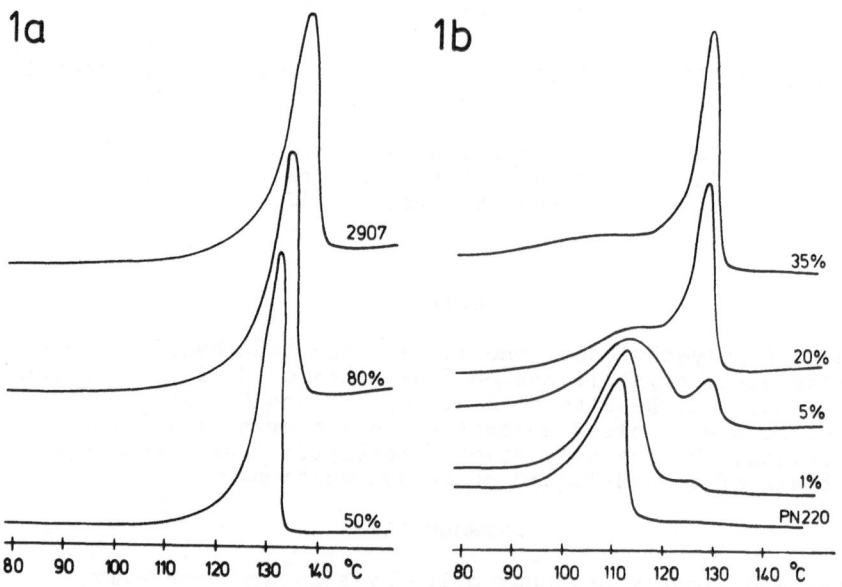

Figure 1. D.S.C. traces obtained by heating blends of HDPE (Sclair 2407) and LDPE (BP PN 220) at 10° C/min after quenching from 160°C into acetone at freezing point. The percentage figure refers to the percentage HDPE in the blend.

We have examined these quenched blends by transmission electron microscopy using replication after permanganic etching. Blends with a single DSC peak exhibit a uniform morphology (figure 2a) and blends with two DSC peaks possess a double morphology (figure 2b). We conclude that those blends with a single DSC peak and morphology crystallised homogeneously from a single phase melt; while the evidence of two distinct morphologies combined with the DSC data showing two peaks whose position is independent of concentration, strongly suggests that such samples crystallised from a biphasic melt.

We can deduce one point on the suggested phase diagram from these results (160°C, 45% HDPE). Further points can be obtained by quenching from different temperatures. The observations thus made indicate a narrowing of the biphasic region both at lower and higher temperatures consistent with the existence of a lower (LCT) and upper (UCT) critical temperature, hence of a closed loop coexistence curve, at least by our interpretation of the observations. While the suggested UCT itself has not been attained so far, the

existence of an LCT is strongly supported by morphological
differences of the crystals formed on cooling, from the
biphasic region or from below the LCT; these are described
below.

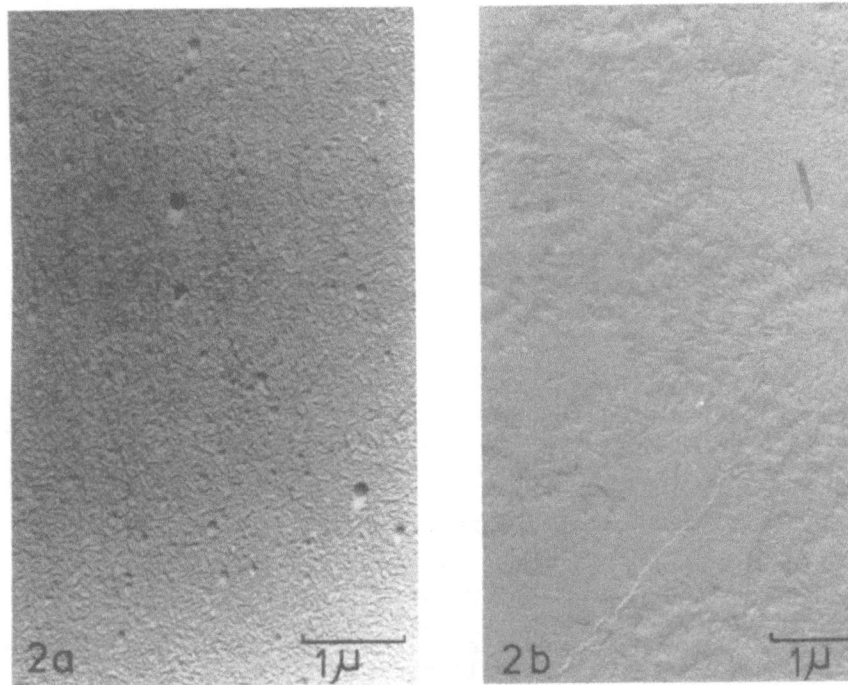

Figure 2. Electron micrographs of 50% (fig. 2a) and 10% (fig. 2b)
blends of HDPE with LDPE quenched from 160° into iced acetone.
Specimen prepared by permanganic etching followed by shadowing and
replication.

EVIDENCE FOR REMIXING

Transmission Electron Microscopy (TEM)

We have crystallised low concentrations of HDPE isothermally.
We have seen two different morphologies. A 0.2% HDPE blend,
crystallised at 128°C gave clusters of crystals isolated from
each other by large volumes of quenched LDPE rich material
(fig.3a). On the other hand the same blend, crystallised at
122°C, gave randomly scattered single (or ocasionally double)
lamellae (fig. 3b), again in a quenched LDPE matrix.

We consider that the clusters grew at 128°C from
approximately spherical zones of the linear-rich component of
the segregated blend. At 122°C, however, the blend appears
to have remixed, thus giving a homogenous distribution of
individual lamellae of HDPE on crystallisation from which LCT
between 128 and 122°C would follow. We have obtained similar
results from 0.2%, 1% and 5% blends.

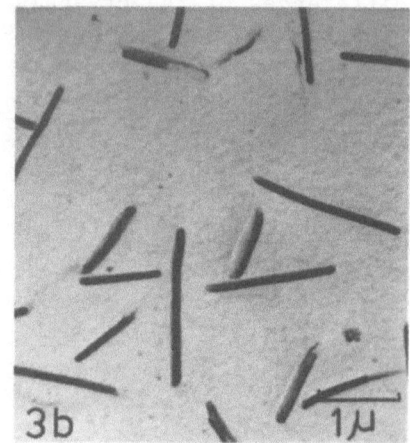

Figure 3. Electron micrographs of replicas made from 0.2% blends of
HDPE with LDPE. Samples were crystallized for one week (in Argon) at
128°C (fig. 3a) or 122°C (fig. 3b) prior to quenching into acetone at
freezing point.

Differential Scanning Calorimetry (DSC)

All blends give double peaks on remelting after complete
isothermal crystallisation, at both 122 or 128°C. New
information can, however, be obtained by quenching the blends
after various times as crystallisation develops. We use here
in fig. 4 the 1% blend for illustration.

After melting at 160°C, cooling to 128°C and leaving for
5 minutes the positions of the two peaks obtained on
reheating are typical of a segregated, quenched melt
(cf. fig. 4a, 5 mins with fig. 1b). Similar peaks are
obtained on quenching after crystallisation for up to 20
hours. After longer crystallisation times we begin to
observe a melting peak at higher temperatures (133°C). We
attribute this peak to a liquid-solid segregation due to
crystallisation of the linear polyethylene within the linear
rich phase. Thus the material melting in the higher
temperature peak observed at up to 20 hours crystallised on
quenching from 128°C, in contrast with the material melting
in the highest peak observed after longer crystallisation
times, which crystallised isothermally at 128°C.

At 122°C two peaks are seen on quenching after 5
minutes, as before (fig. 4b), but over the next 30 minutes
the higher temperature quenched peak completely vanishes, and
only later (seen here well developed at 1 hr) is the peak due
to isothermally crystallised, high melting, material
apparent. We consider that here we are observing the
re-mixing of the blend, at 122°C, over 35 mins. In this
region (128-120°C, 0.2-5%) we have used this technique to
follow segregation and re-mixing with temperature.

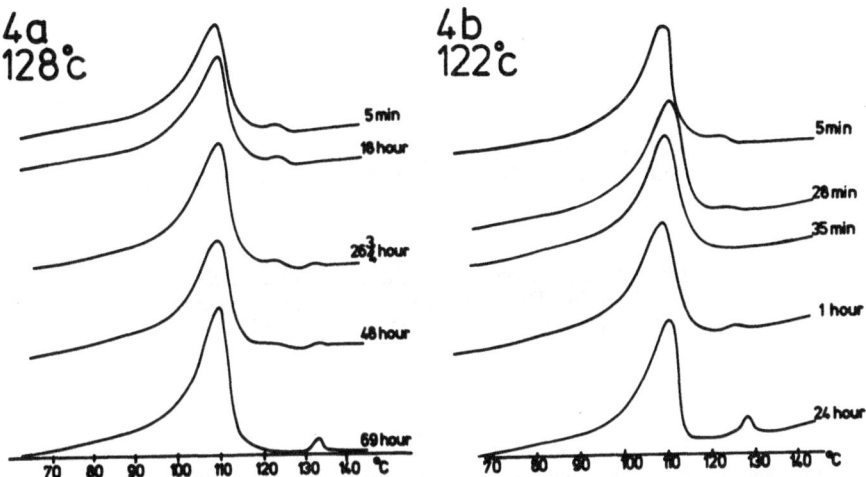

Figure 4. D.S.C. traces of the 1% blend of HDPE with LDPE. Samples
were held at 150°C for 30 mins then held at 128°C (fig. 4a) or 122°C
(fig. 4b) for the times indicated prior to quenching into freezing
acetone (Heating rate 10°C/min).

CONCLUDING REMARKS

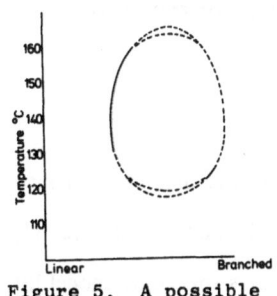

Figure 5. A possible
phase diagram. The
PEs mix outside the
eggshaped region and
separate in it.

We have presented here a small selection
from a much larger set of data. The only
explanation we can conceive, consistent
with this body of work is that liquid-
liquid phase segregation occurs in this
HDPE/LDPE pair in the melt, in the way
illustrated by fig. 5. Experiments such
as those described on re-mixing (figs 3
and 4) demonstrate the presence of an
L.C.T. in the phase diagram. The
narrowing in the liquid-liquid phase
segregated region as the melt temperature
increases leads us to believe that there
may also be an U.C.T. in the phase
diagram (figs 1 and 2 and result on
quenching from 145°).

We concede that we have no observations in the melt
itself and also that even by our indirect indications more
information closer to LCT and UCT would be desirable. Even
so, we consider the effects themselves definitive and
significant in their own right even irrespective of
interpretation. In addition we do offer an interpretation of
possibly wider implications which fits all the facts and to
which we have not been able to conceive an alternative.

Investigation of the Morphology of Spherulitic Crystallized and Solvent Treated Polycarbonate with Electron Microscopy

Heinz Schorn, Robert Kosfeld and Michael Heß
Universität –GH– Duisburg, FB 6/Physikalische Chemie
Lotharstraße 1, D–4100 Duisburg 1, FRG

Summary

The good solubility of amorphous PC in contrast to the high stability of crystalline PC in THF at ambient temperatures enables separation and isolation of PC–spherulites prepared by solvent induced crystallization.

The arrangement of the ordered phases in spherulitic structures is visualized by REM on the surface as well as in inner regions of broken spherulites. Less ordered structures in the sheaf–shaped inner areas can be removed with THF.

Radial compactly grown lamellar sheets stabilize regions apart from the central segments and aggregated spherulites.

The physical properties of semicrystalline polymers depend on these less ordered regions. THF–treated crystalline PC shows melting proceses 30 K higher than untreated semicrystalline material.

Introduction

During the last two decades optical and electron microscopy gave much insight into crystallization in polymers:

Crystallization was observed in the melt and in solution. Lamellar structures [1] were found which aggregate to superstructures, e.g. spherulites.

Electron microscopy on thin films [2–5] led to Peterlin's spherulite model [6] consisting of a sheaf–shaped nucleus growing under radial–symmetric formation of lamellar structures. Within the lamellae Kanig [7] found inclination of chain–segments using a special contrasting procedure.

In any case the analysis of semicrystalline polmyers requires an excellent contrast of the phases.

Experiments

Bisphenol–A Polycarbonate obtained from Bayer AG was dissolved in CH_2Cl_2 and a film cast. The films contained spherulites. THF–treatment yielded a suspension of these superstructures which were further extracted by THF–washing.

The gold sputtered samples were investigated in a Jeol JSM 35 CF raster electron microscope (REM).

The calorimetric experiments were executed in a Perkin Elmer DSC–2 at a heating rate of 20 K/min.

Results and Discussion

The growing of spherulites is observed in swollen amorphous material while the solvent is removed from solution. They are found in groups as well as isolated, embedded in an amorphous matrix.

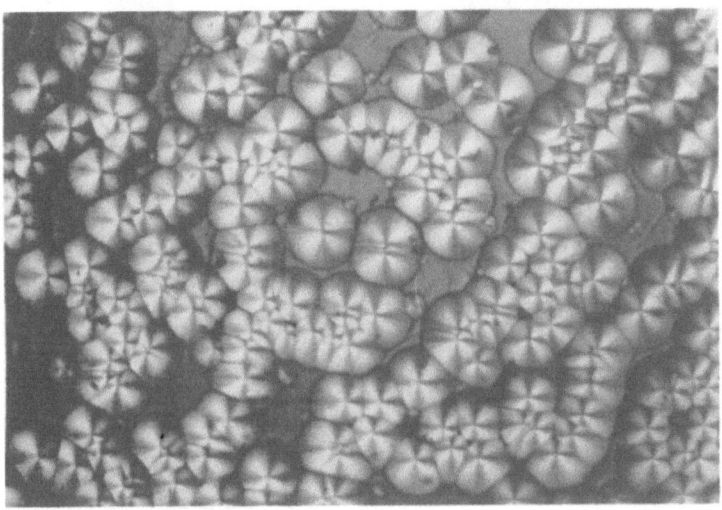

Figure 1.   Optical micrograph of a thin PC–film containing spherulites and spherulite aggregates (magnification 250 x)

Depending on the casting process the resulting films may be totally spherulitic.

Aggregated spherulites in films show plane as well as hyperboloidic contact faces. Plane contact faces are the result of neighbouring spherulites starting their growth simultaneously. Hyperboloidic contact surfaces are the result if neighbouring spherulites start growing at different moments, both cases under the premise of constant rate of radial growth.

THF–treatment of films in the way described above resulted in a separation of spherulites and aggregated spherulites from the amorphous matrix due to the stability of crystalline PC against THF at room temperature. These crystalline structures are ideal objects to study by REM.

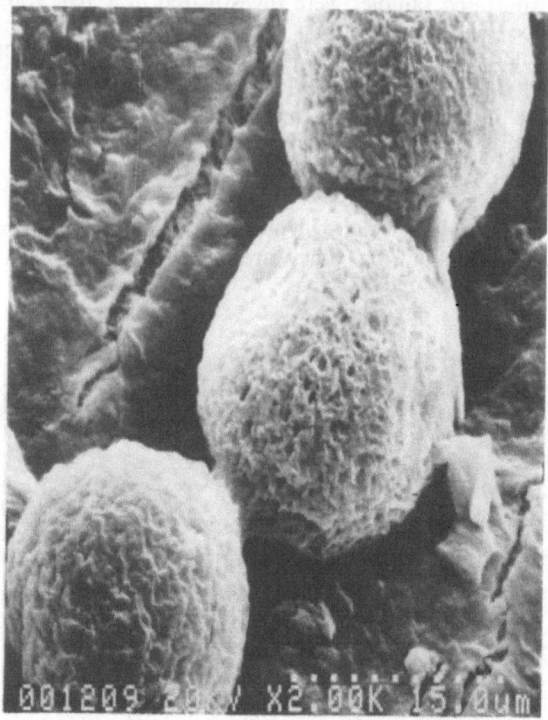

Figure 2.   Electron micrograph of spherulites isolated from the amorphous
matrix with THF

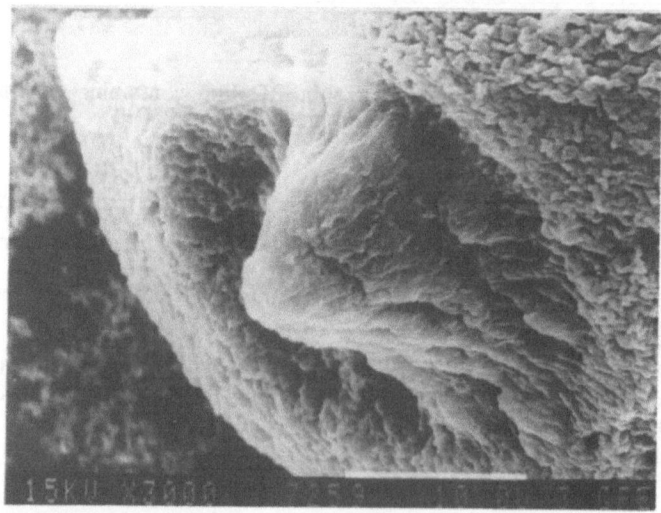

Figure 3.   Electron micrograph of the spherulite surface

The spherulites isolated in the described manner had diameters from 10 μm up to 50 μm. The surface shows a great amount of irregular arranged beads 0,5 μm to 1 μm in diameter. These beads are the ends of the bundles of radially grown lamellare which were proved even in Kanig's investigations [7].

Between these bundles had been a non–crystalline phase which was removed by THF in the outer regions of the spherulite. In the regions localized in the interior of the particle the amorphous interphase is of a less solubility than at the surface due to a greater amount of entanglements. Otherwise the spherulite would collapse due to the THF treatment but this was not observed. Instead of this only a swelling process of this entangled amorphous phase is observed which, if sularge, results in spherulites burst open in a characteristic manner.

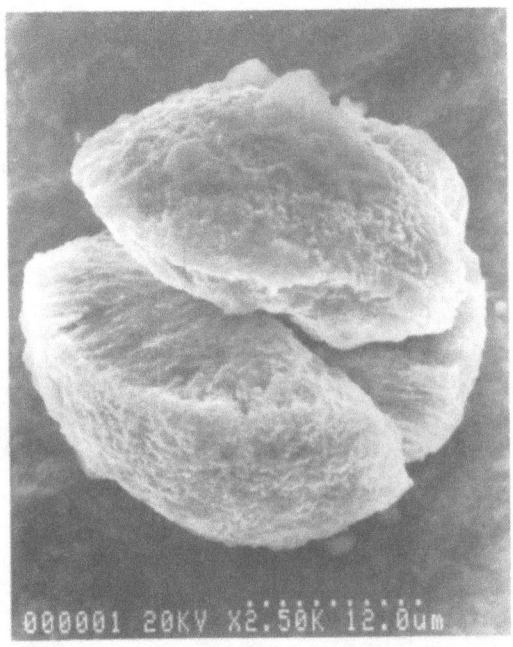

Figure 4.   Electron micrograph of a burst open spherulite

Increasing swelling arches the surface of fracture outwards. The flexibility of the still semicrystalline core in general is able to stand out this stress.

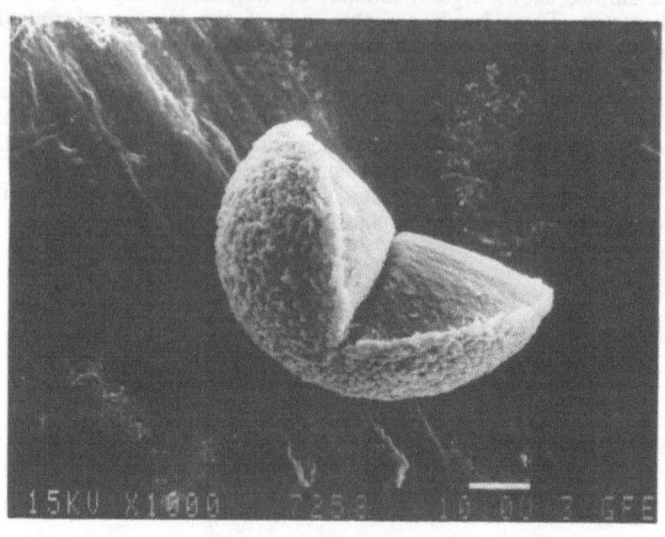

a)

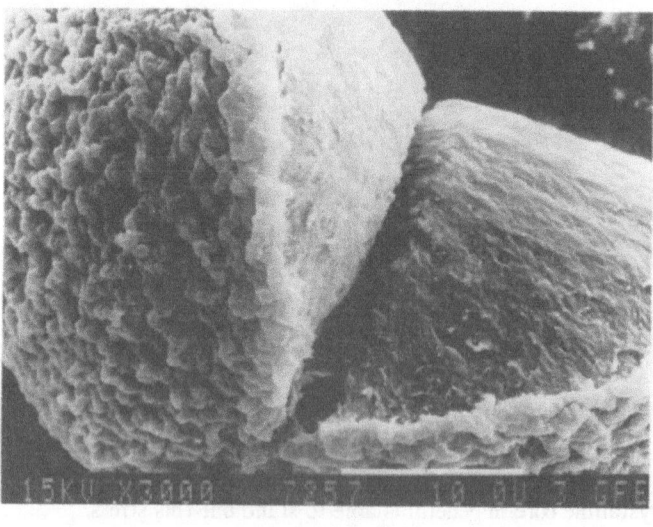

b)

Figure 5.   Electron micrographs of a swollen spherulite

It is an characteristic phenomena that aggregates do not separate at their contact surfaces although they do show fracture on swelling, too.

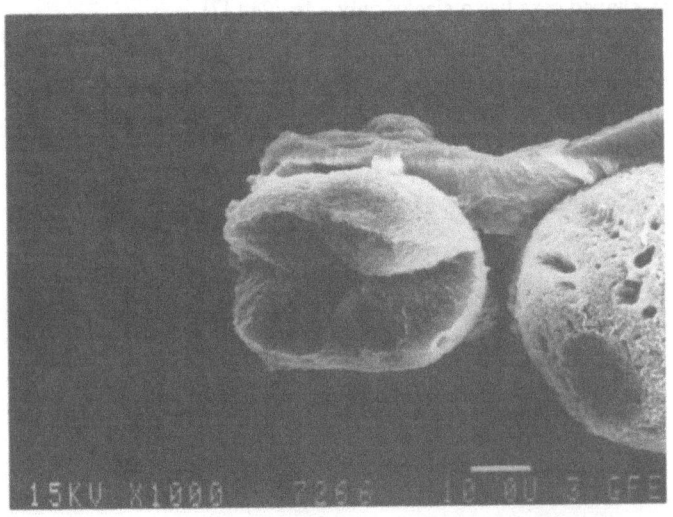

a)

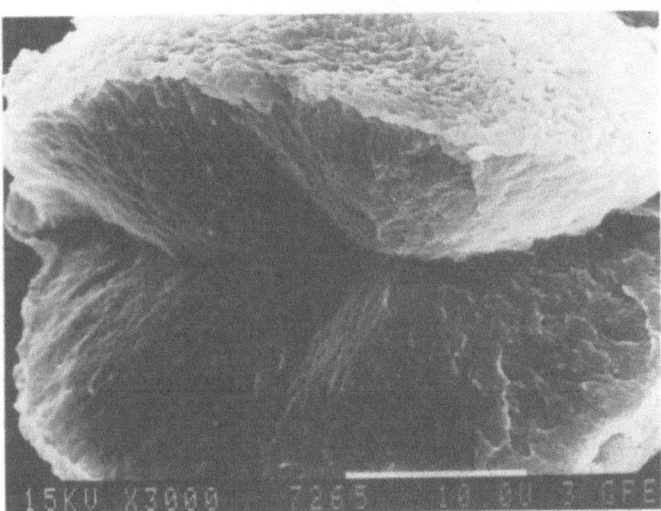

b)

Figure 6.   Electron micorgraphs of a burst aggregate with two spherulite centers

This behaviour is explicable by the sheaf structure of the core:
In the vicinity of the axis of rotation in the centre there is an area where the crystalline lamellae bundles show a lower degree of order (regularity) than in a greater distance where the radial symmetry is much more perfect. This lower ordered, sheaf–shaped core has a size of only a few μm [7].

Are the axis of rotation of neighbouring cores parallel there is a plain perpendicular to them consisting of less ordered bundles of lamellae. It is the plain which cannot stand out the swelling pressure. Here fracture occurs.

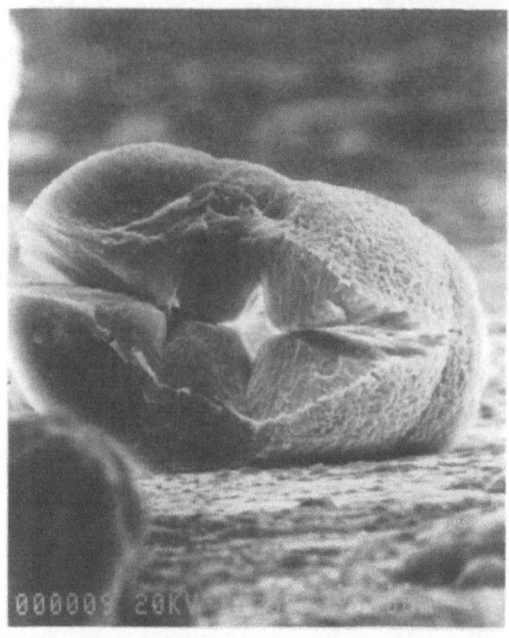

Figure 7.   Electron micrograph of a burst aggregate with four spherulite centers

Between the cores the less ordered bundles are attacked by the solvent and removed while the skeleton distant from the core is – due to its higher organization – more stable against THF–attack.

As the total amount of the THF–treated material contains less amorphous material than the original, an increase of the melting temperature from 225 ^{0}C up to 255 ^{0}C is observed.

LITERATURE

1. Wunderlich B., <u>Macromol. Phys.</u>, Vol. 1 (1973), Vol. 2 (1976) New York
2. Andrews E.H., Owen P.J., Singh A., <u>Proc. Roy. Soc. Lond. A</u>, 324 (1971) 79
3. Philips P.J., Edwards B.C., <u>Polym. Lett.</u>, 14 (1976) 449
4. Philips P.J., <u>Polym. Prepr.</u>, 18 (1977) 612
5. Edwards B.C., Philips P.J., <u>Polymer</u>, 15 (1974) 351
6. Peterlin A., <u>Polym. Sci. Sympos.</u>, 32 (1971) 297
7. Kanig G., <u>Colloid & Polym. Sci.</u>, 260 (1982) 356

INFLUENCE OF DIFFERENT FILLERS ON THE ISOTHERMAL CRYSTALLIZATION AND GLASS TRANSITION TEMPERATURE OF POLYPROPYLENE.

J.L. ACOSTA, E. MORALES, M.C. OJEDA and A. LINARES
Instituto de Plásticos y Caucho (C.S.I.C.)
C/ Juan de la Cierva, 3. 28006 Madrid (Spain)

ABSTRACT

The present work analyze the effect produced by different fillers such as sepiolite, talc and a elastomeric block copolymer on the isothermal crystallization and glass transition temperature of polypropylene, as a way to relate structural and physical properties.

INTRODUCTION

Previous works (1,2) examined the effects produced by the incorporation of an elastomeric block copolymer on the technological properties of polypropylene composites containing sepiolite and talc as fillers.

This work, with the aim of assessing the influence produced by these fillers on the polypropylene microstructure, and hence determining their effect on the physical properties of the resulting composites, reports on a study designed to analyze, by means of an individualized approach, the effect produced by sepiolite, talc and the elastomeric copolymer on the isothermal crystallization kinetics and on the glass transition temperature of the polypropylene. This study permits to correlate more adequately the changes produced in the microstructure of the composites and their physical properties.

EXPERIMENTAL

The materials used were polypropylene ISPLEN (Alcudia S.A.), talc (Ferrocomercial), styrene-butadiene block copolymer Cariflex 4122 (Shell) and sepiolite supplied by Tolsa S.A. in a micronised form. The main physical properties of the sepiolite has been described elsewhere (3).

Samples containing filler loadings of 10, 25 and 40% by weight were made in a Brabender Plasticorder (200°C, 60 rpm,

10 min.)

Isothermal crystallization was followed by means of a differential calorimeter Perkin Elmer DSC-1B. Mechanodynamical test were carried out on a viscoanalyzer Metravib using films 0.5 mm. in thickness and applying the forced vibration technique under tension-compresion.

RESULTS AND DISCUSSION

Fig. 1 shows the crystallization isotherms for the samples crystallized at 392 K. Table 1 compiles the crystallynity grades achieved at this temperature for all the samples and the kinetic parameters inferred from the graphic representation of the Avrami equation.

TABLE 1
Crystallinity and Avrami parameters for the samples crystallized isothermically at 392 K.

Sample	PP wt%	Car wt%	Sep wt%	Talc wt%	Crystal-linity %	Avrami Parameters	
						n	log K
PP	100	–	–	–	38.3	2.78	– 2.57
PP-Car A	90	10	–	–	62.2	3.11	– 1.86
PP-Car B	75	25	–	–	80.3	2.96	– 2.88
PP-Car C	60	40	–	–	9.1	2.12	– 2.53
PP-Sep A	90	–	10	–	41.9	4.31	– 1.20
PP-Sep B	75	–	25	–	44.0	2.68	– 1.37
PP-Sep C	60	–	40	–	–	–	–
PP-Tal A	90	–	–	10	59.1	3.95	1.00
PP-Tal B	75	–	–	25	49.1	3.53	0.45
PP-Tal C	60	–	–	40	5.1	3.53	0.79

Looking at the data assesed, it is clear that the talc and, to a lesser degree, the sepiolite have a marked effect on the crystallization rate of the polypropylene, as deduced from the values of the K constant (Table 1) and by comparison of the conversion-time graphs represented in Fig. 1. This latter fact classify these fillers as nucleating agents, a feature absent in the case of the block copolymer whose impact on the crystallization rate is practically none.

Furthermore can be observed that all the samples with small or medium filler levels achieved crystallinity grades above that of unfilled polypropylene, thus demonstrating that the filler promotes, within the polypropylene matrix, the establishment of conditions which thermodynamically are more favourable to the development of crystalline structures, which in

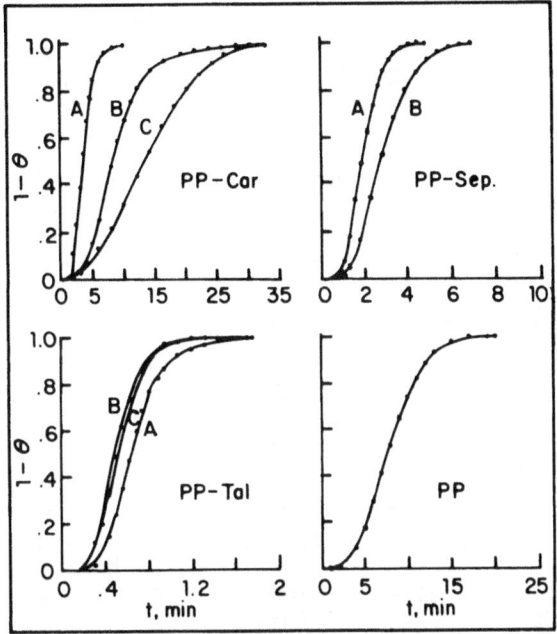

Figure 1.- Crystallization isotherms at 392 K for all the samples.

turn, explains the high crystallinity grades achieved. For higher filler levels (40% by weight) surprisingly lower crystallinities are observed. This could be explained in two ways: either this high filler concentrations inhibits spherulitic growth or else the nucleating effect of the filler causes crystallization to occur at such high rates that they cannot be fully recorder by the technique employed.

When the filler is pressent at polypropylene crystallization, can be expected that this affects its crystalline order, being the Avrami parameter $\underline{n}$ the key to these changes. In our study, whereas in the isothermal crystallization of unfilled polypropylene the value of $\underline{n}$ is near three, i.e. similar to the values reported in the literature (4) corresponding to a tridimensional athermal nucleation, in the composites containing the block copolymer, sepiolite or talc, apart from deviating notably from the value found for unfilled polypropylene, it does not remain constant with filler concentration. The different behaviour observed between the composites containing the different fillers points toward the qualitative fact that different fillers promote different structures in the polypropylene composites containing them, which undoubtlessly has an impact on their physical properties.

In order to prove this postulate a study was carried out
designed to determine, by means of mechanodynamical techniques,
the effect of each filler on the glass transition temperature
and the activation energy (Ea) of the polypropylene. Table 2
shows the Tg values of all the samples as a function of the
frequency as well as the activation energies of the transition.
The first relevant fact is that the Tg values obtained for each
composite family are not equivalent among themselves. In addition, the glass transition for all samples occur at temperatures below that of unfilled polypropylene, decreasing when increasing the level of filler for the composites containing the
block copolymer, while this relation is not so clear in the
case of the composites containing sepiolite or talc. The activation energies follow a similar pattern.

TABLE 2

Glass transition temperature as a function of frequency
for the samples studied.

Sample	Glass Transition Temperature, T_g (ºC)					Energy of activation (KJ/mol)
	11.8 (Hz)	31.8 (Hz)	100 (Hz)	200 (Hz)	250 (Hz)	
PP	12.0	13.0	16.0	17.5	18.0	305.5
PP-Car A	9.0	11.0	14.0	15.0	16.0	269.5
PP-Car B	8.0	9.0	12.0	14.0	15.5	246.8
PP-Car C	2.0	5.0	9.0	12.0	14.0	156.9
PP-Sep A	7.5	8.5	12.0	14.0	14.5	250.0
PP-Sep B	9.0	10.0	13.0	15.0	15.5	273.9
PP-Sep C	7.5	9.5	13.0	–	17.0	197.4
PP-Tal A	11.0	12.0	16.0	18.5	19.5	213.0
PP-Tal B	11.0	13.5	15.5	17.0	18.0	288.6
PP-Tal C	10.0	11.5	15.0	18.0	19.0	206.6

The fact that the behavioural patterns of the different
composite families tested are not comparable with regard to
their glass transition temperatures and activation energy, can
be explained by taking into account the different structural
order developed onto the polypropylene matrix as a function of
the physico-chemical properties of the filler, as well as the
filler loading. This structural order is determined by factors
related to the appearance of new crystalline structures, by
the existence of mesophases surrounding the particles (5) as
well as the result of specific interactions between the phases
present in the composite. These lines of approach are further
being pursued in research in course.

307

REFERENCES

1. A. Linares, M.C. Ojeda and J.L. Acosta, Rev. Plast. Mod., 1987, <u>376</u>, 518.

2. J.L. Acosta, J.P. Vigo, M.C. Ojeda and A. Linares, Rev. Plast. Mod., 1985, <u>354</u>, 735.

3. A. Linares and J.L. Acosta, Angew. Makromol. Chem., 1984, <u>121</u>, 137.

4. L. Mandelkern, "Crystallization of polymers", McGraw Hill, New York, 1964.

5. E. Morales, Ph. D. Thesis, Univ. Complutense Madrid, 1986.

CONFORMATIONAL DISORDER IN POLYDIETHYLSILOXANE (PDES).

G. Kögler, K. Loufakis, R. Bohnert, and M. Möller
Institute für Makromoleculare Chemie
D-7800 Freiburg, West Germany.

ABSTRACT

We examined the phase diagram of PDES using DSC, solid
state NMR, optical microscopy and rheological measurements.
The mobility of the chains in the different phases was
investigated. Changes that occur upon shearing and the
influence of quenching are discussed.

INTRODUCTION

It was recently proposed that between the fully ordered
crystal and the melt there are three mesomorphic states [1].
In addition to the well known liquid and plastic crystals a
third mesophase the "conformational disordered crystal" exist.
This classification results from the recognition that in the
fully ordered crystal three kinds of order exist, namely:
positional, orientational and conformational. Loosing one kind
of order and gaining the corresponding mobility leads
respectively to liquid, plastic and "conformational
disordered" crystals. On quenching in principle, the

corresponding frozen (glassy) states are obtained.

Polydiethylsiloxane (PDES) is a polymer that exist in several polymorphs. This polymorphism along with the fact that the side chains can cause in principle increased mobility within the crystal make it a good candidate for a mesomorphic state forming material. After the first study of its thermal behavior [2] PDES became the subject of several investigations [3] and recently it was proposed that exist in two polymorphs; a and β. Each one of them has a low and a high temperature phase (a_1 and a_2, $β_1$ and $β_2$) [4,5]. In addition prior to the transition to the isotropic melt a so called mesomorphic or viscous crystal state a_m exists.

EXPERIMENTAL

We investigated the thermal behavior and the changes that accompany the phase transitions using Differential Scanning Calorimetry, temperature dependent solid state ^{13}C and ^{29}SI MAS CP NMR and optical microscopy.
The thermal behavior of the material crystallized from the a_m phase is similar to the one observed before [4]. The following transitions were observed: a_1 to a_2 at 212 K, $β_1$ to $β_2$ at 206 K, a_2 to a_m at 279 K, $β_2$ to a_m at 289 K and a_m to the isotropic melt 310 K. Finally a_2 was found to transform to $β_2$ on annealing at temperatures between 220 and 280 K. This indicates that the $β_2$ polymorph is the thermodynamically stable in agreement with the heat of fusion data. The entropies of transition from a_1 or $β_1$ to a_2 or $β_2$ is about 14 $J \cdot mol^{-1} \cdot K^{-1}$ which is the one expected for a fully ordered to "conformationally disordered" crystal transition [1]. The sum of the entropies of the two subsequent transitions is about 7 $J \cdot mol^{-1} \cdot K^{-1}$. It should be noted that the entropy of the transition from the a_m to the melt is very small, about 3% of the total entropy. The melt and the a_m polymorph are thus similar at least

thermodynamically.

We explored the changes during the transitions by solid state NMR. The ^{13}C NMR spectrum shows a splitting of the CH$_3$ and of the CH$_2$ signals in the low temperature phase. Within the time scale of the experiment the side groups are frozen to specific conformations. Above the low temperature transition however the side groups become mobile and the multiple peaks collapse into two: one for the CH$_2$ and the other for the CH$_3$ group. This is also evident on the ^{13}C chemical shift anisotropy (CSA) line shape. It becomes narrower as a result of the side chain conformational motion. The CSA line becomes even narrower at higher temperatures but within the high temperature phase indicating an increase in the exchange rate of the conformations. At the same time the ^{29}Si CSA line shape remains about the same indicating that within the time scale of the experiment the backbone is immobile. The above results showing the conformational mobility and disorder in the side chain permit to classify the a$_2$ and β$_2$ polymorphs of PDES as "conformational disordered crystals".

In both the ^{29}Si and ^{13}C NMR spectra two signals are present below 200 K in a sample that contains a mixture of a and β indicating the presence of a$_1$ and β$_1$ polymorphs. At the high temperature phase however only one signal exist. This is difficult to explain in view of the existence of two polymorphs. We investigated therefore the possibility that the a$_2$ to a$_m$ transition peak is only an annealing peak, thus the possibility that a$_2$ and β$_2$ differ only in the perfection of the crystals. Although this is not a bad assumption in view of the way a$_2$ transforms to β$_2$ (long annealing close to the melting temperature) and the way a$_2$ is obtained (by quenching), subsequent experiments showed that this is unlikely. During these experiments PDES was isothermally crystallized at temperatures ranging from 150 K to 250 K and then submitted to different thermal treatment. The resulting peak was always at 279 K contrary to what one would expect from an annealing peak. These results agree with the X-ray data of a lightly crosslinked sample [5]. The

coincidence of the NMR data should therefore be explained
differently. It seems that in the high temperature phase the
main chains are not influenced by their neighbors. Then if
both polymorphs assume the same conformation they should yield
the same signal. The same conformation for both the polymorphs
was also assumed in the previous X-ray study.

The NMR data indicate that upon the transition of the a_2
and β_2 polymorphs to a_m, additional motion is introduced
in the polymer. As it is evident from the narrowing of the
^{29}Si CSA lines the additional motion is introduced in the
backbone of the polymer. On further heating and after the
final melting the CSA line becomes even narrower indicating a
faster exchange rate of conformations in the melt. The ^{29}Si
chemical shift however remains the same indicating a similar
average conformation in the melt and the mesophase. Recall
that a_m is thermodynamically similar to the melt. X-ray data
indicate that some order exist. This phase was thus
interpreted as a "viscous crystal" being distinctly different
from a liquid crystal [6]. It seems that some type of
preordering of the chains exist in the mesophase. We should
note here that upon shearing even at the lower shear stresses
a transition occurs in the mesomorphic state as it is evident
from a several orders of magnitude increase in the viscosity
of the sample and changes in the texture of the optical
micrographs.

In order to learn more about the a_m polymorph we
crystallized PDES isothermally from the melt. A somewhat
different phase diagram results then. By quenching the
material from the melt to the isothermal crystallization
temperature, bypassing thus the mesomorphic state, the
crystallinity of the sample diminishes and the peaks move to
lower temperature (266 K for the high temperature to
mesomorphic phase transition and 290 K for the low to high
temperature transition). This could be due to smaller
crystals. The position of the peaks however, is not affected
by the crystallization conditions and the subsequent thermal
treatment. The ratio of the heats of fusion of the two peaks
has also changed. These are indications that the phase

obtained during quenching from the melt is different from
that obtained during crystallization from the mesophase. These
results along with the crystallization that occurs upon
shearing indicate that the mesophase promote the
crystallization of PDES. It would be logical to conclude that
this is due to a preordering occurring within the a_m
polymorph.

CONCLUSIONS

Cormational disorder exist in the a_2 and β_2 polymorphs
of PDES. Additional disorder is introduced in the a_m
polymorph, leading to a mesomorphic state that resembles the
isotropic melt. The residual order of this state however
promotes the crystallization of PDES. Upon shearing changes
occur within the a_m polymorph indicating the possibility of
a phase transition.

REFERENCES

1. B. Wunderlich, M. Möller, J. Grebowicz and H. Baur, to be
published, Adv. Pol. Sci., 1988
2. C. L. Lee, O. K. Johanson, O. L. Flaningan and P. Hahn, ACS
Pol. Preprints 10(2), 1319 (1969)
3. C. L. Beaty and F. E. Karasz, J. Pol. Sci., Pol. Phys. Ed.
13, 971 (1975)
4. V. S. Papkov, Y. K. Godovsky, V. S. Svistunov, V. M.
Litvinov and A. A. Zhdanov, J. Pol. Sci., Pol. Chem. Ed. 22,
3617 (1984)
5. D. Y. Tsvankin, V. S. Papkov, V. P. Zhukov, Y. K. Godovsky,
V. S. Svistunov and A. A. Zhdanov, J. Pol. Sci., Pol. Chem.
Ed. 23, 1043 (1985)
6. C. L. Beaty, J. M. Pochan, M. F. Froix, and D. D. Hinnan,
Makromolecules, 8, 547 (1975)

THE EFFECT OF GAMMA IRRADIATION ON THE MORPHOLOGY OF ISOTHERMAL CRYSTALLIZED iPP

D. KOSTOSKI, Z. STOJANOVIĆ AND Z. KAČAREVIĆ-POPOVIĆ
"Boris Kidrič" Institute of Nuclear Sciences-Vinča
P. O. Box 522, 11001 Belgrade, Yugoslavia

ABSTRACT

The effect of low absorbed doses of gamma radiation in the morphology and melting behaviour of isothermal crystallized iPP has been studied using X-ray diffraction, differential scanning calorimetry and gel permeation chromatography.

The changes in the morphology, molecular weight distribution, desrease of melting temperature and crystallinity with absorbed dose are related to the radiation processes in pure iPP.

INTRODUCTION

Isotactic polypropylene (iPP) has been used extensively as a material for medical devices which are sterilized at low-doses gamma radiation (<50 kGy). However, it is well known (1,2) that iPP is vulnerable to degradation induced by irradiation in air, even at low doses. This has been the main reason for studying the effects of low-doses gamma radiation on the morphology of iPP.

By wide-angle X-ray diffraction, four different crystal forms have been identified (3,4). These polymorphs are referred to as the α, β, γ and "smectic" forms and their unit cell structures have been determined to be monoclinic, hexagonal, triclinic and pseudohexagonal, respectively. The relative amounts of these phases are very sensitive to the conditions of crystallization. In the process of slow cooling and isothermal crystallization, the iPP melt crystallizes into the monoclinic, hexagonal and more rarely triclinic lattice. Quenching the melt into the liquid nitrogen produces a "smectic" phase.

The effect of low-doses gamma radiation on the quenched forms of isotactic polypropylene was the subject of our previous work (1). In this work, we have studied the effects of gamma radiation on the monoclinic and hexagonal phase. The macrostructural states of iPP before and after irradiation were followed and the noticed effects were then related to its melting behaviour and molecular weight changes.

EXPERIMENTAL

Material

The isotactic polypropylene used was flake grade Profax 6501 containing no additives, with a number average molecular weight of 4.5×10^5 and a polydispersity 6.5. Specimens of 0.3 mm thickness were prepared by compression of the powder at 190°C and 1.75 MPa pressure. The sheets were melted in a Mettler hot stage in air at 190°C for 10 minutes and then cooled down to 130°C±0.5°C as the apparatus would allow (40 sec) and crystallized for 1 hour at this temperature. After crystallization the samples were quenched in ice-water mixture.

The samples in the form of sheets were irradiated at a dose rate of 15 kGy/hour in the presence of air at room temperature, using a ^{60}Co γ-ray source.

Experimental Methods

Wide-angle X-ray scattering (WAXS): Wide-angle X-ray diffractograms were obtained by using a PW 1010 powder diffractometer (CuK-Ni filtered radiation). Data were collected in the angular range $4° \leqslant 2\theta \leqslant 35°$. The scan rate used was 0.25°/min.

Melting behaviour: A Perkin-Elmer Model DSC-2 differential scanning calorimeter with nitrogen as the purge gas has been used. The samples of 7-8 mg weight cut from the same sheets as in GPC and X-ray measurements were analysed by heating from 310 to 460 K at 10 degree/min rate.

Molecular weight distribution - GPC measurements: A Du Pont HTSEC system with an IR detector and a Zorbax bimodal set of columns was used for GPC measurements; ODCE was solvent and liquid phase, the working temperature 140°C.

RESULTS AND DISCUSSION

Wide-angle X-ray measurements

From the diffractograms (Fig. 1) of the isothermal crystallized samples the monoclinic α and hexagonal β form crystallinity can be identified.

The relative proportion of α and β form present in any specimen was measured by an empirical ratio k, where

$$k = \frac{H_{\beta(300)}}{H_{\beta(300)} + (H_{\alpha(110)} + H_{\alpha(040)} + H_{\alpha(130)})}$$

and $H_{\alpha(1\bar{1}0)}$, $H_{\alpha(040)}$ and $H_{\alpha(130)}$ are the heights of the three strong α-form peaks and $H_{\beta(300)}$ the height of the strong single β peak. The diffractogram made after the irradiation of the same sample in air to the absorbed dose of 50 kGy shows that the total crystallinity decreases.

The effect of low-doses gamma radiation on the relative proportions of α and β forms is evident from Fig. 2, where k decreases with absorbed dose. Namely, k tends to zero when no β form is present and to unity when α form is absent. As it is well known (5), the β form is as stable as the α form, and the β↔α solid-state phase change does not take place. Because of that, the decrease of total crystallinity (Fig. 1) and k (Fig. 2) with absorbed dose can be related with the chain scissions and other changes of the gamma irradiated iPP.

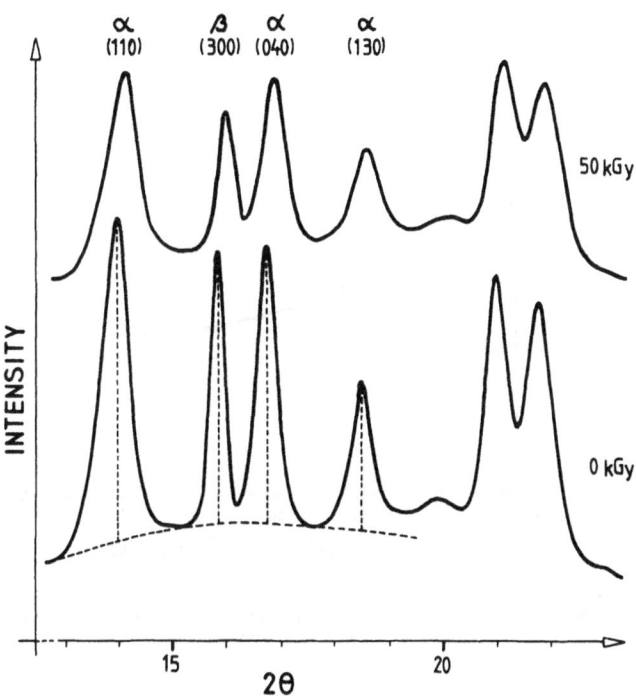

Figure 1. X-ray diffractograms of the isothermal crystallized sample before irradiation   and after irradiation with 50 kGy.

Melting behaviour and GPC Measurements

From the fusion endotherms in Fig. 3 it is evident that the melting temperatures and heats of fusion decrease with an increase of the absorbed dose. Thus the melting behaviour of irradiated samples can be related with the data of GPC measurements, where a strong drop of the molecular weight in the irradiated samples is observed.

TABLE 1

GPC data on molecular weight distribution of isothermally crystallized iPP

	M_n	M_w	M_z	M_w/M_n	M_z/M_n
iPP unirradiated	10300	48600	110000	4.72	10.68
iPP irradiated (50 kGy)	6083	25310	51848	4.16	8.52

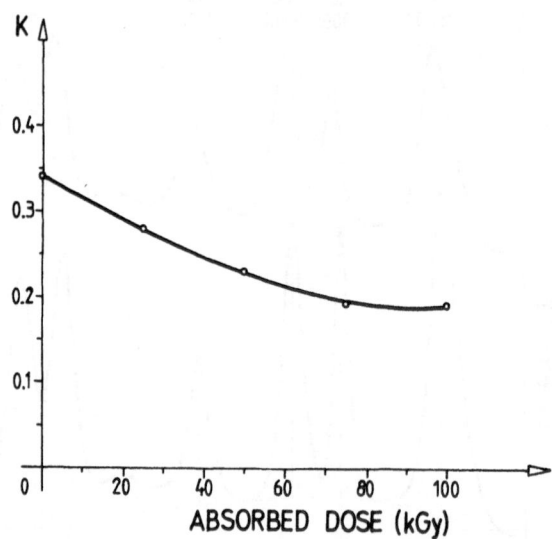

Figure 2. The effect of irradiation on the proportion of α and β form
crystallinity

The lower value of the M_z/M_n for the irradiated sample shows that the bigger molecules are more damaged than the smaller ones. This is confirmed by
a comparison between the changes of M_n representing more the small molecules as well as between the changes of M_z which more strongly reflects the
larger molecules. Namely, a damage of larger molecules is more evident
through an increased presence of small molecules and decreased of bigger
ones, as noticed on the actual MWD curves recorded during the experiments.

CONCLUSIONS

The results of this work show that radiation effects on the morphology,
melting behaviour and molecular weight in isothermal crystallized iPP can
be followed:
1) The melting temperature and crystallinity decrease with an increasing absorbed dose.
2) The content of the β form decreases more rapidly with the absorbed
   dose than the α form of crystallinity.
3) The bigger molecules are more damaged than the smaller ones.

317

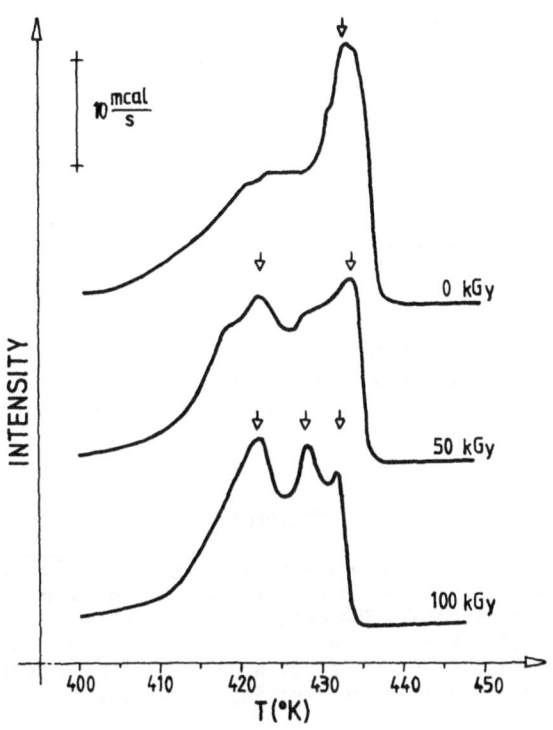

Figure 3. Fusion endotherms of isothermal crystallized iPP. Unirradiated sample and the samples irradiated with 50 and 100 kGy.

REFERENCES

1. Kostoski, D., Stojanović, Z., Gal, O. and Stannett, V. T., <u>Radiat. Phys. Chem.</u>, 1988, <u>00</u>.
2. Kagiya, T., Nishimoto, S., <u>Polymer Degradation and Stability</u>, 1985, <u>12</u>, 261.
3. Nata, G. and Gorradini, P., <u>Nuovo Cim. Supp.</u>, 1960, <u>15</u>, 1.
4. Norton, D. R. and Keller, A., <u>Polymer</u>, 1985, <u>26</u>, 704.
5. Fujiwara, Y., Goto, T. and Yamashita, Y., <u>Polymer</u>, 1987, <u>28</u>, 1253.

SHRINKAGE OF POLYMER MATERIALS AND MEMORY EFFECT

OLGA GAL, DRAGAN BABIČ AND MILOŠ PETRIĆ[*]
"Boris Kidrič" Institute of Nuclear Sciences, 11001 Belgrade, P.O.Box 522
[*]Cable Industry, 35000 Svetozarevo, Yugoslavia

ABSTRACT

Thermomechanical treatment to which polymer materials are subjected during processings: heating, unintended extention, cooling, might cause a deformation of the material. The deformation is reflected on the changes of the macromolecular structure and is manifested as the material shrinkage when reheated. In series of experiments the extent of shrinkage is examined in dependence of the treating conditions.

INTRODUCTION

The end products of polymer materials when subjected to heat shrink sometimes heyond the permissible levels. The origin of such a behaviour has to be search among the conditions of the material processing. The some indication has been shown by series of the preliminary measurements of the present work: The measurements of the length changes have been performed on two groups of specimens made from the same materials by two different procedures. One group presents the specimens cut from the plates made by melting the granules; the specimens of the second group are cut directly from the tubes and wire insulators produced by the routine processing. Their shrinkage differs distinctly: while the samples of the first group shrink 2-3 percent after heating at about the melting temperature of the material, the shrinkage among the second group varies from series to series and goes to few tens of percentage. The measurements have been performed with four commercial polymer material. The larger changes among the specimens from the products than among those cut from the plates strongly indicate that the source of the phenomenon lies among the unsuitable conditions of the processing. The present work is aimed to examine the influence of various parameters of processing and use on the phenomena. The effect of the treatment has been followed by observing the change in dimensions and macromolecular structures of the materials.

MATERIALS AND METHODS

For the most of examination a LDPE (M_w = 110000; M_w/M_n = 4.6, ρ = 0.922 g/cm^3, T_m = 112°C) is used. For comparison some of the examinations have been made with two more different materials. The plates by melting granules have been made and the dumbbell samples cut; the conditions for the plates preparation - melting, cooling etc. - have been used the same throughout the examination. The further treatments of the samples - heating, stretching, cooling, irradiation - have been controlled. Two procedures have been used:

A.　　　　Heating → Stretching → Cooling → Releasing → Reheating
　　　　　10´ at　　at T_{str}　　10´ in　　from the　　10´ at T_h
　　　　　T_{str}　　λ=3　　water　　frame

B.　Irradiation →　　"　　→　　"　　→　　"　　→　　"　　→　　"
　　e⁻($\sim$70% gel)
　　　content

　　　Two techniques have been used: a) the simple measurement of the sample length and b) the DSC method for determination of melting behaviour and calculation of the lamellae thickness distribution (LTD) as a measure of the macromolecular structure. The details on the technique and the calculation are given elsewhere (1)

RESULTS AND DISCUSSION

The full treatment occurs in two phases: I deformation and II recovery. Deformation arises as the consequence of the unsuitable treatment and is taken to be the basic cause of the subsequent shrinkage. The extent of deformation existing at the moment before the recovery starts depends on the used procedure (2), the temperature at which the extention takes place, the stretching itself and the way of cooling. The shrinkage appears in the recovery phase and depends, beside on the extent of deformation aquired in the phase I, on the temperature at which the already deformed material is reheated, T_h. The correlation between the stretching temperature (T_{str}) and the shrinkage (S_h) is presented on the figure 1. The shrinkage, S_h = (L_{str} - L_h)/(L_{str} - L_0), where L_{str}, L_h and L_0 are the lengths after stretching, reheating and before any treatment of the sample respectively, is followed at various heating temperature T_h in the range from the room to about the sample melting. The corresponding residul deformation, R.D. = (L_h - L_0)/L_0, is given in the table 1. The recovery does not necessarily go to the completion. The state when the R.D. becomes zero (S_h → 100%). Both the shrinkage and the residual deformation might differ greatly for two materials treated by the same procedure.
　　　Irradiation does not prevent deformation. In practicle, irradiation takes place after the extrusion process; it means on the already deformed material. However, irradiation might improve the elasticity of the material and thus make easier the completion of the recoveru, table 1.

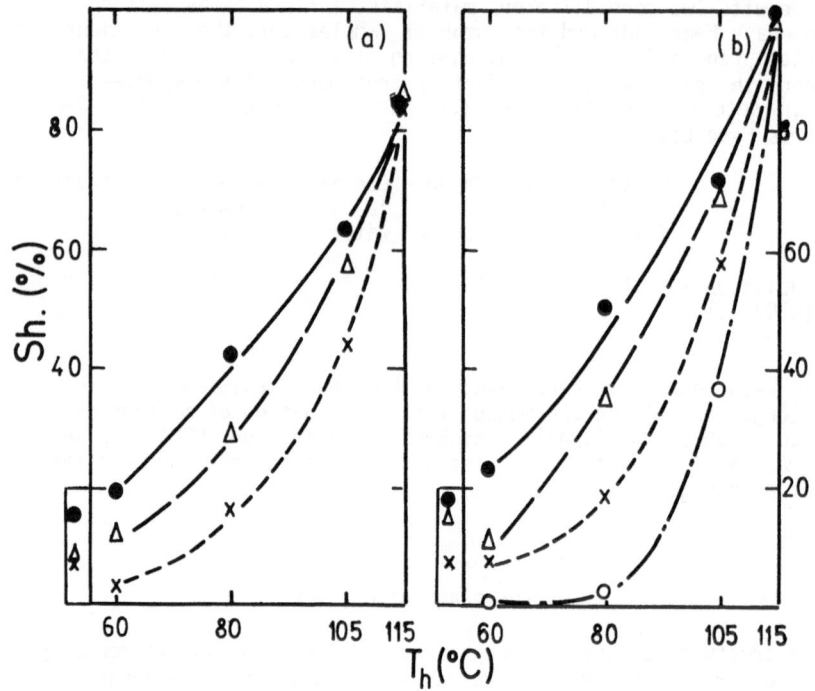

Figure 1. Shrinkage vs heating temperature for various stretching temperatures; on the side columns the shrinkage at the moment of releasing is presented.

TABLE 1

Residual deformation at T_h = 115°C for LDPE samples stretched at various temperatures

Procedure	A				B			
Residual deformation, %	32	28	35	--	0	0	0	0
Stretching temperature, °C	65	85	105	115	65	85	105	115

The shrinkage depends greatly on the time during which the sample is kept at the stretching temperature before cooling. That is seen on the figure 2. The longer time the later recovery. It is clear that the relaxation processes affect the shrinkage behaviour.

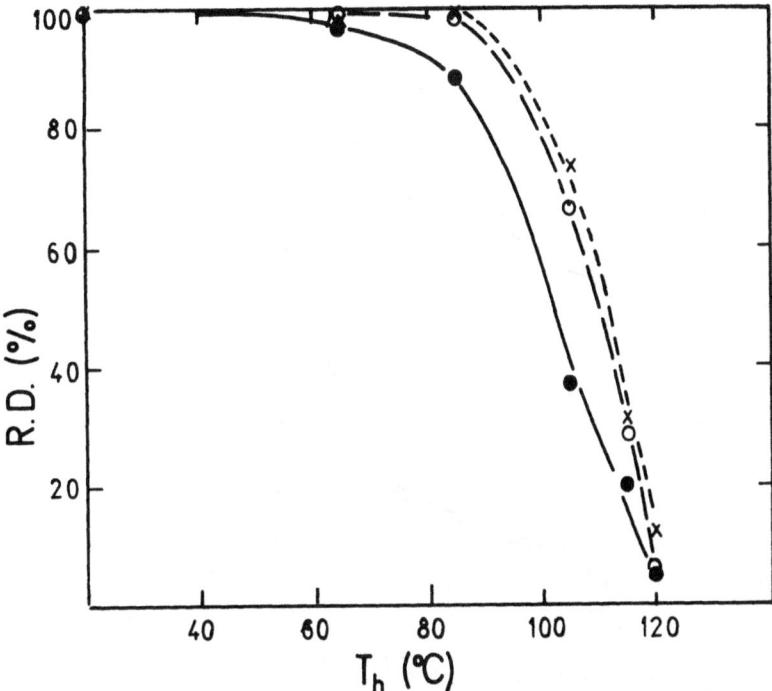

Figure 2. Residual deformation vs heating temperature for the samples
treated by the procedure A, but cooled immediately (●), after
10 (o) and 60 seconds (x) after stretching at 115°C.

By deformation a material aquires, through relaxation mechanisms, the
macromolecular structure in accordance with the applied treatment. The
various structures resulted from the various treatments are observed in
parallel with the measurements of shrinkage on the samples treated
identically. The higher temperature at stretching the more changed
macromolecular structure (2). On the figure 3 the structural change is
presented by the LTD spectra for the sample stretched at the temperature
about the material melting and for the untreated sample. The structural
state of a sample after reheating (recovery) might be compared with that
one obtained by the rescanning during DSC procedure for the treated sample.
Namely, rescanning follows one full run in the DSC procedure: heating and
cooling. Indeed, the LTD spectra of the rescanned samples of both treated
and untreated material are nearly the same and quite similar to the LTD
spectrum of the untreated material taken by the first stanning. It
indicates a tendency of the deformed material to return to a statistically
most probable structure.

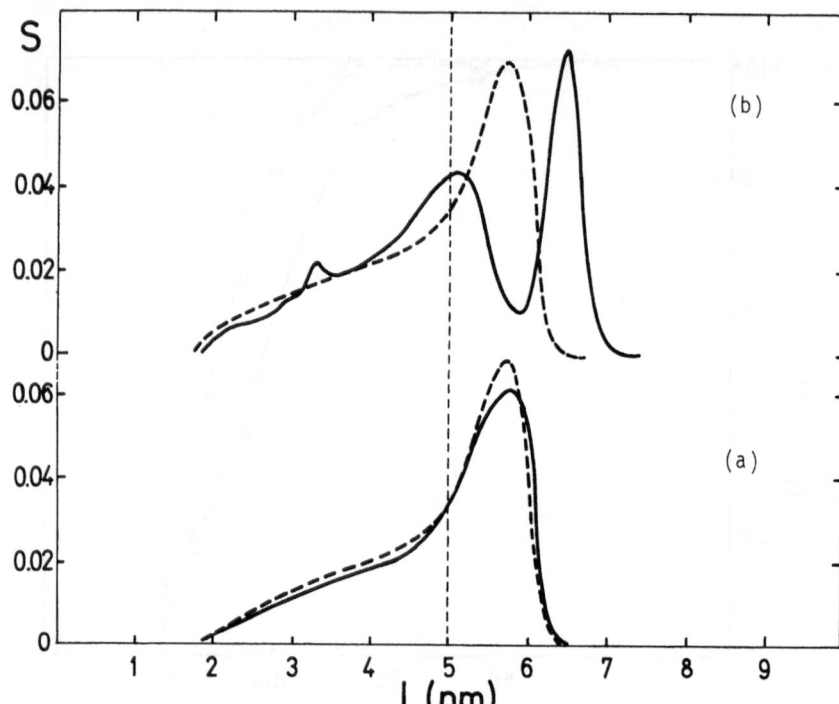

Figure 3. LTD spectra of LDPE untreated (a) and stretched at 115°C (b); —— for the first and ··· for the second scanning.

CONCLUSION

There is mutual correlations among the treatment conditions, supermolecular structure and shrinkage of a polymer material. Residual deformation is the manifestation of the uncompleted recovery to the most probable structural state; it is higher as the temperature of deformation is higher and as the period of keeping the already deformed material at the deformation temperature is longer.

REFERENCES

1. Petrić, M., Gal, O. and Babić, D., Changes of supermolecular structure of polymer materials at various processings, Radiat. Phys. Chem., submitted for publication.

2. Gal, O., Babić, D., Petrić, M., Novaković, Lj. and Charlesby, A., Some additional data on memory effect, Radiat. Phys. Chem., submitted for publication.

OPTICAL AND VISCOELASTIC PROPERTIES OF GELATIN-WATER DURING GELATION

Birgit Burg and Werner Borchard

Angewandte Physikalische Chemie der Universität-GH-Duisburg, FRG

ABSTRACT

The rheological and the optical rotation behaviour of aqueous gelatin solutions during the gelation process for different concentrations and at various gelation temperatures have been studied simultaneously. The change of viscoelastic properties during gel formation was investigated by dynamic measurements in a new oscillating viscometer of a coaxial-cylinder-type in the frequency range from 10 mHz to 3 Hz. The kinetic analysis of rheological and optical properties leads to the result, that no stationary values are reached for long gelation times. The formation of network junctions and ordered helical chain structures during gelation takes place with different reaction rates before and after a critical time for network structurization.

INTRODUCTION

Gelatin is derived from collagen and is built up of long chains of linear poly-(α)-aminoacids with a few hundred monomers. It has a polydisperse distribution of molar masses based on the extreme condition during the denaturation. Gelatin shows - like many other macromolecular components - the ability to form a gel. A gel can be considered as a mixture of at least two components exhibiting elastic properties. If the concentration of the gelatin-water mixture is above the critical concentration for network formation, homogeneous solutions of gelatin are forming a three dimensional polymer network on lowering the temperature. The gelatin/water gels exhibit a thermoreversible sol-gel transition, which is caused by the formation of non-covalent or physical crosslinks.

Beside the large changes in the mechanical properties the sol-gel transition is accompanied by an increase of the amount of the specific optical rotation. The changes of the optical properties can be explained by at least a part of the recovering of the original helical structure of the collagen molecules.

To find out, if there is a correlation between the content of the network junctions and the helical content of the mixture, we have combined kinetic measurements of the mechanical and optical properties. For this reason we have constructed a special measuring cell for the simultaneous detection of the complex shear modulus and the optical rotation of polymer/solvent systems, e.g. gelatin/water up to 4 wt.% of gelatin.

EXPERIMENTAL METHODS

The rheological measurements have been performed in a special dynamic viscometer under oscillatory shear deformation. The viscometer cell consists of two concentric cylinders, where the outer cylinder is fixed and the inner cylinder is suspended between two torsion wires. The polymer solution or the gel is situated in the cylindrical slit between these cylinders. In the slit the sample is forced to sinusoidal oscillating shear deformation. The ratio of the amplitudes and the phase difference between the torsional momentum and the resulting distortion of the oscillating inner cylinder is characteristic for the viscoelastic behaviour of the sample. The analysis of the experimental values leads to the separation of the real and imaginary part of the complex shear modulus G^*, which is consisting of the storage and loss modulus. The storage modulus G' is proportional to part of the reversible stored energy of deformation and the loss modulus G'' describes the dissipated energy during the deformation of one cycle at a fixed oscillating frequency.

The detection of the specific optical rotation $[\alpha]_\lambda$ is possible by an integrated optical channel in the lower part of the viscometer cell. The optical values are registrated continuously in a polarimeter arrangement with crossed nicols and a reference light beam in order to eliminate the increasing turbidity after long gelation times. The source of light is a He-Ne-laser .

RESULTS AND DISCUSSION

The understanding of the physical gelation process of a biopolymer/solvent system is a quite difficult problem, because the gel formation is a non-equilibrium phenomenon. This behaviour leads to an hysteresis effect in the thermal treatment of the sample by cooling and heating below the sol-gel transition temperature. The measured physical values are different for cooling and heating the samples, not only in the mechanical but also in the optical quantities.

This work mainly concerns with the question, if there is a correlation in gelation kinetics between network formation and the increasing helical content of the mixture. For this reason we have examined the time dependence of the mechanical parameters G' and G'' and the specific optical rotation $[\alpha]_\lambda$ during the gelation process, which has been induced by a temperature jump from 313 K to 293 K. The physical values G', G'' and $[\alpha]_\lambda$ are plotted versus the time for a gelatin/water mixture of 0.64 wt/wt% in fig. 1. In this figure it can be seen, that the time dependence of all values namely G', G'' and $[\alpha]_\lambda$ is different. It can be deduced also that at first the specific optical rotation changes its value already within the time necessary for temperature adjustment. After the temperature has reached its lower limit the value of G'' increases nearly stepwize and carries on growing with a small linear slope afterwards. When G''-values are close to the point of inflection of the G''-curve , the parameter G' starts becoming larger noticeably . Finally this process of network formation slows down, because the network has set and all further growth has to take place inside this network. This is also derived from results by G. Rehage and coworkers[1]. But it seems that the network formation has not yet come to an end after several days. Only a quasi-stationary state is reached, because it is not really time independent. This time dependence of the physical parameters are also found by Djabourov, Leblond and Papon[2].

For increasing gelatin concentration the values G' and G'' are higher at 293 K and the time to reach the inflection point in the G'- and G''-curves, which is the induction period of the process, decreases. The final values G' and G'' are approached at longer times. The induction period has been used by Nijenhuis to characterize the transition temperature of the gel formation [3].

The results of the kinetic measurements in this paper are: The final quasi-stationary value of the specific optical rotation is reached much earlier than the values of G' and G''. These kinetic investigations confirm with the results of the time dependence of the reduced viscosity and the specific optical rotation found by Flory and Weaver[4] and from light scattering experiments of very diluted solutions by Doty and Boedtker [5]. The process of network formation is discussed as multiple step mechanism as already proposed by Mercier and coworkers [6]:

1. Helix formation of single and multiple strands by nucleation and growth.
2. Aggregation of helices leading to an increase of the hydrodynamic volume. If the clusters are large enough a weak network is formed.
3. Stacking of helices to crystalline bundles. This process probably is also taking place by steps of nucleation and growth.

During the gelation process a frequency dependence of the mechanical data is observed. In fig. 2 the logarithms of the rheological parameters are plotted versus the logarithm of the frequency of deformation for three different gelation times of a 4.2 wt/wt% gelatin mixture at a temperature of 303.8 K. In comparison with a covalent gelation process, where at the critical degree of polymerisation p_c the infinite network is formed, in a physical gelation process the gel-point is reached at a critical time of gelation t^*. Before a weak network is formed the storage modulus G' is nearly zero but lower than the loss modulus G''. At the gel-point both moduli have the same value. The slope of the lines in fig. 2 are approximately 0.5, following the model of a Rouse-chain[7]. After the crossover of two curves the parameter G' will increase much stronger than G''. If the network formation is advancing the storage modulus is nearly independent of frequency, as has been found by other authors [8].

ACKNOWLEDGEMENT

The financial support by the "**Deutsche Forschungsgemeinschaft**" is greatfully acknowledged.

FIGURES

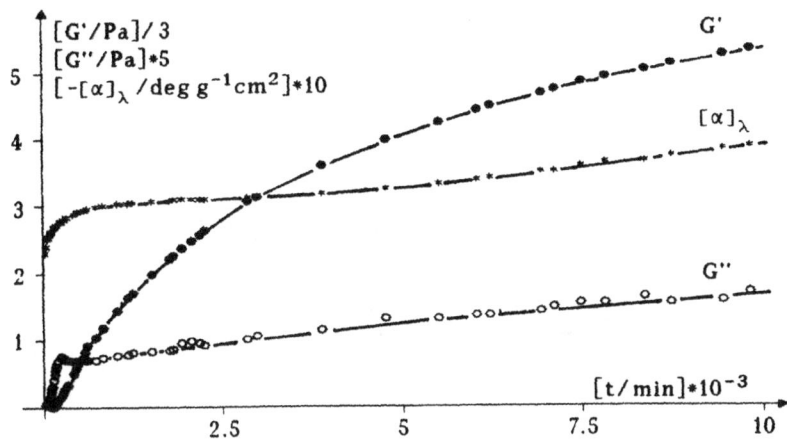

Fig. 1: Time dependence of all physical values during the gelation process induced by a temperature jump from 313K to 293K (0.64 wt/wt %).

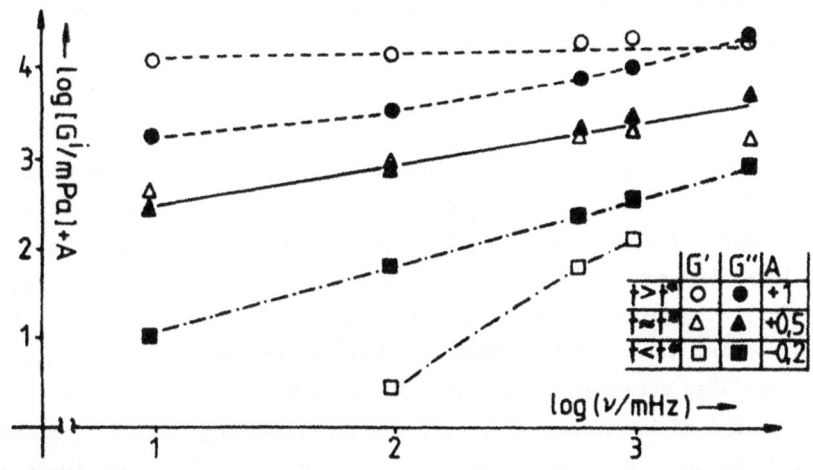

Fig. 2: Frequency dependence of the rheological values of a 4.2 wt/wt % gelatin mixture during the gelation process at 303.8 K.

REFERENCES

[1] G. Rehage, K. Bergmann, W. Borchard. In: "Photographic Gelatin II"; Ed. R.J.Cox; Academic Press; London 1976; p. 57-71
[2] M.Djabourov, J.Leblond, P.Papon, J.Phys.France, 1988, 49, 319-332
[3] K.T. Nijenhuis, Coll.&Polym.Sci., 1981, 259, 522-535
[4] P.J.Flory, E.S.Weaver, J.Amer.Chem.Soc., 1960, 82, 4518
[5] P.M.Doty, A.Boedtker, Ibid., 1954, 58, 968
[6] P. Goddard, J.J. Biebuyck, M. Daumerie, H. Naveau, J.P. Mercier, Polymer Sci., Chem. Ed., 1978, 16, 1817
[7] H.H. Winter, F. Chambon, J. Rheol., 1986, 30(2), 367
[8] M.Djabourov, J.Leblond, P.Papon, J.Phys.France, 1988, 49, 333-343

ON THE STRUCTURE OF HARD ELASTIC POLYPROPYLENE FIBRE

TONGYIN YU, QIANGUO DU AND JIACONG HU
Department of Material Science
Fudan University
Shanghai, China

ABSTRACT

The superstructure of hard elastic polypropylene (HEPP) fibre was investigated. By means of SAXS studies, stress in various media, and birefringence determinations, a more precise model is suggested.

INTRODUCTION

Hard elastic polypropylene (HEPP) fibre is a kind of fibre with crystalline lamellae aligned normal to the fiber direction. It has high elasticity when extended. The elasticity of this crystalline polymer is primarily not rubber-like, but is rather controlled by the energy effect. Recently many models were proposed to explain its elasticity. The simplest one is the leaf-spring model which is based on reversible bending of lamellae. Some models also consider the contribution of lamellae rotation, the entropy change of interlamellar layer, the surface energy effect of microvoids and the surface energy effect of microfibrils (1-6). Each structure model explains certain aspects of hard elasticity.

Formerly, hard elastic behavior was thought to be acquired only from crystalline polymers by a combination of melt extrusion and stress crystallization followed by annealing under tensile stress. Recently, Xu et al. (7) worked out high elastic polyethylene by cold-drawine and followed by constrained annealing from non-elastic fibres. Further, Bair and co-workers (8) reported the development of hard elastic material from high-impact polystyrene, an amorphous glassy polymer, by the induction of profuse crazing stacked perpendicular to the applied uniaxial tension. Thus, a more precise view must be introduced to account for the hard elastic property.

MATERIALS AND METHODS

HEPP fibre (SM-45-A) was kindly supplied supplied by Dr. H. D. Noether of Celanese Research Company. The liquids used in experiment are all of anlytical grade except liquid paraffin (chemical grade).

The tensometer used is DL-1000 made by the Second Material Testing Factory, Changchun. The extension experiments were carried out at the temperature range of 20-25° unless otherwise specified, with sample length of 50 mm, and draw speed 50 mm/min. The sample and clamps can be both immersed in liquids during experiment.

The measurements of contact angles between polypropylene fibre and liquids were performed on WDS-I contact angle meter made by the East China Institute of Chemical Technology.

A Rigaku 2203 E1 X-ray diffractometer was used for the small angle X-ray scattering measurements with experiment conditions of copper target, K_α ray, 40 kV and 15 mA.

The birefringence was measured on polarizing microscope made by Shanghai Optical Instrument Factory.

RESULTS AND DISCUSSION

SAXS studies

It is well known that the stacked lamellae would open to form micro-voids during extending. This creates more surface area and is bond to enhance its surface energy. Figure 1 is the SAXS curve of HEPP at meridian direction. The long period increases at 20% elongation, which accounts for the spraying apart of lamellae and the formation of micro-voids (9).

Surface energy

In order to semi-quantitatively estimate the surface-energy-associated retractive stress, the stress relaxation curves were studied, curve 1 in Figure 2 is the stress relazation curve of HEPP at 50% elongation. It tends to level off after 6 min. When samples were soaked in liquids which can wet polypropylene surface, the stress dropped immediately as shown by curves 2 and 3 in Figure 2. After liquids were removed and evaporated from the fibre, the stress relaxation curves went back to their previous height.

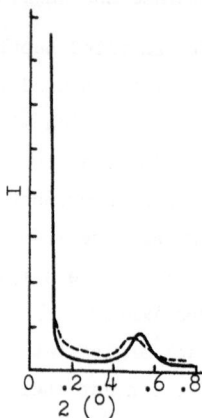

Figure 1. SAXS curves of HEPP fibre at meridian direction. ___ before
elongation; --- at 20% elongation

By measuring the contact angles between the polymer and various
liquids, it is possible to calculate the proportion of surface energy
effect in the initial retractive stress using Young-Dupre's equation, the
value is around 0.3. After six minutes of relaxation, it increases to 0.5.

Orientation of polymer chains in amorphous region

The calculation of orientation function in amorphous region from the data
of birefringence shows that it is close to zero (10). It means that this
part of polymer chains is more or less random in nature. A rough estimation from Figure 2 of its contribution to elasticity is less than 20%, it of
course will be lost at low temperature.

Model

The morphology of HEPP fibre is a composite of alternative layers of
stacked crystals and amorphous region. Under tension, micro-fibrils are
formed in the latter. The large surfaces created manifest the greater
part of hard elastic behavior.

CONCLUSIONS

The superstructure of HEPP is a manifestation of bulk-microfibril composites. The elastic force at room temperature consists of three parts:
entropic elasticity (~20%), surface energy (~50%) and deformation of
lamellae (~30%).

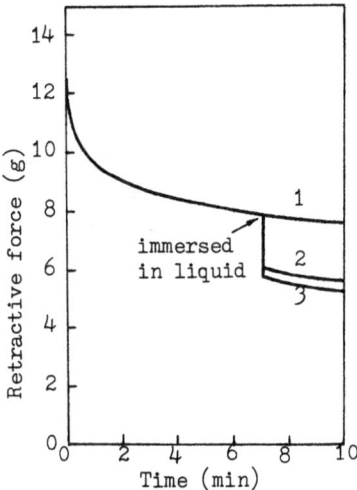

Figure 2. The stress relaxation curves of HEPP fibre. 1. air; 2. ethanol; 3. butanone.

REFERENCES

1. Sprague, B.S., Relationship of structure and morphology to properties of "hard" elastic fibers and films. J. Macromol. Sci.-Phys., 1973, B 8(1), 157.

2. Cannon, S.L., McKenna, G.B. and Statton, W.O., Hard elastic fibers, J. Polym. Sci., Macromol. Rev., 1976, 11, 209.

3. Miles, M., Petermann, J. and Gleiter, H., Structure and deformation of polyethylene hard elastic fibers. J. Macromol. Sci.-Phys., 1976, B12(4), 523.

4. Samuels, R.J., High strength elastic polypropylene, J. Polym. Sci., Polym. Phys. Ed., 1979, 17(4), 535.

5. Hosemann, R. and Cackovic, H., Adhäsive Reibungselastizität von hart-elastischem polypropylen (HEPP). Colloid Polym. Sci., 1981, 259, 15.

6. Chou, C.J., Hiltner, A. and Baer, E., The role of surfaces stresses in the deformation of hard elastic polypropylene. Polymer, 1986, 27, 369.

7. Xu, G., Du, Q. and Wang L., Hard elasticity of polyethylene formed by cold-drawing and annealing. Makromol. Chem., Rapid Commun., 1987, 8, 539.

8. Moet, A., Palley I. and Baer, E., Development of hard-elastic solids from glassy polymers. J. Appl. Phys., 1980, 51(10), 5175.

9. Yu, T., Du, Q. and Hu, J., Elasticity of hard elastic polypropylene fibre and surface energy. Polym. Commun.,(Beijing), 1985(1), 35.

10. Du, Q., Lin M. and Yu, T., Character of the structure of hard elastic polypropylene fibers. J. Fudan Univ.(Natural Sci.), 1986, 25(4),375. (in Chinese); Chem. Abstract, 1987, 106(22), 177975e.

Part 5

SPECIALITY POLYMERS

POLYMERS FOR OPTICAL RECORDING MEDIA

R.G.GOSSINK, G.H.WERUMEUS BUNING, R.WIMBERGER-FRIEDL
Philips Research Laboratories
P.O.Box 80,000, 5600JA Eindhoven, The Netherlands

ABSTRACT

Three roles of polymers in optical recording media can be distinguished: the recording layer, the protective coating, and the transparent substrate. This review focusses on the last role, and in particular on the problem of birefringence induced by polymer chain orientation and how to minimize this. Emphasis is placed on reducing the stress optical coefficient of polycarbonate type polymers by modifying the polymer structure or by blending polycarbonate with a second polymer.

INTRODUCTION

Polymers play three different roles in optical media: as the recording layer, as the protective coating on this recording layer, and as the transparent substrate. Since the last role is by far the most important, the emphasis in this review will be placed on substrate materials. In particular the results of recent work done at our laboratories on the problem of birefringence in optical disc substrates and how this is related to processing and to polymer chain structure will be discussed.

As an example of an optical recording medium, Figure 1 shows schematically the construction of a Compact Disc (CD).

It consists of a transparent substrate, for which polycarbonate is used, with a thickness of 1.2 mm. This substrate has a structure of depressions on its inner surface, with a depth of 0.12 μm. The length of and distance between these pits varies; each transition into or out of a pit represents a 1, a flat region represents

a (series of) 0 in the digital signal. Other important parts in the disc structure are the reflective metal layer (Al or Ag) and the polymeric protective coating.

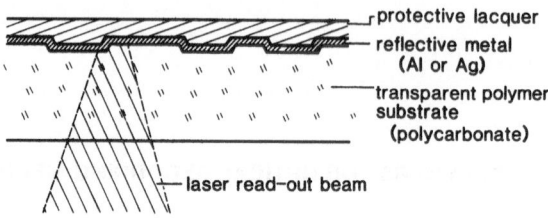

Figure 1. Construction of a Compact Disc

The information contained in the pit pattern is read out with a laser beam with a wavelength of about 800 nm, emitted from a solid-state laser and focussed to a spot of about 1 μm in diameter. When this spot overlaps a pit, there is destructive interference between the light reflected from the bottom and the surroundings of this pit, thus reducing the amount of light reflected as compared to no pit.
There are three types of optical recording media (1):

1. Prerecorded media, such as the CD and the LaserVision (LV) disc. The polymer substrates containing the prerecorded pit structure are usually made by injection moulding.

2. Write-once media, in most cases based on burning pits in an organic or inorganic layer which melts, evaporates or decomposes under the influence of the laser beam. Another possibility is inducing a local phase transition, e.g. from polycrystalline to amorphous. Read-out is done at a lower intensity.

3. Reversible media, in which information can be recorded and erased repeatedly. Here again, phase change is a possibility, but the more prominent recording principle is magneto-optic recording. In this recording mode (1) the sensitive layer consists of a ferro- or ferrimagnetic alloy, e.g. GdFe, with a strong tendency for magnetization perpendicular to its surface. When a ferromagnetic alloy is heated locally above its Curie temperature using a focussed laser beam in the presence of an external magnetic field, the initial direction of magnetization can be reversed. This can be read out optically, since linearly polarized light in these materials undergoes a rotation of its polarization plane depending on the direction of magnetization. This (Kerr) effect is small, typically $< 1°$, and is therefore easily disturbed by optical imperfections in the substrate, such as birefringence.
Although the tendency in reversible media is strongly towards inorganic layers, below also some reversible organic layers will be discussed.

Several types of write-once and rewritable media are currently on the market or are at an advanced stage of development; they are in first instance intended for professional or semi-professional use.

ROLE OF POLYMERS IN OPTICAL MEDIA

Reminding of the structure of a CD (Figure 1), we can distinguish between (i) the transparent substrate, which has the function of a mechanical support, but also takes care that scratches and dust at the surface of the disc are out of focus of the laser beam, (ii) the layer in which the information is recorded, and (iii) the top coating, which should protect the recording layer against mechanical and chemical damage.
Polymers are generally used for the protective coating, because they can be easily applied. Suitable coatings spun on from solution as well as solvent-free photocurable coatings have been developed (2).

The information layer of prerecorded media is inherently based on a polymer, since the information is moulded in the substrate surface. Write-once and reversible media have a separate recording layer, in which polymers play a less dominant role. The media on the market or in development largely rely on inorganic recording layers. Nevertheless, a number of interesting organic concepts have been described : write-once by thermally forming pits in dye or dye-containing polymer systems, and also the formation of vesicles under the action of decomposing polymers, e.g. nitrocellulose or PMMA (3,4).
A possible reversible organic layer (see Figure 2) may be built up of an elastomer containing dye I, and a thermoplastic polymer containing dye II (5).

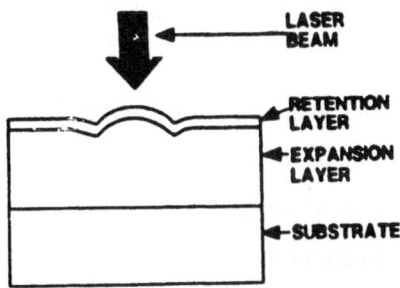

Figure 2. Construction of reversible organic bilayer (5).

These dyes have their absorption maxima at two different wavelengths. In the writing step the wavelength I at which dye I absorbs is used to heat the elastomeric layer, which will expand as soon as thermal conduction brings the thermoplastic layer above its glass transition temperature. This expansion is fixed by the rapid temperature drop, as soon as the laser beam leaves the irradiated area. In the erasing step, wavelength II is used to directly heat the thermoplastic layer, which flattens again under the influence of the stressed elastomer.
Another proposal for a reversible recording process involving a polymer as the recording layer is based on a liquid-crystalline side-group polymer (6). Apart from the mesogenic side-groups, the polymer contains dichroitic dye groups which are ordered along with the mesogenic groups. An example of such a polymer is shown in Figure 3.

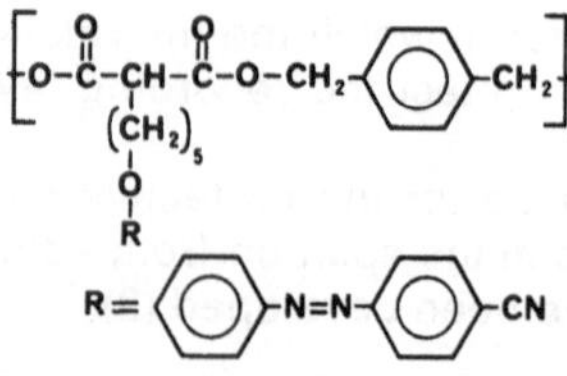

Figure 3. Mesogenic polyester for reversible optical recording (6).

The recording process relies on thermally disturbing the initial ordering in the polymer, giving rise to local light scattering. In the erasing process the ordering is restored.

POLYMERS FOR SUBSTRATES

Since the major role that polymers play in optical recording media is in substrates, the remaining part of this review will focus on this aspect. The most imporant requirements on polymers for optical disc substrates are, in a qualitative sense, the following:

- high optical transparency at the write and read wavelengths.

- high purity (freedom from inclusions).

- high dimensional stability with regard to moisture.

- high heat deflection temperature.

- low optical anisotropy.

- good processing properties.

- good mechanical properties.

- good adhesive properties vs. adjacent layers.

- low cost.

Although thermoset and photoset materials are certainly among the candidates for specific optical media substrates, such as large recordable discs, the discussion will be restricted here to thermoplastics. Suitable candidates are then found among amorphous polymers with excellent optical transparency and an acceptable mix of other properties.

Hennig (6) is one of the authors who have made a comparison between the few serious candidates available. Table 1 is based on his comparison.

TABLE 1
Qualitative comparison of a few substrate materials for optical discs.

property	PMMA	PC	glass	c-MMA-S	TPX	PC-PS
birefring.	+	0	+	0	+	+
H$_2$O -abs.	–	+	+	+	+	+
permeabil.	–	–	+	–	–	–
HDT	0	+	+	0	0	+
toughness	+	+	–	0	+	+
E-modulus	+	0	+	+	–	0
processing	+	0	–	+	–	0
cost	+	0	–	+	0	0

permeabil. = permeation coefficient for oxygen and water vapour; HDT = heat distortion temperature.

The three main substrate materials used in the optical media presently on the market or in development are PMMA, polycarbonate (PC) and glass. From the table it is seen that PMMA has a high water absorption and a medium HDT, PC has a medium tendency to birefringence and a high permeability, whereas glass is brittle and expensive. The present situation is that PMMA is used for the 30 cm diameter LV discs where because of their sandwich structure (two discs glued back-to-back) water uptake does not lead to high dimensional instability. PC is used for the CD and is also a good candidate for the smaller write-once and reversible types, whereas glass is still the preferred material for the 30 cm recordable media.

Since there is no ideal substrate material available yet, considerable R&D efforts have been and are being made to develop improved materials. In the right half of Table 1 some possibilities are listed: the copolymer of MMA and styrene, with a reduced water absorption when compared to PMMA at the cost of a higher birefringence; an olefinic polymer, poly(4-methyl 1-pentene) (TPX), with favourable water absorption and birefringence but poorer processibility; and

finally the possibility of a blend of PC and polystyrene is indicated. This would offer a reduced birefringence with respect to the PC homopolymer by compensation of the stress-optical coefficient, as we will see later.

BIREFRINGENCE IN POLYCARBONATE SUBSTRATES

PC thus is the material used in today's highest volume application, the CD, as well as some other optical media, and its tendency to birefringence is something to be improved. Birefringence in the substrate leads to optical retardation, i.e. the birefringence Δn integrated over the thickness of the substrate. A high retardation leads to changes in the state of polarization of the laser beam, giving unwanted feedback of part of the reflected light into the laser. The retardation in optical disc substrates has to remain below a level of 50 nm (single pass; focussed beam) for a CD substrate, and below 20 nm for LV and magneto-optic reversible media substrates.

In a stress-free amorphous polymer there is no birefringence. However, application of stress above or below T_g introduces birefringence, i.e. an anisotropic refractive index. The contribution induced above T_g is due to frozen-in polymer orientation, and that induced below T_g to frozen-in thermal stresses. A linear proportionality exists between stress $\Delta\sigma$ and birefringence Δn with the stress optical coefficient C as the proportionality constant. C has a different value above and below T_g (C_{or} and C_{th}, respectively). This linear relationship is valid except for the region closely around T_g.

$$\Delta n = C_{or} \times \Delta\sigma_{or} + C_{th} \times \Delta\sigma_{th}$$

The equation above shows that the birefringence is a function of both the material and the process. The stress optical coefficients are material constants, whereas the induced stresses are influenced by both the material and the process. Which of both contributions, that by orientation and that by thermal stresses, has the largest influence in the case of injection moulded CD substrates made of PC ? Wimberger-Friedl et al. (8) have made a careful analysis of the distributions of Δn in such substrates. The material used was Makrolon CD-2000 from Bayer, and the moulding machine used was of the Meiki Dynamelter type.

In the analysis cross sections of the substrate, both in radial and tangential direction, were studied under a polarization microscope. From these Δn distributions, the birefringence over the thickness of the disc can be calculated. Figure 4 gives a characteristic result.

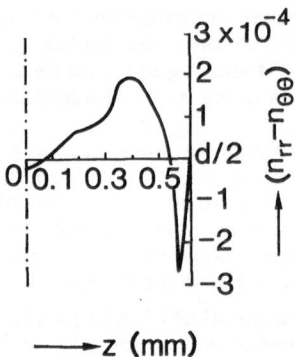

$\longrightarrow$ z (mm)

Figure 4. Birefringence distribution over the thickness of an injection moulded CD substrate at r = 35 mm (8).

A large negative value is observed just below the surface, and a positive value further down. At the centre the birefringence is almost zero. The overall retardation is thus determined by two large contributions of opposite sign, which makes it no wonder that the retardation level is extremely dependent on processing conditions, in particular on finding a good balance between the filling and compression stages, which can be related with the negative and positive regions, respectively.

An important result of the analysis of Wimberger-Friedl et al. has been, finally, that the main birefringence contribution that must be taken into consideration is that caused by orientation, and not by thermal stresses, since the latter works out in a similar way in both radial and tangential directions.

We will therefore concentrate the discussion on birefringence brought about by orientation stresses. One measure to reduce this is to bring down these stresses as far as possible. From the material suppliers side this has been done by developing a polycarbonate grade especially for optical disc substrate production. This has an improved flowability due to a considerable reduction of the molecular weight, however to such a value that the thermal and mechanical properties still remain acceptable (9). Other necessary modifications have been an improvement of heat stability and optical purity.

From the above birefringence analysis it can further be expected that by finding a good balance between the many injection moulding parameters, such as melt temperature, injection speed, mould temperature, and packing pressure profiles, the birefringence distribution can be optimized and thereby the optical retardation reduced. Indeed by using well-controlled moulding machines, especially modified for disc substrate production, the retardation in polycarbonate CD-size substrates can be brought back to a value not far above 20 nm. These moulding machines are generally equipped with plastification units giving high melt temperature but short residence time to prevent degradation of the polymer, and a good thermal homogeneity (10).

Balancing the injection moulding conditions so as to reduce the birefringence would be less critical, if the stress optical coefficient C_{or} of PC would be smaller. C_{or} of a polymer can be related to Boltzmann's constant k, to the temperature T, to the average refractive index $\bar{n}$, and is directly proportional to the difference in polarizability parallel and perpendicular to the polymer backbone ($\alpha_1 - \alpha_2$):

$$C_{or} = \frac{2\pi}{45kT} \; \frac{(\bar{n}^2 + 2)^2}{\bar{n}} \quad (\alpha_1 - \alpha_2)$$

In the structure of PC the presence of the aromatic rings is observed (Figure 5); these rings have a high in-plane polarizability compared to the direction perpendicular to the ring.

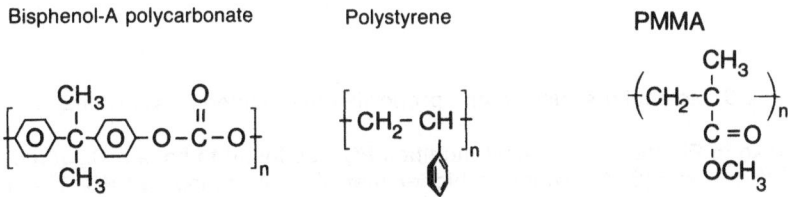

Figure 5. Molecular structure of PC, PS and PMMA.

Although the chain conformation is more complicated than depicted in the figure, PC proves to have a high positive value of the stress optical coefficient (11). In PS the opposite situation is encountered: the aromatic ring is positioned perpendicularly to the polymer backbone, leading to a high negative value of C_{or}. PMMA on the other hand has a low and not very different polarizability in both directions, yielding a low value of C_{or}.

The question arises whether the stress optical coefficient of a polycarbonate-type polymer can be reduced by modifying the material whilst maintaining its otherwise attractive properties. This could in principle be done in two ways:

1. changing the polymer chain structure.

2. blending PC with a polymer with opposite sign of C_{or}, such as PS.

An attempt along the first route was undertaken by Werumeus Buning et al. (11), who synthesized a benzyl-modified PC. Depending on the orientation of the aromatic side-group, this could lead to a lowering but also to an increase of the stress optical coefficient.

C_{or} of this polymer as a function of temperature was determined using a flow birefringence apparatus developed by Janeschitz-Kriegl and co-workers (see ref. 11). In this method an independent determination of the melt viscosity at the same temperature is needed.

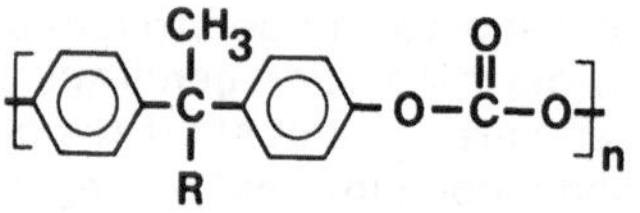

R	C_{or} $(10^{-9} Pa^{-1})$	T_g (°C)
CH_3	3.5 - 3.7 (170 - 230 °C)	145
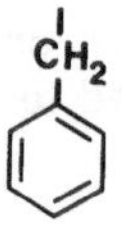 CH_2 (benzyl)	4.0 (216 °C)	165
 (phenyl)	1.8 - 2.1 (239 - 253 °C)	188

Figure 6. Molecular structure and properties of modified polycarbonates.

As shown in Figure 6, the benzyl-modified PC was found to have a C_{or} value of $4.0 \times 10^{-9} Pa^{-1}$ (at 216 °C), which is higher than that of unmodified PC. This increased C_{or} value suggests that the aromatic side-ring is oriented largely parallel to the polymer backbone. Molecular mechanics calculations, in which the lowest-energy conformation of the benzyl-modified polymer was determined, clearly support this explanation (11).

Another modified PC, in this case with a phenyl side-group in the bisphenol moiety, was made available by Bayer (see Figure 6). Molecular mechanics calculations predicted that for this polymer the aromatic side-group in its lowest-energy conformation is in the desired perpendicular position. And indeed the stress optical coefficient resulting from the flow birefringence measurement for this polymer has a value of $1.8 \times 10^{-9} Pa^{-1}$ (at 253°C), considerably lower than that of unmodified PC (11).

A recent result concerns the possibility of compensating the stress optical coefficient of PC by blending it with a polymer with a C_{or} value of opposite sign (12).

A candidate blend mentioned earlier is PC-PS, but this is not a practical proposition since the components of this blend are known to be incompletely miscible. However, the polycarbonate modified by incorporation of four methyl groups in the bisphenol-A moiety (ortho to the carbonate oxygen) is known to be compatible with polystyrene, and both polymers have approximately the same absolute value of C_{or}. The tetra-methyl polycarbonate used in the study by Werumeus Buning and Gijsen (12) was again made available by Bayer.

Different compositional ratios of this blend were analyzed for their stress optical coefficients, applying a method in which an 80 μm thick foil of the blend, obtained from solution, is stretched about 15°C above T_g of that composition in an Instron tensile tester at different stress values, after which the foil is quenched. The birefringence is then analyzed in a polarization microscope. From the measured stress versus elongation and the birefringence versus elongation, the ratio between birefringence and stress can be calculated, yielding C_{or}.

Figure 7 gives the dependency of stress optical coefficient on blend composition.

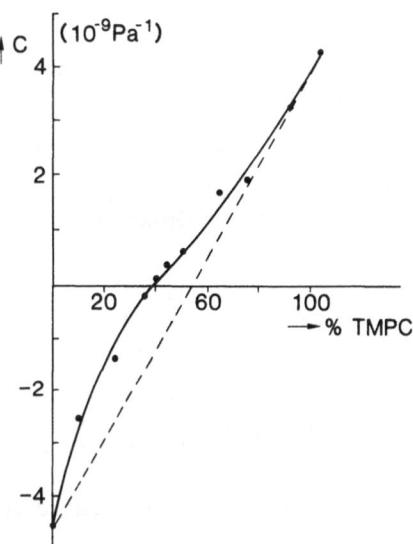

Figure 7. Stress optical coefficient as a function of composition in the blend tetra-methyl polycarbonate / polystyrene.

Indeed compensation of C_{or} is found, but interestingly not at 50% tetra-methyl PC. This is provisionally ascribed to the fact that, although completely miscible, the two polymers are oriented in the blend to a different extent.

CONCLUSIONS

Polymers play a vital part in optical recording media, mainly as the substrate material. Polycarbonate is the material of choice for Compact Disc substrates. However, in particular for magneto-optic rewritable disc substrates it could be attractive to reduce the birefringence not only by optimizing the moulding process, but also by modifying the material towards a lower orientational stress optical coefficient. Work at our laboratories has shown that this can be achieved both by changing the polymer chain structure itself and by composing a suitable blend.

ACKNOWLEDGEMENTS

The authors thank Drs. K.Idel and U.Grigo (Bayer AG, Uerdingen, FR Germany) for providing the modified polycarbonate samples, and Dr.W.G.O'Brien (Philips and DuPont Optical, Wilmington, DE, USA) for helpful discussions.

REFERENCES

1. Bouwhuis, G. et al., Principles of Optical Disc Systems, Adam Hilger Ltd., Bristol, 1985.

2. Lamberts, J.J.M., UV-haertende Materialien als Schutzlack und Replikationsmedium, Proc. Fachtagung Polymerwerkstoffe fuer die Elektronik. SKI, Wuerzburg, FR Germany, October 14-16, 1987, pp. 177-96.

3. Huijser, A., Materials for on-line optical recording. In ref. (1), pp. 210-27.

4. Broer, D.J., Application of organic coatings in the electronics industry, Proc. Thirteenth International Conference in Organic Coatings Science and Technology, Athens, Greece, July 8-12, 1987.

5. Hartman, J.S. and Lind, M.A., Erasable bilayer dye-polymer optical recording medium. Proc. Topical Meeting on Optical Data Storage, Lake Tahoe, Nevada, USA, March 11-13, 1987, pp. 155-8.

6. Eich, M. Wendorff, J.H. Reck, B. and Ringsdorf, H., Reversible digital and holographic optical storage in polymeric liquid crystals, Proc. 16. Arbeitstagung Fluessigkristalle, Freiburg, FR Germany, March 19-21, 1986, paper 6.

7. Hennig, J., Polymer substrates for optical discs, Jap. J. Appl. Phys., 1987, 26, Suppl. 26-4, 9.

8. Wimberger-Friedl, R. and Janeschitz-Kriegl, H., Birefringence in injection-moulded optical components. In Integration of Fundamental Polymer Science and Technology, Vol.2, Kleintjens, L.A. and Lemstra, P.J., eds., Elsevier, London, 1988.

9. Siebourg, W., Polycarbonat - ein Werkstoff fuer optische Speichermedien, Kunststoffe, 1986, 76, 917-9.

10. Anders, S. and Hardt, B., Spritzgieszen von Compact Discs, ibidem, 1987, 77, 21-6.

11. Werumeus Buning, G.H. Wimberger-Friedl, R. Janeschitz-Kriegl, H. and Ford, T.M., Optical anisotropy of polycarbonates. In Integration of Fundamental Polymer Science and Technology, Vol. 2, see ref. (8)

12. Werumeus Buning, G.H. and Gijsen, R.M.R., Stress-optical study of the tetramethyl-polycarbonate / polystyrene blend, Polymer Preprints, 1988, 29, 2 (Fall), in print.

NEW MODEL AND STARTING COMPOUNDS FOR LIQUID
CRYSTALLINE POLYMERS

GÜNTER LATTERMANN, GÜNTER STAUFER UND GERHARD RENNER
Makromolekulare Chemie I, Universität Bayreuth
Postfach 101251, D-8580 Bayreuth

ABSTRACT

It seems to be important for the synthesis of discotic liquid
crystalline polymers to find simple and clear synthetic path-
ways for related monomers.
In the course of our work, we found that the reaction of
3,4,5-tris(alkoxy)benzoic acid chloride with e.g. phloroglucin
or cis,cis-phloroglucit leads to the formation of tristar-
discogens, a new class of discotic liquid crystalline compounds. By means of blocking reactions selective substitutions
at the central core e.g. the cyclohexane core becomes possible.
The resulting compounds, possessing two hydroxy groups per
molecule are liquid crystalline themselves, among them a new
class of discotic associates.
The phase behaviour was examined by polarization microscopy
and by means of differential scanning calorimetry.
Different substituents show a drastic influence on the liquid
crystalline properties.
The formation of polymers of this intermediate compounds in
the synthetic pathway to discotic polymers have been examined,
their phase behaviour is discussed.

INTRODUCTION

Till now only few examples of discotic liquid crystalline
polymers are known. In the first example described in
literature (1),hexasubstituted triphenylene cores as discotic
sidegroups, are linked to a polysiloxane backbone via a
flexible alkyl spacer. Discotic main chain polymers (2,3)
posses as disclike core benzene or tripenylene derivatives
with the same high degree of substitution, i.e. six.
An alternative should be the synthesis of discotic monomers
with only tetra or trisubstituted central cores.
Several attempts have been made to realize this idea, first

with model compounds. We succeeded in finding a new class of
discotic liquid crystalline compounds (4), which show three
common structural principles: a) the central core is trisubstituted, b) the overall number of lateral chains is nine and
c) the side chains do not start from the central core but from
regional cores.

RESULTS

As an example   3,4,5-tris(alkoxy)benzoic acid chloride can be
reacted with cis,cis-phloroglucit. The resulting tristar-
discogen 1

$$R = -O-C_{10}H_{21}$$

1

exhibits an enantiotropic discotic mesomorphism with a hexagonal columnar phase. It shows the following transition scheme:

$$K \xrightarrow{\text{31.3 (146.5)}} D \underset{\text{57.9 (16.9)}}{\overset{\text{61.5 (17.6)}}{\rightleftharpoons}} I$$

The basic requirement for a controlled synthesis of discotic
polymers is the possibility of introducing different substituents in the trisubstituted central core.
Following the reaction scheme in figure 1, we reacted cis,cis-
phloroglucit 2 with phenylboronoic acid 3, leading to a
protection of two functional groups of the cyclohexane core.
After the reaction with 3,4,5-tris(alkoxy)benzoic acid chloride
and the subsequent deblocking reaction we obtained product 4
in good yield.

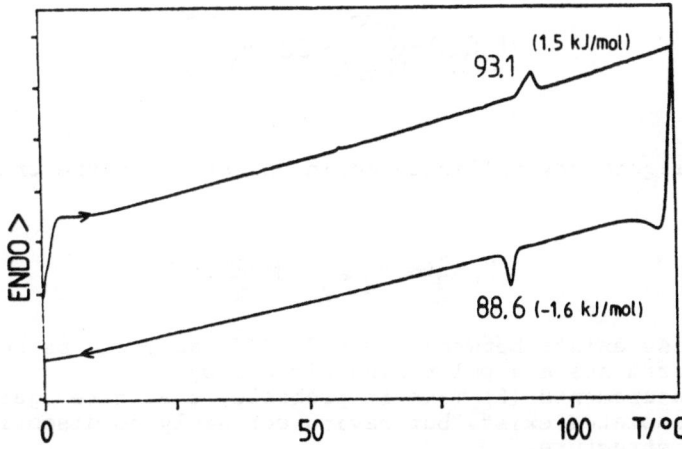

The diol-compound <u>4</u> representsnot only one of the key-compounds along the controlled way to related discotic polymers, but is already a liquid crystalline compound itself.
DSC-measurements (cf. Figure 1) reveals on heating and cooling only one peak at 93.1°C, resp.88.6°C.

Figure 1. DSC-thermogram of diol-compound <u>4</u>. 10 K/min.

The low transition enthalpies of about 1.5 kJ/mol indicate
mesophase transitions. This was confirmed by polarizing microscopy, which show,beside homeotropic domains,also focal conic
textures.
Preliminary X-ray measurements (5) exhibit a hexagonal columnar
structure of the discotic phase. Moreover the intercolumnar
distance of 40.8 Å corresponds to nearly the 1.5 fold length
of one diol-molecule 4. Together with the existence of a strong
absorption band at 3378 cm, typical for hydrogen bridges, this
shows, that the diol-compound 4, though being not disclike in
its chemical structure, is capable to form discotic liquid
crystalline phases. This should be due to the possibility,
that presumably more than three molecules can arrange to disc-
like associates.
It should be evident, that a perfect peripheric space filling
of the lateral alkylchains, which is regarded as an essential
criterion for the ability of conventional disc-like molecules
to form disophases (6), determinates also the behaviour of this
novel discotic associates.
Diminishing the lateral space filling changes drastically the
behaviour of the related compounds.
So the 3.5-disubstituted analogon

don't exhibit any mesophase.
The only 4-monosubstituted diol-derivative

reveals liquid crystallinity again. As shown in the transition
scheme

$$K_1 \xrightleftharpoons[75]{85} K_2 \xrightleftharpoons[92]{100} M \xrightleftharpoons[100]{104} I$$

a mesophase exists between 100 and 104°C as could be seen by
DSC-measurements and polarizing microscopy.
X-ray measurements (5) make it probable, that here again mesogenic associates exist, but having certainly no discotic
columnar structure.
To demonstrate the importance of the hydroxy groups in forming
discotic associates compound 4 was esterified with decanoyl-

chloride. By means of DSC-measurements and polarizing microscopy the following transition scheme was established:

The very low temperature range of the detectable monotropic
liquid crystalline phase characterizes a strong decrease of the
mesophase stability. This emphasizes on the other hand the
importance of the hydroxy groups, e.g. in compound 4, concerning the formation of discotic associates.
It is widely accepted, that in general mesophases in polymers
are more stable than in related monomers.
So we synthesized the following product by addition polymerization of diol-compound 4 with decanedicarboxylic acid chloride:

$R = -OC_{10}H_{21}$

Despite all attempts of annealing this polymeric material
over longer periods of time no liquid crystallinity could be
detected.

CONCLUSIONS

The controlled synthesis of discotic polymers requires a functionalized tristar-discogen-monomer. The described reaction of 3,4,5-tris(alkoxy)benzoic acid chloride with protected cis,cis-phloroglucit and the subsequent deprotection leads to a first key-compound in this synthetic pathway: the diol-compound 4, member of a novel class of discotic associates. A controlled functionalization of the 3,4,5-tris(alkoxy)derivatives should lead to the second key-compound in the stepwise synthesis of discotic polymers.

REFERENCES

1. Kreuder, W. and Ringsdorf, H., Makromol.Chem.Rapid Commun., 1983, 4, 1794

2. Kreuder, W. and Ringsdorf, H., Makromol.Chem.Rapid Commun., 1985, 6, 376

3. Wenz, G., Makromol.Chem.Rapid.Commun., 1985, 6, 577

4. Lattermann, G., Liquid Crystals, 1987, 2, 723

5. Ebert, M. and Wendorff, J.H., personal communication

6. Kohne, B. and Praefcke, K., Chem.Ztg., 1985, 109, 121

A NOVEL HIGH TEMPERATURE THERMOPLASTIC

D J Sikkema

Akzo Research Laboratories Arnhem

P O Box 60

6800 AB Arnhem, the Netherlands

ABSTRACT

Amorphous fully aromatic polyamides can be prepared by multicomponent copolymerization. Injection mouldable products were polymerized via an acidolytic process, i.e. in the melt. The products show $T_g > 250\ ^oC$, respectable mechanical properties and they do not absorb common organic solvents.

INTRODUCTION

The search for thermoplastic materials that may be used at elevated temperatures has received much attention in recent years. Materials that are relatively difficult to prepare like polyetherimide and polyetheretherketone have been proposed as High Temperature Plastics. Our aim has been the development of materials that would be fairly inexpensive by virtue of simple raw materials and a relatively simple polymerization process.

The fact that the glass transition temperature defines a significant drop in the mechanical properties in all polymers except the exceedingly crystalline, suggests a high T_g as an all-important goal in the search for new HTPs. If one confines oneself to amorphous polymers, T_g may be targeted as high as about 100 ^oC below the highest temperatures that one could contemplate for thermoplastic forming work with the polymer.

The traditional drawback of amorphous polymers is their ready dissolution or at least swelling in various solvents.

RESULTS AND DISCUSSION

Our approach has been to look for amorphous fully aromatic polyamides, to optimize both T_g (by virtue of their rigid chains) and solvent resistance (by virtue of their high polarity and chain-chain interaction). It turned out that random copolymerization employing at least four monomers (in the four-comonomer case: in close to equimolar amounts) was necessary and sufficient to render the aramid polyemer amorphous – the monomers had to be selected to an extent of at least half the total mass, from the group that would lead to some form of kinking in the chain.

Most of the work has been performed with the system 3-aminobenzoic acid, 4-aminobenzoic acid, isophthalic acid, bis(4-aminophenyl)methane. These monomers, in about equimolar quantities (and of course the diamine(s) and diacid(s) in stoichiometric quantities with respect to each other) were polymerized by an acidolytic scheme. Normally they were treated with acetic anhydride in slight excess relative to the amines present and – after the exothermic amine acetylation subsided – heated to remove acetic acid, ultimately in a high vacuum. Other anhydrides can of course be used. This means a melt polycondensation approach; although this is not without its pitfalls (high temperatures, high viscosities, high vacuum) we envision economic benefits, compared with polymerizations requiring work in solution. Whereas it is difficult to satisfactorily copolymerize aminoacids in solution (this requires the cumbersome preparation of the hydrochlorides of amino acid chlorides, or using the expensive phosphorylation polymeriza-tion procedure [2] rather than acid chlorides), the choice of monomers expands to include aminoacids by the acidolysis process. On the other hand, the high reaction temperature at relatively high acidity limits the monomer choice to exclude monomers that are much activated for Friedel Crafts chemistry, such as m-phenylene diamine. Surprisingly, aliphatic diamines such as hexamethylene diamine could be included without serious polymeriza-tion problems to lower T_g. Such aliphatic-modified products, in our (limited) experience, did show problems during processing due to their reduced thermo-oxidative stabiity.

The properties of the aromatic products, which so far have been prepared at a fairly small scale only, i.e. laboratory and limited semi-technical scale, are promising. T_g values over 250 $^{\circ}$C, excellent solvent resistance (in terms of swelling, or rather non-swelling) to e.g. ASTM oils, toluene, alcohols, chlorinated hydrocarbons, ethers and esters and attractive (bending) modulus and strength were recorded. It came as no surprise, for rigid materials like the present polymers, that tensile evaluations were difficult (clamp-induced failure). On failure, the unrein-forced glassy materials break without crazing or fibrillating.

Naturally, we are investigating the possibilities to combine these HTPs with reinforcing fibres.

LITERATURE

1. D J Sikkema, NL 8600965
2. N Yamazaki, M Matsumoto and F Higashi, J Pol Sci, Chem 13, 1373 (1975)
   W R Krigbaum, R Kotek and Y Minara, J Pol Sci, Chem 22, 4045 (1984)

Properties of high-temperature plastics as determined in injection moulded specimens made in our laboratory

	Experim. aramid	Ultem 1000	PEEK after annealing, 60 min, 100 $^{\circ}$C to crystallize
T inj. moulding, $^{\circ}$C	365	400	400
T mould	140	90	140
HDT, method A	213	186	161
T_g start	235	203	150
peak tan d	265	230	165
E-modulus, GPa	4.0	3.0	3.5
bending strength, MPa	110	94	101
elongation, %	3.2	11	5.5/46[*)]
relative creep rate, 200°C	1	8	?

[*)] strain to start plastic flow and breaking strain, respectively

STUDY ON THE SYNTHESIS AND PROPERTIES OF A
NEW POLYIMIDE END-CAPPED WITH MALEIC ANHYDRIDE
I.Synthesis of Amic Acid and Its Kinetics of Imidization

Wang Xueqiu, Bian Jinhua and Li Shijin
Dept. of Polymer Sci. & Techn., East China Univ. of Chem.
Techn., 130 Meilong Road, Shanghai 200237, P. R. China

ABSTRACT

In order to synthesize a new polyimide, an amic acid(MPE) was synthesized.

The kinetics of imidization of the amic acid (MPE) was studied with a new method by computerized DSC experimental data. The order of imidization reaction of MPE amic acid is about equal to 3, and it is not variable in the whole imidization. The new method is substantiated feasible by the experimental results of TGA and IR.

INTRODUCTION

In order to obtain a new polyimide with good properties and cost cheaply, a new amic acid capped with maleic anhydride was synthesized. According to this model compound, its kinetics of imidization was studied.

Generally, the kinetics of imidization of polyamic acid were studied by means of infra-red spectra (1) and thermal analysis (TGA) (2). These methods are more complex and with more errors. In order to overcome above shortcomings, a new method by means of DSC spectra analysis was proposed.

EXPERIMENTAL

Preparation of MPE amic acid

The MPE amic acid was prepared shown as following scheme:

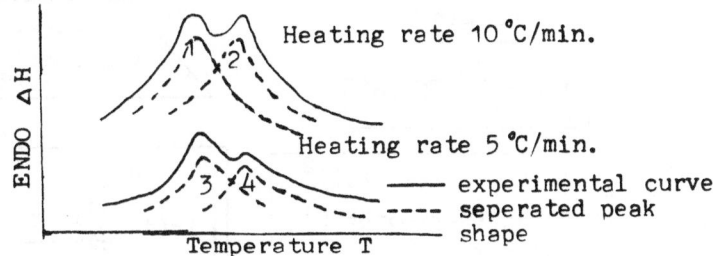

In a 250 ml flask, 2.98 g of bis(p-diaminophenyl)ether was dissolved in 35 ml DMF using mechanical stirring. Then added dropwise 2.416 g of 3,3′4,4′-benzophenonetetracarboxylic dianhydride (dissolved in 40 ml DMF) over a period of 100 minutes. Then 1.47 g of maleic anhydride was slowly added, maintained not over 20 °C. The resulting MPE amic acid was precipitated in water, filtered and washed with water, dried. MPE amic acid was characterized by Infra-red, DSC and Elemental analysis.

The kinetics of imidization

DSC analysis: Because of two types of imidize group in MPE amic acid molecule, i.e. and , there are two peaks indicated by imidization in DSC spectra.
But the two peaks are overlaped (Fig.1)

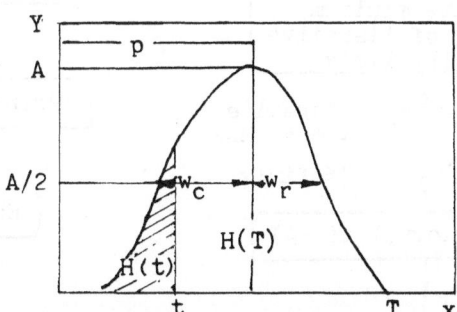

Figure 1. DSC specra of MPE amic acid

The peak shape function is described as follows (3);

$$Y = A / (1 + ((x-p)/(w_c \cdot w_r))^2 \cdot |((w_c - w_r) \cdot x + (w_c - w_r - p)^2 - (p - w_c)(p + w_r))|)$$

Where A, p, w_c, w_r are peak shape parameters(as shown in Fig.2).

Figure 2. Peak shape in DSC spectra.

The kinetic equation may be written as follows:

$$\frac{dp}{dt} = K(1-p)^n \qquad (1)$$

Where p is the reaction degree, K the rate constant, n the order of the reaction.

Because the heat effect $H(t)= p(t)$, $H(T)= p(T)$, where t is the reaction time, T the total time of the reaction. $H(t)/H(T)= p(t)$. The heat effect H is proportional to the peak area S,

Then $1/S_T$. $ds/dt = K(1-S/S_T)^n$, S_T is the total peak area at T time, given by $\int_0^T Ydt$. Due to $ds = Ydt$, $K=Ae^{-E/RT_m}$, Then

$$\ln Y+\frac{E}{RT_m} -n \ln(1 - \int_0^t Ydt/ \int_0^T Ydt) = \ln(A . \int_0^T Ydt) \qquad (2)$$

It can be solved by computerized the DSC experimental data, the program is shown as follows:

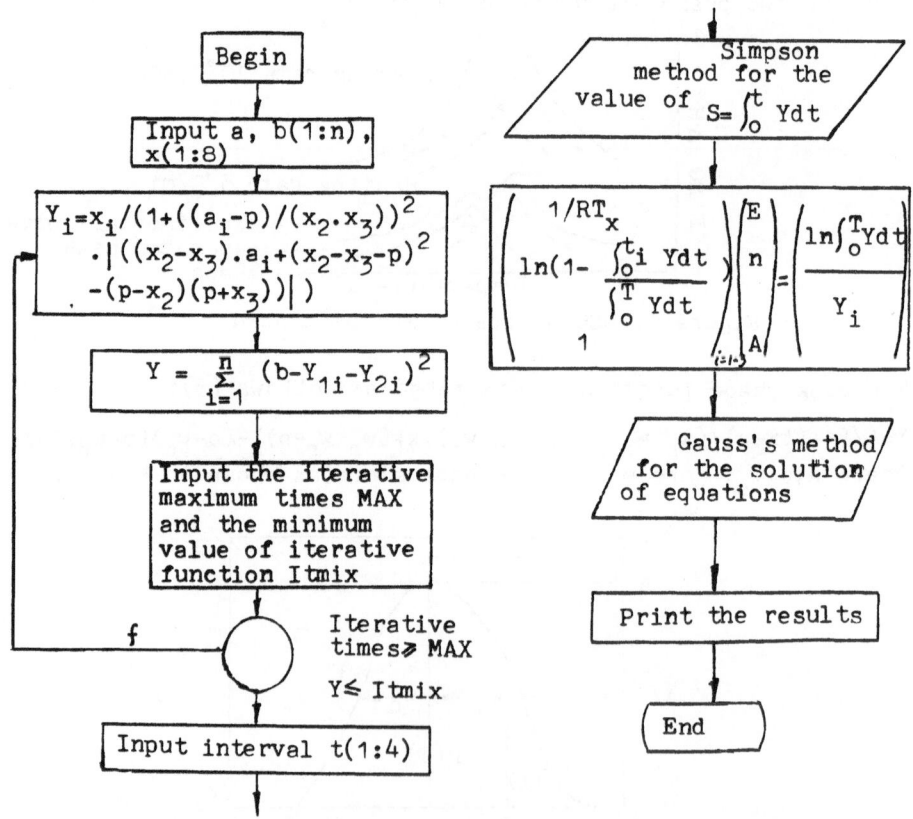

IR analysis and TGA analysis: Study the kinetics of imidization by means of IR and TGA experiments by the method (1) and (2) respectively.

RESULTS AND DISCUSSION

The DSC analytic spectra was shown in Fig.1, the special data of the kinetics were tabulated in table 1.

TABLE 1

The kinetics data of MPE amic acid imidization reaction

Heating rate	Peak number	Frequency factor $A \times 10^{-8}$	Activation energy E (kcal./mol.)	The order of reaction n
5 °C/min.	1	6.35	18.7	2.94
	2	6.02	21.4	3.00
10 °C/min.	3	6.78	18.8	2.98
	4	6.95	21.7	3.01
average	first stage	6.57	18.8	2.96
	second stage	6.49	21.6	3.01

The order of imidization reaction can be regarded as 3, and not variable in the whole imidization reaction. The kinetic equations are expressed as follows:

$$\frac{dp_1}{dt} = 6.57 \times 10^{-8} \cdot e^{-18.8/RT_m} (1 - P_1)^3$$

$$\frac{dp_2}{dt} = 6.49 \times 10^{-8} \cdot e^{-21.6/RT_m} (1 - p_2)^3$$

where P_1, P_2 are the imidization reaction degree of imidizing groups

respectively.

The order of imidization reaction of MPE amic acid obtained by IR method is 3.05, obtained by TGA analysis is 3.

The resulting kinetics data obtained by above three methods are almost the same. But more tests must be done and the error is always included by IR analysis. TGA analysis is easy to understand, however the precision is limited because the error by drawing is larger. The new method by computerized DSC data is an ideal method,it only needs to do one test for obtaining the whole imidization reaction kinetics data.

CONCLUSIONS

1. A new amic acid (MPE) end-capped with maleic anhydride has been synthesized and characterized.
2. For studying the kinetics of amic acid imidization reaction, a new method by computerized DSC data was proposed. The method is substantiated feasible by the experimental results of TGA and IR analysis.
3. The order of imidization reaction of MPE amic acid is about equal to 3, and is not variable in the whole imidization reaction.

REFERENCES

1. N.A.Adrova, M.I.Bessonov, L.A.Laius and A.P.Rudakov, POLYIMIDES, TECHINOMIC Publishing Co.,Inc., 1970, pp.27-44.

2. H.E.Kissinger, Reaction Kinetics in Differential Thermal Analysis. Analytical Chemistry, 1957, 29, pp. 1702-1706.

3. 鄧云鶴, 梁学润, 高分子通讯 1980,4, pp.231.

SYNTHESIS OF POLYMERS WITH LONG SIDE-CHAIN OF N-ALKYL ESTERS AND THEIR EFFECTS ON POUR-POINT DEPRESSION OF OIL

LI ZHUOMEI ZHANG WEIBANG ZENG HUIYANG
Institute of Polymer, Zhongshan University, Guangzhou, PRC

ABSTRACT

Three poly(n-alkyl acrylates) PA-14, PA-16, PA-18 and two poly-(vinyl alcohol aliphatic acid esters) PVA-14 and PVA-16 have been synthesized. They are white crystalline polymers at room temperature, characterized with IR, DSC, SALLS-photometer and polarizing microscope. Only PA-14 and PVA-14 are effective for O# diesel oil. The influences of molecular weight, molecular weight distribution and amount of addition of PA-14 and PVA-14 upon the pour-point depression of oil are very alike. These results reveal the length of n-alkyl side-chain of the polyesters plays the most important role in depressing the pour-point of petrolic oil and the order of linking in the ester group seems in no relation to this effect.

INTRODUCTION

Although poly(n-alkyl acrylates)(PA-esters) with long side-chain have been used early as depressant for pour-point of petrolic oil(1), no studies have been reported on the poly-(vinyl alcohol aliphatic acid esters)(PVA-esters) in this application. We suppose there would exist such possibility as the structures of PA-esters and PVA-esters are very similar to each other except that their forms of ester linking are in reverse order. Therefore it is valuable to investigate the structural features of PA-esters and PVA-esters, and their effects on pour-point depression of oil, not only for scientific interest but also for practical use.

MATERIALS AND METHODS

Monomers C_{14}-acrylate(A-14), C_{16}-acrylate(A-16) and C_{18}-acrylate(A-18) were prepared by alcoholysis of methyl methacrylate

identified with IR and elementary analysis, and polymerized in
benzene at 60-70°C to synthesize the corresponding polymers
PA-14, PA-16 and PA-18. Two poly(vinyl alcohol aliphatic acid
esters) PVA-14 and PVA-16 were synthesized in DMF at 60-70°C
by the reaction of PVA with excess acid chloride, of which the
degree of esterification was determined by acetylation method
(2). The structural features of both kinds of polyesters were
examined by 5DX-FTIR spectrophotometer, by Perkin-Elmer DSC
apparatus and by Ying-Kou LS-1 small angle laser light-scat-
tering photometer or Jiang- Nan XPT-6 polarizing microscope.
The molecular weight of polymers was determined by Chromatix
KMX-6 and dn/dc by Chromatix-16 at 30°C with heptane as solvent for PA-esters and dioxane for PVA-esters. The pour-point
of oil was detected with Tian-Jin NGB2 oil freezing point
tester.

RESULTS AND DISCUSSION

Characterization of PA-esters

All three PA-esters are white solids at room temperature.
Their IR spectra show strong absorption band at 1730 cm^{-1} due
to C=O stretching vibration in ester group and no absorption
band at 1630 cm^{-1} due to C=C stretching vibration. This fact
indicates that the PA-esters have not been contaminated by the
monomer residues.

DSC thermograms are with rather sharp endothermic peaks,
showing the PA-esters are crystalline polymers. Their melting
points and heats of fusion are listed in Table 1.

TABLE 1
Melting points, heats of fusion and radii of
spherulites for PA-esters

PA-ester	M.P.($^{\circ}$C)	ΔH_f(cal g^{-1})	R(μm)
14	29	12.1	2.55
16	41	20.1	3.05
18	49	24.0	3.52

The sizes of the spherulites of PA-esters are so small
that they are not easily observed by the polarizing microscope as shown in Figure 1, but can be examined by SALLS
photometer. The radii of the spherulites of PA-esters are
also listed in Table 1. It can be seen that the longer the
alkyl side-chain, the larger is the spherulite with higher
melting point and heat of fusion.

50μm

Figure 1. PLM of PA-16

Characterization of PVA-esters

PVA-14 and PVA-16 are also white solids at room temperature.
Their IR spectra are quite similar to those of PA-esters,
except that the intensity of absorption band at 3300 cm⁻¹ due
to the stretching vibration of -OH would increase as the degree of esterification decreases.

DSC thermograms with sharp endothermic peaks also show
that the melting point and heat of fusion are higher for PVA-
ester of longer side-chain as PA-ester does (see Table 2). And
for PVA-16 the sample of lower degree of esterification has
higher melting point and heat of fusion owing to the hydrogen
bonding between -OH's as expected.

TABLE 2
Degrees of esterification, melting points and
heats of fusion for PVA-esters

PVA-ester	D.E.(%)	M.P.($^{\circ}$C)	ΔH_f(cal g^{-1})
14(N-1)	98.6	50.0	23.8
16(N-2)	96.5	58.0	34.7
16(N-3)	87.1	65.7	43.7

The sizes of the spherulites of PVA-esters are large
enough to be observed very clearly by polarizing microscope
as shown in Figure 2.

The presence of -OH, even in small amount, in PVA-esters
affects their thermal behaviors considerably. In comparison of

the data in Table 1 and Table 2 we find the melting point and heat of fusion of PVA-14(N-1) are much higher than the corresponding PA-14, but can be comparable with PA-18. These facts are consistent with the result mentioned above that the spherulites of PVA-esters are larger than PA-esters.

50μm

Figure 2. PLM of PVA-16(N-2)

Depression of Pour-point of petrolic oil

All the PA-esters and PVA-esters used in this research are ineffective for pour-point depression of 10# transformer oil and 10# engine oil. Only PA-14 and PVA-14 are effective for 0# diesel oil while PA-16, PA-18 and PVA-16 are still ineffective yet. The influences of molecular weight, molecular weight distribution and amount of addition of PA-14 and PVA-14 on the depression of pour-point of 0# diesel oil are quite similar to each other as seen in Table 3 and Table 4.

TABLE 3
Influences of mol. wt. and mol. wt. distribution on
the pour-point depression of 0# diesel oil

Fraction		1	2	3	4	5	mixed
PA-14	$MwX10^{-4}$	235.3	195.6	145.3	69.6	27.9	
	$-\Delta t(^{\circ}C)$	11	12	12	12	12	12
PVA-14	$MwX10^{-4}$	184.1	101.4	35.3	18.6	4.14	
	$-\Delta t(^{\circ}C)$	12	11	12	12	12	12

Amount of polyester added= 0.1%

TABLE 4
Influences of amount added on the pour-point
depression of O# diesel oil

Amount added(%)	0.01	0.02	0.05	0.07	0.10	0.50
PA-14($-\Delta t^\circ$C)	4		8	10	12	12
PVA-14($-\Delta t^\circ$C)	2	2	5	8	10	11

All these facts reveal that the length of n-alkyl side-chain of the polyesters plays the most important role in depressing the pour-point of oil and the order of linking of the ester-group to the backbone seems in no relation to this effect.

CONCLUSIONS

The polyesters PA-14, PA-16, PA-18 and PVA-14, PVA-16 are crystalline polymers at room temperature. The melting points and heats of fusion of PVA-esters are higher than the corresponding PA-esters. Of all these polyesters only PA-14 and PVA-14 are effective for pour-point depression of O# diesel oil.

As pointed out by Gavlin(1) the pour-point depression of lubricating oil by alkyl polymethacrylates was attributed to the adsorption of the depressants on a growing wax crystal face preventing the formation of a three-dimensional wax network. Our previous work(3) supported this viewpoint. The further study in this paper show even more distinctly that the adsorption of polyesters occurs to the long alkyl side-chain only and the ester linking part, owing to its polar character, is excluded from the wax crystal face. This is why PA-esters and PVA-esters have the same effect of pour-point depression of oil regardless of the difference in their ester-linking.

REFERENCES

1. Gavlin, G., Swire, E. A. and Jones, S. P., Jr., Pour Point Depression of Lubricating Oils. Ind. Eng. Chem., 1953, 45, 2327-35.

2. Zhang Weibang, Zeng Huiyang and Li Zhuomei, Syntheses and Structures of Poly(long chain aliphatic ethannl esters). (to be published)

3. Zhang Weibang, Feng Ke and Li Zhuomei, Synthesis of Poly(n-alkyl methacrylates) and Their Effects on the Pour-point Depression of Oil. Polymeric Materials Science and Engineer-ing. 1987, 3(4), 16-21.

Part 6

TECHNIQUES

DEFORMATION INDUCED EMISSION OF ELECTRONS
AND POSITIVE IONS

JÜRGEN FUHRMANN, LEO NICK

Institut für Physikalische Chemie, TU Clausthal

Arnold-Sommerfeld-Str. 4, 3392 Clausthal-Zellerfeld, FRG

GÜNTER H. SCHERER

BASF Ludwigshafen, FRG

INTRODUCTION AND HISTORICAL BACKGROUND

Although the mechanically stimulated particle emission from polymer surfaces is a rather new field of investigation, similar effects at surfaces of various materials have been known for a long time.

Particle emission from metal surfaces

The earliest report on mechanically stimulated particle emission known to us dates from 1897 when W. J. Russel [1] found that a zinc plate, which is freshly rubbed with sand paper, could produce an image of the scratches when it was covered by a photographic plate after the stimulating treatment. His understanding of the phenomenon enabled him already to question if it is *a form of energy (possibly what has been called dark light) that these bodies emit ?'*.

Radiation induced particle emission lead to disturbing counting rates of Geiger-Müller counting tubes in the 1940ies, e. g. in X-ray experiments at the Van-de-Graaff generator at M.I.T. [2], but no attempt was made to explain the effect at that time. The first systematical investigations of such emission processes were done by Kramer [3], who supposed that these emission effects were causally determined by exothermic processes, e. g. crystallisation of woods metal. This hypothesis of 'exothermic electron emission' gave a name to a whole class of particle emission phenomena, i. e. 'exoemission'. This name is used up to now for all emission penomena which are not to be explained by standard theories like field emission, thermionic emission etc. despite the fact that the correlation with exothermic processes is often not found.

Particle emission from semiconductor surfaces

Thermally stimulated exoemission of silicon turned out to be no material property but to depend on the existence of a silicon oxide surface layer [4]. Exoemission of ZnO was shown to depend on the adsorption kinetics of oxygene at the surface [5]. Emission phenomena which result from external factors like adsorbate layers may be classified by the term *extrinsic* exoemission, while other surface and subsurface emission effects that are due to factors inherent to the material should accordingly be called *intrinsic* exoemission [6].

Particle emission from polymer surfaces

In 1963 Polyakov and Krotova [7] were the first to observe mechanically stimulated electron emissions of a macromolecular substance. Stepwise straining of gutta-percha lead to emission maxima at each step. This result seemed to have been forgotten for ten years until Zakrevskii and Pakhotin [8-10] showed that a number of polymers (poly(ethylene), poly(caprolactam), poly(ethyleneterephthalate), poly(urethane)) exhibited the effect of mechanically stimulated electron emission during straining. A first suggestion was that electrons might be emitted by strong electrical fields at the surface, but the expected influence of surface charges could not be proved. A second model for the emission of electrons proposed by the same authors will be discussed later.

Radiation activated and optically stimulated electron emission of polymers were not found [11], except for the activation by electron bombardement.

Advances in the measurement technics in the last 15 years especially the development of channel electron multipliers (CEM's) render now the possibility to detect besides electrons (or in general: negative particles) also positive ions. Dickinson et al. [12-22] examined particle emissions and corresponding emissions of photons, radio-waves and accustic waves accompanying macroscopic fracture of polymers (fractoemission). They also proposed a physical emission model based on a high degree of charge separation and gaseous breakdown. The present paper deals with the particle emission during homogeneous deformation.

MECHANICALLY INDUCED PARTICLE EMISSION DURING DEFORMATION OF POLYMERS

Before we present some experimental results, we want to have a brief look on the model of Zakrevskii and Pakhotin [10]. Because their experimental equipment allowed only the detection of negatively charged particles, the model exclusively gives an explanation for the situation of electrons. By now there is up to our knowledge no reasonable model capable to deal with the mechanically stimulated emission of positive ions by polymers.

Fig. 1 shows a diagram of energy states [10].

The principle idea is that the ionisation energy of the macromolecules I_{gas} is locally lowered by mechanical activation and thermofluctuations to ~ 50 %. From this situation (energy E_{loc}) a tunnel transition, which leads to either two macroions or a trapped electron and a macroion, is possible.

These processes are:

a) $R - R \xrightarrow{\sigma, T} R^+ + e^-_{loc} \xrightarrow{R^{\cdot}} R^- + R^+$

The energy is given by

$E_1 = I_{gas} - P_{e-h}(r) - A - Q(r)$

with

I_{gas} : (hypothetical) ionisation energy of the macromolecule in the gaseous phase

$P_{e-h}(r)$: polarisation energy of the polymer by a charge pair

$Q(r)$: coulomb energy

A : electron affinity of the macroradical R

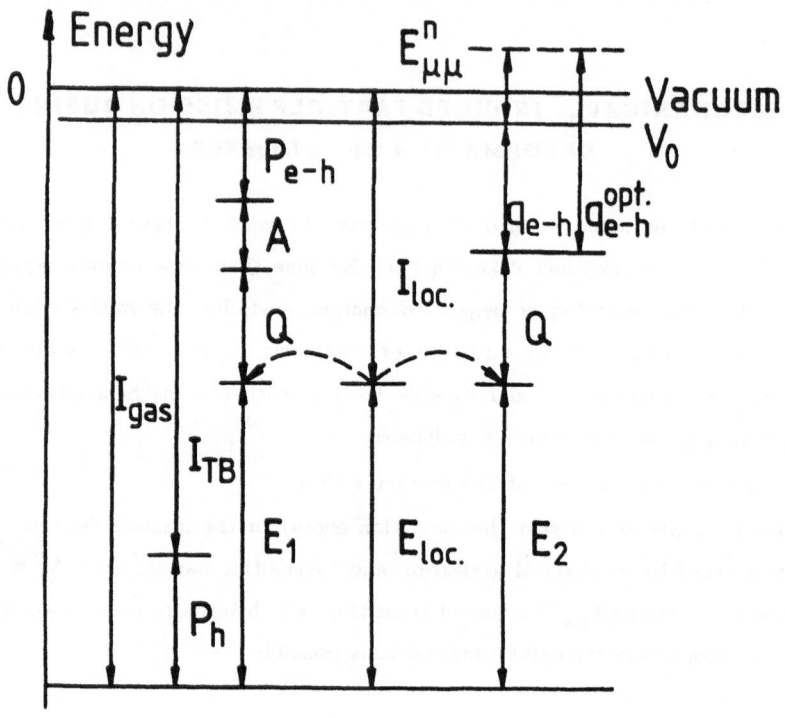

Figure 1.   Energy states of electrons in polymers [10]

I_{gas} Ionization energy of a macromolecule in the gas phase

I_{loc} Ionization energy of a stressed macromolecule

P_h Energy of polarization of the polymer by a single positive charge

$P_{e-h}(r)$ Energy of polarization of the polymer by a charge pair in distance r

A   Electron affinity of the macroradical R·

$Q(r)$ Coulomb energy

q_{e-h} Depth of an electron trap interacting with a positive ion

V_o Energy of quasi free, electrons

E_{loc} Localized electron state in a stressed macromolecule

E_1 Energy of charge separation (charge pair formation), when the electron is captured by a macromolecule

E_2 Energy of charge separation, when the electron is trapped

Broken arrows: tunnel transitions

This process results in effects like the *anomalous decay of mechano-radicals* observed in EPR-investigations on the mechanical degradation of polymers [23]. If polypropylene is ball-milled in presence of a strong electron scavenger (Tetracyano-ethylene. TCNE) a concurrent reaction for the capture of electrons is induced. Instead of being trapped by macro-radicals, which would lead to anions which are *invisible* by EPR-means. the electrons are captured by TCNE and result in radical-anions which can be examined by EPR-methods. Sakaguchi et al. [24] estimated the proportion of heterolytical to homolytical bond cleavage in this experiment to be at least 37%.

b) $\quad R - R \xrightarrow{\text{σ. T}} R^+ + e^-_{loc} \longrightarrow e^-_{trap} + R^+ \;\text{-----}\overset{?}{\text{--->}}\;$ Exoemission
$\qquad\qquad\qquad\qquad\qquad \diagdown \text{----------------}\overset{?}{\text{--->}}$

The energy is given by

$E^l_2 = I_{gas} - P_h - q + V_o$

$P_h \qquad$: polarisationenergy of the polymer by a single trapped hole

$q \qquad$: depth of an electron trap

$V_o \qquad$: energy level of quasi mobile states (conduction band)

This energy E^l_2 is the energy of the system assuming that the electron is able to separate from the macroion via quasi mobile states before being trapped.

The energy E_2 shown in Fig. 1, results of the same case when a charge pair is formed. In that case. instead of the energies of the polarisation P_h of the polymer by a single trapped hole and the depth q of an electron trap, the Coulomb energy $Q(r)$ of the charge pair (distance r) and the depth of the 'electron-hole-trap' q_{e-h} have to be taken into account.

$E_2 = I_{gas} - q_{e-h} + V_o - Q(r)$

While Zakrevskii and Pakhotin merely wanted to show that autoionisation processes in polymers under combined mechanical and thermal activation are quite reasonable to discuss. we like to draw the attention to some extending aspects.

In all cases discussed above the charged fragments of bond cleavage - electrons and positively or negatively charged ions - remain in the material. In order to be emitted the electrons or ions must get an additional energy. that enables them to reach the vacuum. We suggest an Auger-like recombination mechanism of electrons and/or ions involved in the bond cleavage reaction to be able to provide this energy. The energy distribution of the emitted negatively charged particles of polyethylene (fig. 3) supports a recombination model.

So the questions stimulating our former and current work are:

How large is the escape probability (escape length) of an electron or a positive ion ?

Are recombination processes possible, that provide the additional energy for particles to leave the polymer ?

Is the only effect of mechanical stimulation that it populates traps or is it needed to create a sufficient number of suitable traps ?

To what extend do bulk effects interfere with the emission out of the surface ?

Cartier and Pfluger [25] and Racz and Ohki [26] measured the mean escape length of hot electrons with kinetic energies of 5 to 6 eV to be in the range of 3 to 9 nm in polyethylene and model substances for polyethylene.

Recently Cartier and Pfluger [27] reported that hot electrons with a threshold energy of about 4 eV induce a radiation damage in a polyethylene model substance. This means that there exists a coupling of local electronic states with quasi mobile states, as supposed by Zakrevskii and Pakhotin.

EXPERIMENTAL SETUP

The emission of positively and negatively charged particles is registered simultaneously during deformation in high vacuum. The principal equipment consists of two channel electron multipliers as detectors, appropriately biased to detect the differently charged particles and a two channel single particle counting system. A detailed description of the equipment is given in [28].

Example I : High Density polyethylene (HDPE)

We discuss briefly the results of some investigations on HDPE. For a detailed discussion see [29].

The samples (Lupolen 6041^{TM} , provided by BASF, Ludwigshafen) consisted of blown films and had a lamellar morphology with the normal direction of the lamellae parallel to the drawing direction (uniaxial strain).

Fig. 2 (Lit. [29]) shows the emission rate of positive ions (PIE) and negative particles (EE) as function of the deformation ratio λ. The emission is seen to rise immediately with beginning deformation, to yield a maximum in the vicinity of the mechanical yield point and to drop almost to zero at further elongation. At the moment of macroscopic fracture an emission maximum occurs again.

A (bulk) deformation model is able to explain the emission characteristics, if we consider the correlation of the emission processes with molecular fracture. The HDPE is biaxially preoriented, due to the manufacturing process. Defining the correlation length l as the distance in which the mechanical force Ψ acting on a point is reduced to a portion α, we may characterize the lamellae and the amorphous regions by their correlation length distributions.

In the crystal lamellae the correlation length in normal direction $l_{\parallel}$ and in direction of the lamellae $l_{\perp}$ is rather high, i. e. the lamellae are capable to transfer a relatively great amount of mechanical energy. In the amorphous phase the correlation length in deformation direction $l_{\parallel}$ is large for tie-molecules, i. e. for those molecules which are established firmly in adjacent lamellae. These molecules are not able to distribute the mechanical energy over a greater 'area', because the lateral correlation length $l_{\perp}$ is low.

The correlation length gradient at the boundary lamellae/amorphous phase renders it possible to transfer a force Ψ on single tie-molecules that enables a combined mechanical/thermofluctuation activation of bond cleavages (Zhurkov mechanism).

At the mechanical yield point, the lateral dimensions of the lamellae decrease dramatically [30], i. e. the correlation lengths in lateral lamellae direction decrease. The lamellar morphology changes gradually to a fibrillar one. The crystalline and amorphous regions approach each other in their correlation length distributions. In this more uniform situation the mechanical straining of single tie-mole-

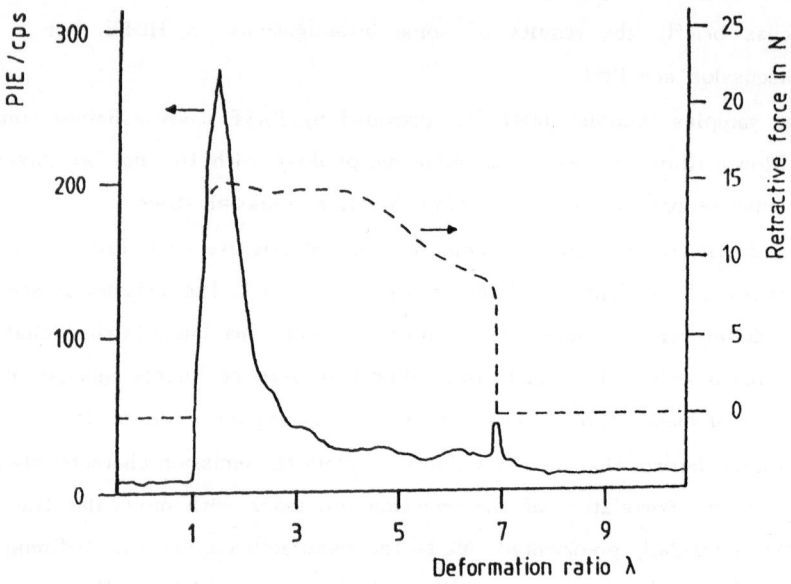

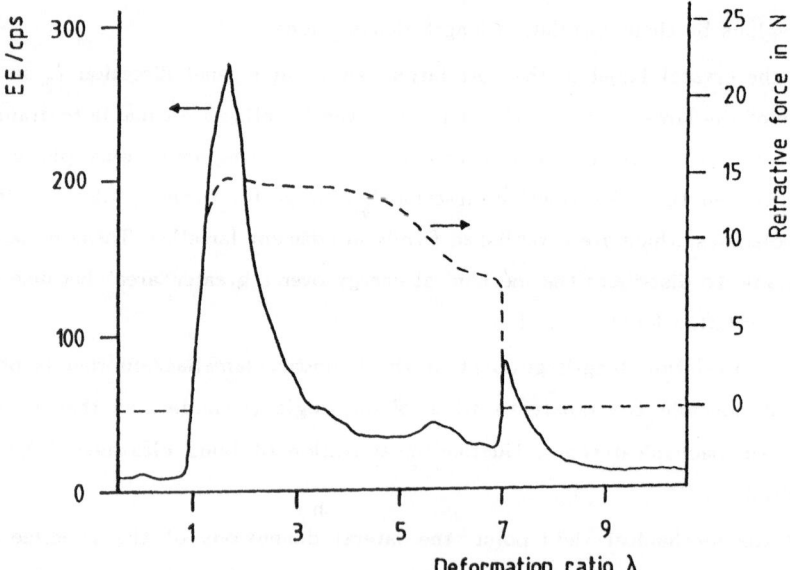

Figure 2. Positive ion emission (PIE), electron emission (EE) – solid lines – and the retractive force of the sample – broken lines – during deformation of high density polyethylene (HDPE) [29].

Experimental conditions: rate of deformation 10 %/s; p= 10^{-5} Pa; T = 298 K

sample dimensions 20 x 10 x 0.098 mm

cules up to rupture becomes rare; accordingly the exoemission rates decrease.

The extreme inhomogenity of correlation lengths in the region of macroscopic fracture enables again macromolecular rupture and a corresponding emission maximum.

In order to support or abandon this emission model, we investigated further aspects.

A central issue was whether the emission is an extrinsic or an intrinsic one. The latter would be expected according to the deformation and bond rupture model outlined above. For this purpose the surrounding gas in the vaccum recipient was varied (air, dry air, nitrogen, argon) and its concentration was changed (accessible pressure interval: $< 10^{-3}$ Pa for linearity of the CEM's), but no significant variation of the emission could be detected. To exclude the influence of eventually existing surface charges the samples were washed with methanol prior to the experiment. Although no influence of this treatment was found (see also [9]), it was applied in all experiments.

Variations of the deformation rate (5 to 50%/s) gave no deviations in neither the emission rate nor the total (integral) emission, so that effects of elastically stored deformation energy seem to be of no influence in this deformation rate interval.

Stepwise elongation (0 to 50%, 50 to 300%, 300% to fracture) with 5 minutes of relaxation time between the steps produces a superposed emission curve which is very similar to that of the transient experiment, indicating that the emission originating processes are instantaneous and not resulting of or superimposed by a relaxation of perturbations created in the very first stages of deformation [29].

Experiments with cyclic deformation (up to 10% elongation) exhibit an emission signal only in the first cycle. Subsequent deformations to greater elongations lead to emissions only if the deformation ratio exceeds that of previous cycles. This shows the irreversible nature of the emission generating processes.

All these experimental results are readily to be explained in terms of the deformation and emission model outlined above.

The energy distribution determined by retarding field methods (Fig. 3) of the emitted negatively charged particles shows a rather narrow maximum, which can, for some experimental uncertainties, only be said to lie between 5 and 15 eV. The narrowness of the distribution prohibits field emissions or thermionic emissions, whereas Auger-like recombination mechanisms of excited species and trap-sites are favourable, because the first ionisation energies of polyethylene observed in x-ray photoelectron spectra (XPS) are 7.7 eV, designated to C-H bond electrons, and 13.2 and 18.8 eV, designated to C-C bond electrons [31].

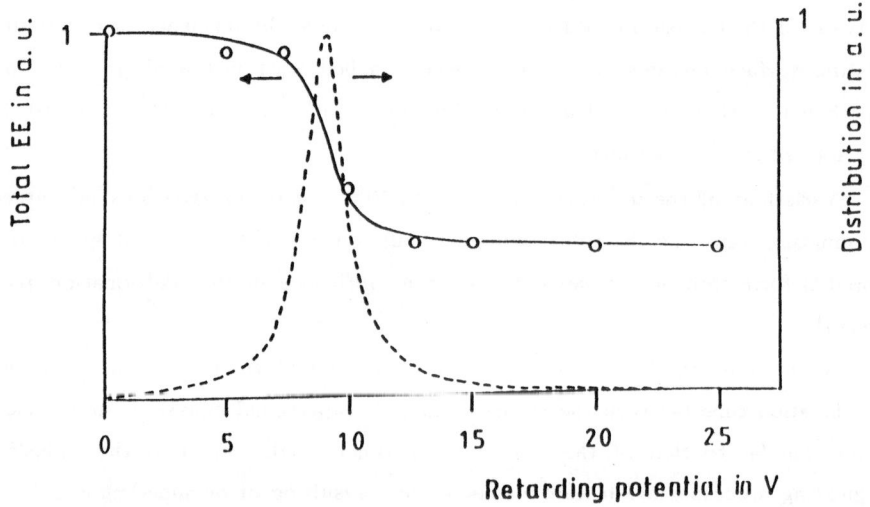

Figure 3. Energy of electron emission (EE) [29] of high density polyethylene film. Solid line: total emission as function of the applied retarding field voltage. Each point (circles) is determined by five measurements at least. Broken line: energy distribution function; the derivative of the solid line function
Rate of deformation 10 %/s; p= 10^{-5} Pa; T = 298 K;
sample dimensions 20 x 10 x 0.098 mm

Example II : Polyamide 6

The investigated polyamide 6 [32] (Ultramid B4^{TM}, provided by BASF. Ludwigs-hafen) consisted of films of thickness 100 μm. These films had a crystallinity of 34 % (DSC) and were slightly oriented in calandration direction.

Fig. 4 shows the exoemission rates (PIE and EE) of Polyamide 6 as function of the deformation ratio λ. The deformation rate was 5 %/s and the samples which were cut from the foil had a dumbbell form with an efficient length of 15 mm and were 10 mm wide. The deformation direction was perpendicular to the machining direction of the manufacturing process.

The emission of both PIE and EE exceeds the level of background noise only in the deformation range between the mechanical yield point and $\lambda = 2 - 3$. Although the emission rate is very low, it is reproducible to ~20%. An emission maximum at the moment of macroscopic fracture is only observable in some experiments. Due to the great deformation ratios the macroscopic fracture often occurs in a region beyond the efficient detection area of the CEM's.

Comparing the morphology of polyamide 6 and HDPE we can readily explain the differences in the emission behaviour. Because the crystallites in polyamide 6 are rather small. they are not able to transfer as much mechanical energy to tie-molecules as the lamellae of HDPE can do. Furthermore the crystallinity α_c is relatively low (35%. DSC). In terms of correlation lengths this means that the lateral correlation length $l_\perp$ of the crystallites is relatively small. Simultaneously the lateral correlation length in the amorphous regions is - due to carbonamide H-bond-bridges - greater than in HDPE, so that mechanical energy is distributed from single molecules on to their neighbouring ones.

Molecular fracture can therefor only be expected, when large scale reorientation processes occur. In the region before the mechanical yield the deformation is merely elastical and homogeneous. When the material starts to flow and to develop fibrillar structures after yielding large relative displacements of molecules occur. So the conditions outlined above for molecular fracture are given. At higher deformation rations. when the *natural* deformation ratio is surpassed, a fibrillar structure has been reached. In this deformation range single microfibrils slide against each other without molecular reorientations and almost without molecular fracture.

378

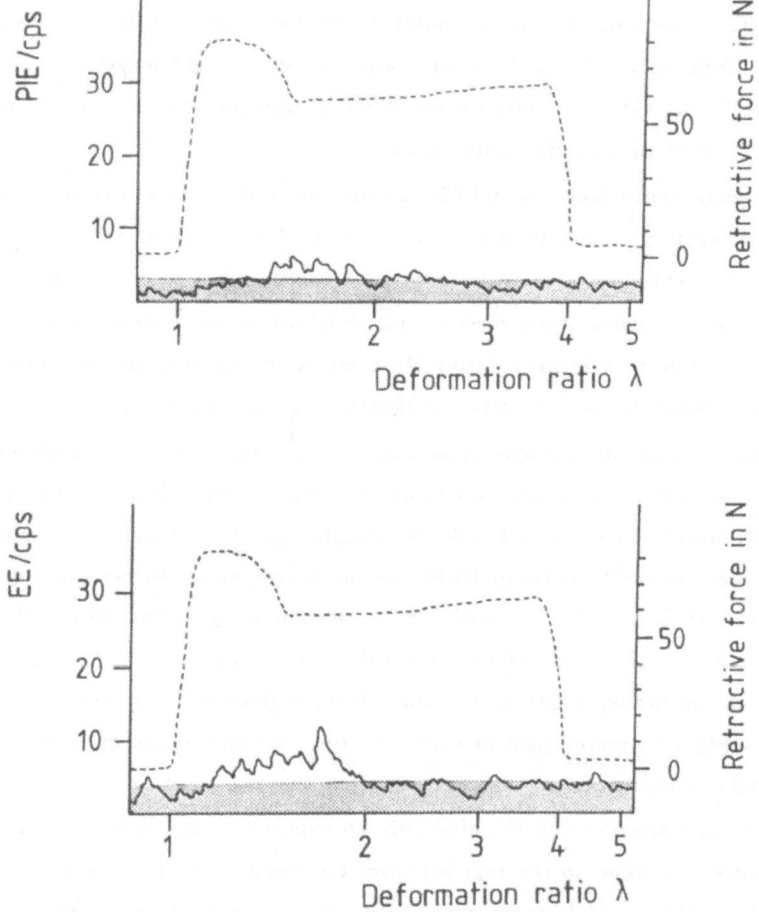

Figure 4. Positive ion emission (PIE), electron emission (EE) - solid lines - and retractive force of the sample - broken lines - during deformation of polyamide 6 [32].
Rate of deformation 5 %/s; p= 10⁻⁵ Pa; T = 298 K;
sample dimensions 15 × 10 × 0.1 mm

379

REFERENCES

1. Russel. W. J.. Proc. Roy. Soc. [London] **61**, 424 (1897)

2. Van Atta. L. C.. Northrop, D. L.. Van de Graaff, R. P.. Van Atta, C. M.. Rev. Sci. Instr. **12**, 534 (1941)

3. Kramer, J.. Z. Physik **125**, 729 (1949)

4. Drenckhan. J.. Gross, H.. Glaefeke, H.. Phys. Stat. Sol. **(a) 2**, K201 (1970)

5. Menhold. R.. Z. Physik **157**, 499 (1960)

6. Ramsay, I. A.. "Exoelectric Emission" in: Progress in Surface and Membrane Science **XI** (1976)

7. Polyakov. A. M., Krotova, A. N., Dokl. Akad. Nauk. SSSR **151**, 130 (1963)

8. Zakrevskii. V. A.. Pakhotin, V. A., Polym. Sci. USSR **17**, 653 (1975)
   Zakrevskii. V. A.. Pakhotin, V. A., Vysokomol Soyed **A17**, 468 (1975)

9. Zakrevskii. V. A.. Pakhotin, V. A., Sov. Phys., Solid State **20**, 214 (1978)
   Zakrevskii. V. A.. Pakhotin, V. A., Fiz. Tverd. Tela **20**, 371 (1978)

10. Zakrevskii. V. A.. Pakhotin, V. A., Polym. Sci. USSR **23**, 741 (1981)
    Zakrevskii. V. A.. Pakhotin, V. A.. Vysokomol. soyed. **A23**, 658 (1981)

11. Arabin. G.. Kriegseis, W.. Scharmann, A.. Z. Naturforsch. **27A**, 1378 (1972)

12. Dickinson. J. T., Donaldson, E. E., Park, M. K., J. Mater. Sci. **16**, 2897 (1981)

13. Dickinson. J. T., Donaldson, E. E., Park, M. K., J. Mater. Sci. **17**, 3173 (1982)

14. Dickinson. J. T., Jensen, L. C., J. Polym. Sci., Polym. Phys. Ed. **20**, 1925 (1982)

15. Dickinson. J. T., Jensen, L. C., Park, M. K., Appl. Phys. Lett. **41**, 443 (1982)

16. Dickinson. J. T.. Park, M. K., Donaldson, E. E., Jensen, L. C., J. Vac. Sci. Tech. **20**, 436 (1982)

17. Dickinson. J. T., Jensen, L. C.. Jahan-Latibari, Rubber Chem. Technol. **56**, 927 (1983)

18. Dickinson. J. T., J. Mater. Sci. **19**, 2426 (1984)

19. Dickinson. J. T., Jahan-Latibari, A., Jensen, L. C., J. Mater. Sci. **20**, 229 (1985)

20. Dickinson. J. T., Jensen, L. C., J. Polym. Sci., Polym. Phys. Ed. **23**, 873 (1985)

21. Dickinson. J. T., Jahan-Latibari, A., Jensen, L. C., in "Molecular Characterisation of Composite Interfaces", Ed. H. Ishida, G. Kumar, Plenum Press New York, 1985, p. 111

22. Dickinson, J. T., in "Adhesive Chemistry - Developments and Trends", Ed. C. H. Lee, Plenum Press New York, 1984, p. 193

23. Sohma, J., Sakaguchi, M., Adv. Polym. Sci. **20**, 111 (1976)

24. Sakaguchi, M., Kinpara, M., Hori, Y., Shimada, S., Kashiwabara, H. Polymer **25**, 944 (1984)

25. Cartier, E., Pfluger, P., Phys. Rev. B **34**, 8822 (1986)

26. Racz, I., Ohki, Y., Jpn. J. Appl. Phys. **26**, 209 (1987)

27. Cartier, E., Pfluger, P., IEEE Trans El. Ins., **EI-22**, (1987)

28. Fuhrmann, J., Scherer, G. H., Hofmann, R., Polym. Comm. **27**, 164 (1987)

29. Fuhrmann, J., Scherer, G. H., Nick, L., Makromol. Chem. **188**, 2241 (1987)

30. Heise, B., Kilian, H.-G., Wulff, W., Progr. Coll. Polym. Sci. **67**, 143 (1980)

31. Pireaux, J. J., Riga, J., Caudano, R., Verbist, J., ACS Symp. Ser. **162**, 169 (1981)

32. Nick, L., Diplomarbeit, Kaiserslautern 1986

MEASUREMENT OF VOLUME CHANGES DURING TORSION OF UNPLASTICIZED PVC BY MEANS OF A TORSIONAL DILATOMETER: PHYSICAL AGING BEHAVIOUR

RALF PIXA, VERONIQUE LE DU, CONSTANT WIPPLER
Institut Charles Sadron (EAHP-CRM)
4 rue Boussingault, F - 67000 Strasbourg

ABSTRACT

A new dilatometer has been constructed which, during torsion, allows the simultaneous measurement of shear deformation, twisting moment and volume variation. This apparatus has been utilized principally for the study of volume changes during deformation of unplasticized PVC near ambient temperature. Complex volume changes have been observed: with increasing twist angle, volume first decreased, reached a minimum value and eventually increased again, for sufficient strain levels. In the latter case, after coming back to zero strain, net volume increase and subsequent recovery have been observed. This behaviour has been interpreted in terms of deformation induced rejuvenation, followed by physical aging.

INTRODUCTION

Glassy amorphous polymers are generally not in thermodynamic equilibrium. Consequently, their structure is not stable and evolves continually towards the equilibrium configuration. Due to highly reduced molecular mobility, this results in a long-term volume decrease if it takes place at a temperature much lower than the glass transition temperature. This phenomenon has been termed structural or volume recovery. It is accompanied by changes of many related physical properties which have been referred to as physical aging (1).

The kinetics of this behaviour have been described by various theories. In particular, Kovacs et al. (2) have proposed a multiparameter model which describes most of the experimentally observed effects. It is based on the free volume concept and assumes that the overall free volume influences the contribution of the structural parameters on the recovery behaviour.

Volume recovery and aging after cooling through the glass transition and its subsequent erasure by heating above T_g have been extensively studied (1,2). It has however been shown (1) that large deformations also erase previous aging, giving rise to a reactivation of the aging process.

Principally two hypotheses exist to explain this mechanical rejuvenation: by increase of free volume, analogous to the behaviour observed after thermal rejuvenation (1), or by reactivation of molecular mobility without any volume effect (3). We have undertaken several experiments in order to discriminate between these two possibilities.

In order to measure the volume changes accompanying deformations, we have tried to minimize the influence of purely elastic volume changes. For this reason, deformation in torsion has been chosen, for which only small elastic volume changes are expected (4,5).

MATERIALS AND METHODS

Cylindrical samples from rigid PVC were machined from the wall of an extruded pressure pipe manufactured by the SHELL company about 2 years earlier. Accordingly, the material structure could be considered as reasonably stable in order to show no measurable physical aging prior to deformation.

Construction of the torsional dilatometer was initiated by Kovacs and coworkers and will be described in detail elsewhere (6). It has been designed to measure simultaneously twist angle, torque, volume variation, length change and normal force of cylindrical samples, subjected to torsional movement at constant angular speed.

All measurements were made at temperatures near 30°C (maintained constant to within $\pm$ 0.01 °C) and at an angular speed of about 0.1 rad/min.

RESULTS

Figure 1 shows the results obtained during successive cycles of increasing shear strain (as measured at the outer cylinder surface). Roughly three stages in deformation behaviour can be distinguished: In the quasi linear part of the stress-strain curves, the sample volume contracted approximately proportionally to the applied strain. Then, the absolute slopes of both stress and relative volume as function of strain decreased simultaneously. Stress remained virtually constant for the largest strains whereas volume increased again.

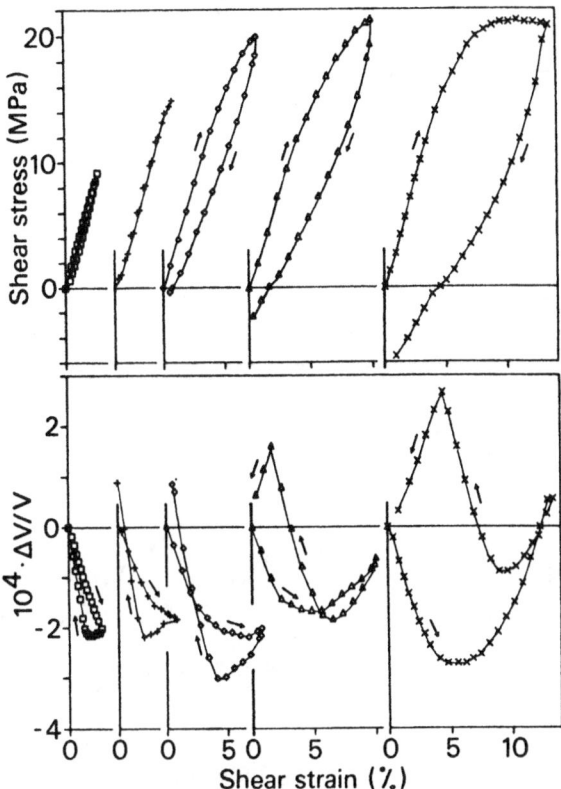

Figure 1. Shear stress and volume variation during cyclic straining experiments.

The curves show, for small strains, that the volume recovered to its initial value after the return to zero strain. This domain corresponds to the linear part in the stress-strain diagrams. At the largest strains, volume changes were more complicated and plastic deformation took place. Therefore, when going back in torsion, zero stress occurred before zero strain was attained. As seen in figure 1, volume reached its maximum value at zero stress and went down again at zero strain, remaining, however, larger than before torsion.

At zero strain, after each strain cycle, we have measured the variation of residual volume dilation and of residual stress as function of time elapsed after deformation (figure 2): After the smallest strain maxima, residual dilation remained constant. At larger strains, however, volume recovery was observed. Rate of recovery of residual volume increased with increasing strain level.

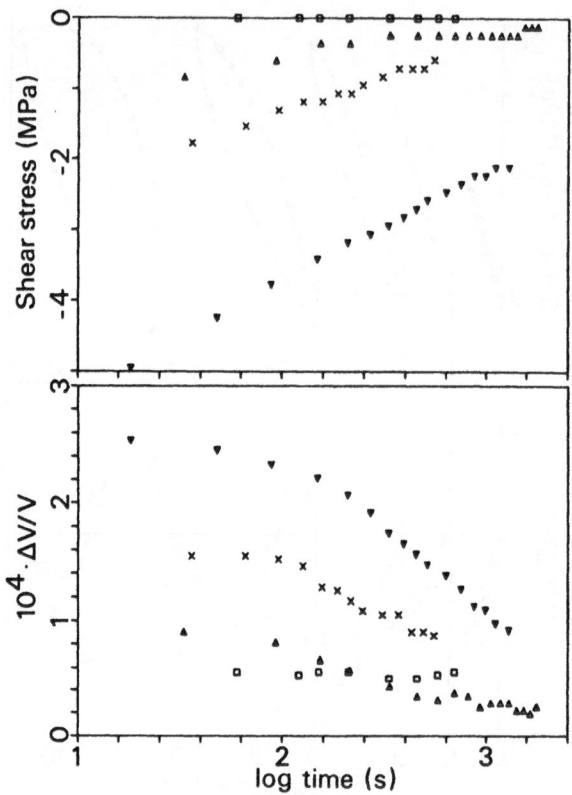

Figure 2. Stress relaxation (top) and recovery of residual dilation (bottom) at zero strain (preceding strain levels: □: 3.7 %; △: 9.2 %; x: 10.1 %; ▽: 13.4 %).

DISCUSSION

It seems from our measurements that volume variation during torsion resulted from two superposed phenomena: contraction at small strains, followed by dilation (figure 1). The small strain contraction may be attributed to nonlinear elasticity (4,5), as already shown for other polymers (7). Concerning the dilation at larger strains, which has been observed for the first time in our experiments, it may be explained by Struik's hypothesis (1) that a sufficiently large deformation should produce free volume, rejuvenating the material. This appears to invalidate other theories (3) for deformation induced rejuvenation.

The volume recovery curves of figure 2 show, after a first stage of slow evolution, a fairly constant slope. The recovery is the faster, the larger the created volume, that means, the larger the preceding strain maximum. Consequently, it seems that residual dilation and volume recovery are directly related; the recovery rate being roughly proportional to the amount of created free volume.

CONCLUSIONS

The experiments described above have shown that, during torsion, PVC underwent first elastic volume contraction, as postulated by some nonlinear elasticity theories (4,5). At larger strains, however, volume dilation occurred which seems not to be due to craze formation. After returning to zero strain a greater volume than before torsion was observed. This volume excess seems to confirm Struik's hypothesis concerning free volume creation (1). As supposed by this concept, volume recovery has been observed immediately after stopping torsion. The larger the preceding strain maximum was, the faster the volume recovery. These results seem to show that physical aging was reactivated by mechanical destabilization.

REFERENCES

1. Struik, L. C., Physical Aging in Amorphous Polymers and Other Materials, Elsevier, Amsterdam, 1978.

2. Kovacs, A. J., Aklonis, J. J., Hutchinson, J. M. and Ramos, A. R., Isobaric Volume and Enthalpy Recovery of Glasses. II. A Transparent Multiparameter Theory. J. Polym. Sci., Phys., 1979, 17, 1097-1163.

3. Myers, F. A., Cama, F. C. and Sternstein, S. S., Mechanically Enhanced Ageing of Glassy Polymers. Ann. N. Y. Acad. Sci., 1976, 279, 94-99.

4. Treloar, L. R. G., The Physics of Rubber Elasticity, Clarendon Press, Oxford, 1975.

5. Wack, B., Second and third-order effects in the torsion of circular tubes and rods. J. Mécan., 1981, 20, 737-787.

6. Pixa, R., Le Dû, V. and Wippler, C., Dilatometric study of deformation induced volume increase and recovery in rigid PVC. Submitted for publication to Colloid & Polymer Sci.

7. Wang, T. T., Zupko, H. M., Wyndon, L. A. and Matsuoka, S., Dimensional and volumetric changes in cylindrical rods of polymers subjected to a twist moment. Polymer, 1982, 23, 1407-1409.

13C SOLID STATE NMR STUDY OF
ETHYLENE-VINYL ALCOHOL COPOLYMERS

H. Ketels[1], J. de Haan[2], A. Aerdts[2]
[1] Laboratory of Polymer Technology
[2] Laboratory of Instrumental Analysis
Eindhoven University of Technology
P.O. Box 513, 5600 MB Eindhoven
The Netherlands

and

G. v.d. Velden
DSM-Research BV
P.O. Box 18, 6160 MD Geleen
The Netherlands

ABSTRACT

In the present study ethylene-vinyl alcohol copolymers (E-VOH) with an increasing content of ethylene were synthesized and investigated using high-resolution solid state ^{13}C NMR. A splitting of the methine carbon resonance into three peaks could be observed. This splitting could be explained taking into account tacticity <u>and</u> sequence effects.

INTRODUCTION

Recently, high-resolution ^{13}C NMR spectra in solids have become readily measurable by using the combined techniques (CP/MAS) of high polar proton decoupling, cross polarization, and magic angle sample spinning [1]. In solid polymers, however, line broadening, due to chemical shift dispersion, smears out the splitting resulting from differences in tacticity. Splitting due to differences in tacticity was observed for the first time in the methine carbon region of solid poly(vinyl alcohol) (PVOH) by Terao [2]. This splitting was ascribed to enhanced downfield shifts by formation of intramolecular hydrogen bonds. In this paper we are concerned with the ^{13}C NMR spectra of solid E-VOH copolymers. The spectra of E-VOH copolymers with different contents of ethylene will be presented.

MATERIALS AND METHODS

Five samples of E-VOH copolymers (A-E, powders) were prepared as recently described [3]. The degree of hydrolysis was confirmed to be higher than 99 mol% (^{1}H NMR). Three commercially available E-VOH copolymers (pellets) have been obtained from Kuraray, coded EPL (F, $Mn = 32$ kg mol^{-1}), EPF (G, $Mn = 21$ kg mol^{-1}) and EPG (H, $Mn = 20$ kg mol^{-1}). The E-VOH copolymer (powder) with an ethylene content of 90% (I, $Mn = 25$ kg mol^{-1}) was obtained by hydrolizing the corresponding E-VA copolymer ELVAX-260 (Dupont). PVOH powder (J) has been obtained from Hoechst (Mowiol 66-100) and is characterised by a viscosity of 66 ± 4 centipoise (4% water solution at 20°C).

The NMR data were acquired at room temperature on a Bruker CXP 200 spectrometer operating at 50.3 MHz. Samples were loaded into an air driven two component Beams-Andrew BN-POM rotor with a polyoxymethylene (POM) cap. The samples were spun at 3.0-3.5 kHz. The chemical shifts were referenced to the external chemical shift of the crystalline and amorphous POM-resonances (both at 88.8 ppm).

RESULTS AND DISCUSSION

Figure 1 shows the 50 MHz ^{13}C CP/MAS-NMR spectra of PVOH and the E-VOH copolymers A-I.

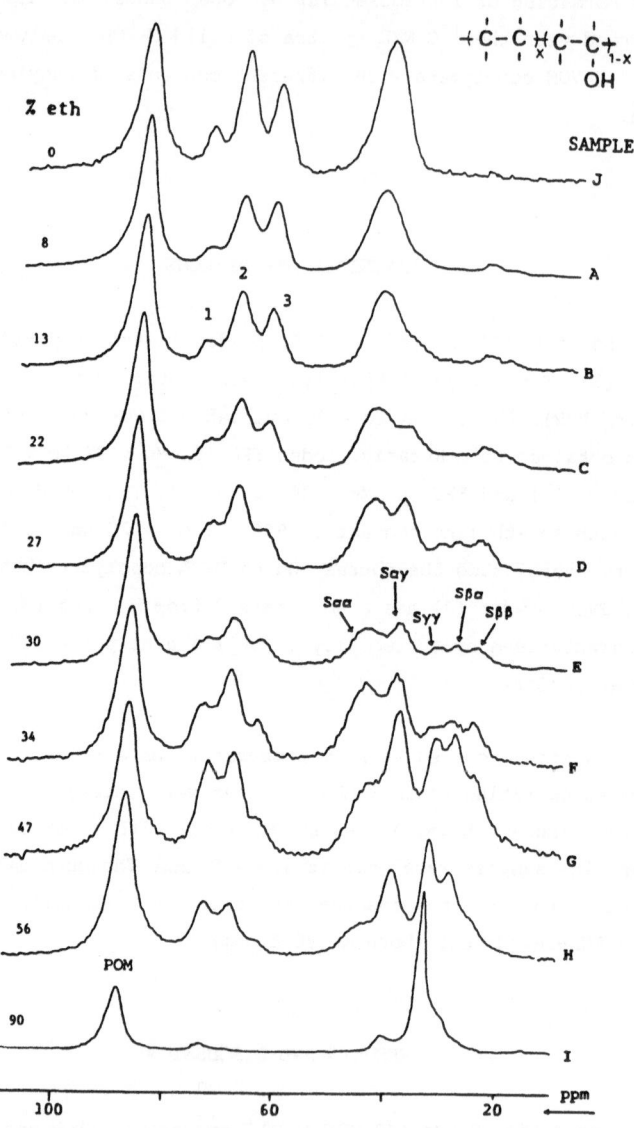

Figure 1. ^{13}C CP/MAS-NMR spectra of PVOH and E-VOH copolymers A-I

The complete assignment of the chemical shifts together with the relative
area intensities for the methine carbon resonances are given in Table 1
and will be discussed now in some detail.

TABLE 1

Chemical shifts and relative area intensities of the methine carbon
resonances

	% eth	1	2	3	A_{i-1}	A_{i-2}	A_{i-3}	$S_{\alpha\alpha}$	$S_{\alpha\gamma}$	$S_{\gamma\gamma}$	$S_{\varepsilon\alpha}$	$S_{\beta\beta}$
PVOH(J)	0	77.3	71.2	65.4	0.14	0.5	0.36	45.7	-	-	-	-
A	8	76.9	71.1	65.4	0.13	0.48	0.39	46.0	-	-	-	27.3
B	13	76.8	71.2	65.4	0.17	0.51	0.32	45.9	-	-	-	27.2
C	22	75.6	70.5	65.2	0.18	0.52	0.30	45.9	42.2	33.1	29.8	27.1
D	27	74.6	70.4	65.5	0.29	0.52	0.19	45.3	40.5	33.6	31.0	26.3
E	30	74.9	70.4	65.0	0.31	0.50	0.20	46.3	40.7	33.9	29.8	26.3
F	34	74.8	70.4	65.4	0.30	0.51	0.19	46.1	40.7	33.7	30.9	26.9
G	47	74.0	69.5	65.2	0.40	0.43	0.17	49.7	39.7	33.2	29.9	26.2
H	56	73.7	69.0	65.1	0.43	0.42	0.15	45.2	40.0	33.5	29.9	25.9
I	90	73.3	69.3	-	0.79	0.21	-	-	40.4	33.5	-	-

A_i = Area intensity

The spectra can be subdivided into a low field region (65-80 ppm) which
is assigned to all methine carbon resonances and a high field region
(25-50 ppm) for all methylene resonances. In the high field region of the
spectra five resonance patterns can be observed for the methylene carbons
(see Figure 1). In the nomenclature used in Figure 1 each of the
methylenes or secondary (s) carbon atoms in the sequences is identified
by a number of Greek letters denoting the nearest methine carbon atoms
[3,4].
In the methylene carbon region no significant additional resonances can
be observed in comparison with the solution NMR results [3], so we have
focussed our attention to the methine carbon region, because hydrogen
bonding certainly has more effect on the chemical shifts of these
resonances.
The CP/MAS spectra are a superposition of the spectra of the crystalline
and the amorphous material. However the smaller linewidths and the CP

behaviour of the crystalline fraction cause it to dominate the CP/MAS spectrum, and the lines assigned here are therefore essentially due to the crystalline fraction.

The low field region shows three methine carbon resonances, peaks 1-3. Referring to Terao [2], and concentrating for the moment only at PVOH, peaks 1-3 in the solid state cannot simply be assigned to the tacticity induced splitting of the mm, mr and rr triads respectively: the relative area intensities of the three peaks in the solid state are not consistent with the triad tacticity observed in solution [2]. Moreover a significant downfield shift of the resonances 1 and 2 occurs compared with the observed chemical shifts for mm and mr triad resonances in solution, which results in much larger mutual chemical shift differences between the methine carbon resonances in the solid.

This downfield shift is tentatively assigned to the action of intramolecular hydrogen bonds [2]. This effect is larger in solid than in solution because of the fact that the hydrogen bonds are stronger in solids. In the case of the mm triad, the oxygen atom bonded to the central methine carbon can form two, one or zero intramolecular hydrogen bond (s) and for the mr triad one or zero intramolecular hydrogen bond. In the rr triad no intramolecular hydrogen bond is postulated to be present. By assuming that the line position shifts downfield by about 6 ppm per intramolecular hydrogen bond, Terao assigned peak 1 in PVOH to the mm triad with two intramolecular hydrogen bonds, peak 2 to the mm and mr triads with one intramolecular hydrogen bond each and peak 3 to the mm, mr and rr triads with no intramolecular hydrogen bonds. In the spectra of the E-VOH copolymers peaks 1-3 can also be observed. see Figure 1. In Table 1 the chemical shifts and the relative intensities of peaks 1-3 of the E-VOH copolymers are shown with increasing ethylene content of the E-VOH copolymers. An upfield shift of 4 ppm is observed for carbon resonance 1, a lesser upfield shift for resonance 2, while no effect has been observed for peak 3. The area intensities of the three peaks also change with different ethylene contents: the area intensity of peak 1 increases, the area intensities of peaks 2 and 3 decrease with increasing ethylene content. At an ethylene content of 90 mol % peak 3 has alsmost disappeared. The area intensities of the three peaks of the methine carbon resonances in the CP/MAS spectra showed similar contact time dependences in the relative signal responses for two different E-VOH copolymers (A and H).

In the assignment of the methine carbon resonances in the E-VOH
copolymers we have to take in account both tacticity effects, as in PVOH,
and sequence distribution effects. For the calculation of the [13]C
chemical shifts in the sequence triads of solid E-VOH copolymer, the [13]C
chemical shift additivity rules for aliphatic alcohols, as calculated by
Ovenall [5] from solution NMR measurements, have been used:

$$R - YCOH - {}^{\beta}CH_2 - {}^{\alpha}CHOH - {}^{\beta}CH_2 - YCHOH - R^1$$

$$\alpha = 40.8 \text{ ppm} \qquad \beta = 7.7 \text{ ppm} \qquad \gamma = -3.4 \text{ ppm}$$

Assuming no conformation dependent chemical shift effects to occur and
using the chemical shift of orthorhombic polyethylene (33 ppm) [6] we can
now calculate the chemical shifts of the methine carbon atoms in the
three triads of the solid crystalline E-VOH copolymer, respectively OOO
(67 ppm), OOE (70.4 ppm) and EOE (73.8 ppm) where OOO, OOE, EOE are
abbreviations for (VOH, VOH, VOH), (VOH, VOH, E) and (E, VOH, E) triads.
The chemical shift values presented above are only meant to yield useful
assignments of the several methine carbon ^{13}C NMR signals of E-VOH
copolymers. These assignments are necessary because Ovenall [5] did not
report dependable estimates for all three types of methines sustained by
experimental results. We are aware of chemical shift differences between
liquids and solids. Moreover, the choice of orthorhombic polyethylene as
a basis for the shift calculations is rather arbitrary but this will only
cause the same uncertainty in each of the three shifts. Of more
importance is the known sensitivity of substituent-induced shifts towards
different conformational equilibria. From results obtained by Cantow [7]
for different poly (1,2-dimethylbutane) polymers it can be estimated that
the uncertainties in our estimations amount to ca. 2 ppm. It is, however,
improbable that the order of the three methine carbon signals will be
misjudged.

Besides these sequence effects, also effects similar as observed by Terao
for PVOH (i.e. hydrogen bond induced tacticity shifts), have to be taken
into account. In Table 2 the theoretically calculated chemical shifts of
the methine carbon atoms are presented, adopting the same assumption as
made by Terao, i.e. a down field shift of approximately 6 ppm per formed
intramolecular hydrogen bond. The chemical shift of OOO m(2) can e.g. be
calculated as $\delta_E + \alpha + 2\gamma + 2(6) = 79$ ppm.

TABLE 2

Theoretically calculated chemical shifts of the methine carbon atoms

$$OOO = 67.0 \text{ ppm} \quad mm \text{ ———— } 2 \text{ H-bond } = 79.0 \text{ ppm}$$

$$mr + rm- \; 1 \text{ H-bond } = 73.0 \text{ ppm}$$

$$rr \text{ ———— } 0 \text{ H-bond } = 67.0 \text{ ppm}$$

$$OOE = 70.4 \text{ ppm} \quad m \text{ ———— } 1 \text{ H-bond } = 76.4 \text{ ppm}$$

$$r \text{ ———— } 0 \text{ H-bond } = 70.4 \text{ ppm}$$

$$EOE = 73.8 \text{ ppm} \qquad\qquad = 73.8 \text{ ppm}$$

Due to chemical shift dispersion, considerable overlap occurs, as is evident via an inspection of the results in Table 2 and the experimentally observed chemical shift patterns (Figure 1). Therefore only three peak regions can be discerned and subsequently assigned (the number between brackets indicates the level of hydrogen bonds):

Peak 1 = OOO mm (2) + OOE m (1) + EOE

Peak 2 = OOO mm (1) + OOO mr (1) + OOE m (0) + OOEr (0)　　　　(1)

Peak 3 = OOO mm (0) + OOO mr (0) + OOO rr (0)

From equation 1 it is apparent that this resonance assignment includes the model proposed by Terao for PVOH (one extreme, Figure 1; spectrum J). For ethylene rich copolymer (the other extreme) this model predicts the observation of one major resonance situated at 74 ppm. From an inspection of Figure 1 (spectrum I) this appears to be true.

REFERENCES

1. Schaefer, J., Stejskal, E.O., J. Am. Chem. Soc., 1976, **98**, 1031–1032.
2. Terao, T., Maeda, S., Saika, A., Macromolecules, 1983, **16**, 1535–1538.
3. Ketels, H., Beulen, J., v.d. Velden, G., Macromolecules, 1988, accepted for publication.
4. Carman, C.J., Wilkes, C.E., Rubber Chem. Techn., 1971, **44**, 781.
5. Ovenall, D.W., Macromolecules, 1984, **17**, 1458.
6. Kitamaru R., Horii, F., Murayama, K., Macromolecules, 1986, **19**, 636.
7. Möller, M., Cantow, H.-J., Polymer Bulletin, 1981, **5**, 119.

INDEX OF CONTRIBUTORS

Subject Index